Benjamin Fine, Anja Moldenhauer, Gerhard Rosenberger, Annika Schürenberg,
Leonard Wienke
Geometry and Discrete Mathematics

Also of Interest

Algebra and Number Theory. A Selection of Highlights
Benjamin Fine, Anthony Gaglione, Anja Moldenhauer, Gerhard
Rosenberger, Dennis Spellman, 2017
ISBN 978-3-11-051584-8, e-ISBN (PDF) 978-3-11-051614-2,
e-ISBN (EPUB) 978-3-11-051626-5
2nd edition is planned!

General Topology. An Introduction
Tom Richmond, 2020
ISBN 978-3-11-068656-2, e-ISBN (PDF) 978-3-11-068657-9,
e-ISBN (EPUB) 978-3-11-068672-2

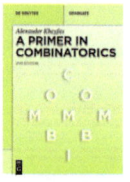

A Primer in Combinatorics
Alexander Kheyfits, 2021
ISBN 978-3-11-075117-8, e-ISBN (PDF) 978-3-11-075118-5,
e-ISBN (EPUB) 978-3-11-075124-6

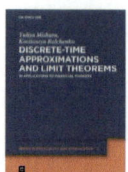

*Discrete-Time Approximations and Limit Theorems. In Applications to
Financial Markets*
Yuliya Mishura, Kostiantyn Ralchenko, 2021
ISBN 978-3-11-065279-6, e-ISBN (PDF) 978-3-11-065424-0,
e-ISBN (EPUB) 978-3-11-065299-4

*Discrete Algebraic Methods. Arithmetic, Cryptography, Automata and
Groups*
Volker Diekert, Manfred Kufleitner, Gerhard Rosenberger, Ulrich
Hertrampf, 2016
ISBN 978-3-11-041332-8, e-ISBN (PDF) 978-3-11-041333-5,
e-ISBN (EPUB) 978-3-11-041632-9

Benjamin Fine, Anja Moldenhauer,
Gerhard Rosenberger, Annika Schürenberg,
Leonard Wienke

Geometry and Discrete Mathematics

—

A Selection of Highlights

2nd edition

DE GRUYTER

Mathematics Subject Classification 2010
Primary: 00-01, 05-01, 06-01, 51-01, 60-01; Secondary: 03G05, 05A10, 05A15, 06A06, 20H10, 20H15, 51F15, 53A04, 55U05, 60A10, 94C15

Authors
Prof. Dr. Benjamin Fine
Fairfield University
Department of Mathematics
1073 North Benson Road
Fairfield, CT 06430
USA

Dr. Anja Moldenhauer
20535 Hamburg
Germany

Prof. Dr. Gerhard Rosenberger
University of Hamburg
Department of Mathematics
Bundesstr. 55
20146 Hamburg
Germany

Annika Schürenberg
Grundschule Hoheluft
Wrangelstr. 80
20253 Hamburg
Germany

Leonard Wienke
University of Bremen
Department of Mathematics
Bibliothekstr. 5
28359 Bremen
Germany

ISBN 978-3-11-074077-6
e-ISBN (PDF) 978-3-11-074078-3
e-ISBN (EPUB) 978-3-11-074093-6

Library of Congress Control Number: 2022934378

Bibliographic information published by the Deutsche Nationalbibliothek
The Deutsche Nationalbibliothek lists this publication in the Deutsche Nationalbibliografie; detailed bibliographic data are available on the Internet at http://dnb.dnb.de.

Preface

To many students, as well as to many teachers, mathematics may seem like a mundane discipline, filled with rules and algorithms and devoid of beauty, culture and art. However, the world of mathematics is populated with true gems and results that astound. In our series *Highlights in Mathematics* we introduce and examine many of these mathematical highlights, thoroughly developing whatever mathematical results and techniques we need along the way.

We regard our *Highlights* as books for graduate studies and planned it to be used in courses for teachers and for general mathematically interested, so it is somewhat between a textbook and a collection of results. We assume that the reader is familiar with basic knowledge in algebra, geometry and calculus, as well as some knowledge of matrices and linear equations. Beyond these the book is self-contained. The chapters of the book are largely independent, and we invite the reader to choose areas to concentrate on.

We structure our book in 11 chapters that are arranged in three parts: In the first seven chapters we examine results which are related to geometry. In Chapter 8 we give a connection of geometric ideas and combinatorically defined objects. In the last three chapters we further investigate topics in combinatorics, discuss a glimpse of finite probability theory and end our book with Boolean algebras and Boolean lattices.

In Chapter 1 we look at general geometric ideas and techniques. In the second edition we added a primer on curves in the real space $\mathbb{R}^3$ to this chapter to give a little insight into the richness of differential geometry.

In Chapter 2 we discuss the isometries in Euclidean vector spaces and their classification in $\mathbb{R}^n$. We realize that the study of planar Euclidean geometry depends upon the knowledge of the group of all isometries of the Euclidean plane and hence devote a section to them. The study of geometry using isometries and groups of isometries was developed by F. Klein, and this approach is fundamental in the modern application to geometry. A first application is in Chapter 3 where we give a classification and a geometric description of the conic sections.

In Chapter 4 we describe certain special groups of planar isometries, more precisely, we describe the fixed point groups and classify the frieze groups and the planar crystallographic groups. This especially leads to a classification of the regular tessellations of the plane. In this second edition we included a beautiful non-periodic tessellation of the real plane $\mathbb{R}^2$, the Penrose tiling which gets along with only two prototiles.

In Chapter 5 we present graph theory and graph theoretical problems. In particular, we discuss colorings, matchings, Euler lines and Hamiltonian lines along with their rich and current applications such as the marriage problem and the travelling salesman problem. In contrast to the first edition of this book the chapter on graph theory is now an extended stand-alone chapter and the discussion of spherical geometry and the Platonic solids takes place in a new Chapter 6.

https://doi.org/10.1515/9783110740783-201

There, we discuss the Platonic solids which historically have played an outsider's role in our view of the universe. For the description and the classification of the Platonic solids we use Euler's formula for planar, connected graphs and the spherical geometry of the sphere S^2. In Chapter 7 we complete the discussion of planar geometries with the introduction of a model for a hyperbolic plane and a look at the development and properties of hyperbolic geometry.

In the second edition we added a new Chapter 8 on simplicial complexes and topological data analysis – two important concepts from the emerging field of applied topology.

Chapter 9 gives a detailed path through combinatorics, combinatorial problems and generating functions. Finite probability theory is heavily dependent on combinatorics and combinatorial techniques. Hence in Chapter 10 we examine finite probability theory with a special focus on the Bayesian analysis.

Finally, in Chapter 11 we consider Boolean algebras and Boolean lattices and give a proof of the celebrated theorem of M. Stone which says that a Boolean lattice is lattice isomorphic to a Boolean set lattice. Hence, Boolean algebras and Boolean lattices are crucial in both pure mathematics, especially discrete mathematics, and digital computing.

We would like to thank the many people who were involved in the preparation of the manuscript as well as those who have used the first edition in classes and seminars for their helpful suggestions. In particular, we have to mention Anja Rosenberger for her dedicated participation in translating and proofreading. We thank Yannick Lilie for providing us with excellent diagrams and pictures. Those mathematical, stylistic, and orthographic errors that undoubtedly remain shall be charged to the authors. Last but not least, we thank De Gruyter for publishing our book. We hope that our readers, old and new, will find pleasure in this reviewed and extended edition.

Benjamin Fine
Anja Moldenhauer
Gerhard Rosenberger
Annika Schürenberg
Leonard Wienke

Contents

1 Geometry and Geometric Ideas

1.1 Geometric Notions, Models and Geometric Spaces

Geometry roughly is that branch of mathematics which deals with the properties of space: points, lines, angles, shapes, etc. However, the definitions and implications of these concepts will depend on the chosen axioms as we now explain.

To understand modern geometry we must look at the historical development. *Geometry* means *earth measure* in Greek. Historically geometry was developed by the ancient Egyptians, Chinese, Babylonians, Greeks, Hindus and others into an ad-hoc system of formulas and observations to find areas and volumes of plane and solid figures. The ancient Greeks were the first to try to arrange this material in a rigorous fashion. This work culminated in *Euclid's Elements* from about 300 BC.

This famous set of books attempted to give a completely rigorous treatment to planar geometry. It sets the pattern for not only the study of geometry but the study of mathematics in general. After the Bible it is the most widely read book in the Western world.

The basic plan of Euclid was that geometry consisted of certain *undefined* notions: points and lines, together with certain assumptions, the *axioms*. From these all other geometrical facts could be proven as *theorems* in a sequential manner using rules of logic. A geometrical "fact" is not a fact unless it can be deduced in the above manner.

Euclid used certain *common notions* on equality together with five basic axioms.

Euclid's Axiom 1. For every point P and for every point Q with $P \neq Q$ there exists a *unique* line PQ passing through P and Q. (Two distinct points determine a unique line.)

Euclid's Axiom 2. For every line segment $\overline{AB}$ and for every line segment $\overline{CD}$ there exists a unique point E such that B is between A and E and the line segment $\overline{CD}$ is congruent to $\overline{BE}$. It follows that any line segment $\overline{AB}$ can be extended by a line segment $\overline{BE}$ congruent to a given line segment $\overline{CD}$, see Figure 1.1.

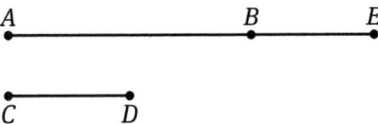

Figure 1.1: Line Extension.

Euclid's Axiom 3. For every point O and every point $A \neq O$ there exists a circle with center O and radius $\overline{OA}$.

Euclid's Axiom 4. All right angles are congruent.

https://doi.org/10.1515/9783110740783-001

Note. In Axiom 4 a *right angle* is defined as an angle which is congruent to its supplementary angle.

Euclid's Axiom 5. In a plane, for every line ℓ and for every point P not on ℓ there exists a unique line M through P and parallel to ℓ, that is, $\ell \cap M = \emptyset$.

Note. Axiom 5 is called the *Euclidean Parallel Postulate*. Euclid's original formulation was somewhat different and we will describe it later in this chapter.

As mathematicians over the centuries looked at Euclid's method, they noticed that there were many shortcomings in Euclid's Elements. First, there was the problem of *diagrams*. Euclid relied extensively on diagrams or ideas from diagrams. When working on the foundations of geometry, nineteenth century geometers discovered that the use of these diagrams did *not* all follow from the five basic axioms. Therefore more axioms are necessary to justify Euclid's Theorems. There are now several sets of (equivalent) formulations of complete sets of axioms for Euclidean geometry. Ironically these deficiencies in the axioms for Euclidean geometry were only discovered after the discovery of non-Euclidean geometry. We will discuss complete sets of axioms for Euclidean geometry in Section 1.3.

The next and perhaps more serious problem was the problem with the parallel postulate. Axioms 1 through 4 are concepts that can be "verified" by straightedge and compass construction. Axiom 5 is qualitatively different. Euclid himself seemed to have a problem with Axiom 5 and did not use it until rather late in his theorems. There were many attempts to prove Axiom 5 in terms of Axioms 1 through 4. Actually, what these attempts led to were alternative equivalent forms of the parallel postulate. We mention several of these and discuss them later.

(1) If two lines are cut by a transversal in such a way that the sum of 2 interior angles on one side of the transversal is less than 180° then the lines will meet on that side of the transversal (Euclid's original form).
(2) Parallel lines are everywhere equidistant.
(3) The sum of the angles of a triangle is 180°.
(4) There exists a rectangle.

In the nineteenth century, N. I. Lobachevsky (1792–1856), J. Bolyai (1802–1860) and C. F. Gauss (1777–1855) working independently discovered that by assuming other parallel postulates a geometry can be developed, that is, every bit is logically consistent as Euclidean geometry. Their work was the basis for *non-Euclidean geometry*.

The discovery of non-Euclidean geometry had a profound effect on human thinking. Essentially it said that geometry is dependent on the axioms chosen and nothing is a priori true. This type of thinking led almost directly to the discoveries in modern physics highlighted by relativity theory.

The proper modern approach to the study of geometry is that Euclidean geometry is only one geometry among infinitely many. Further there is some evidence that our universe is *non-Euclidean*! In addition, geometry, through transformation groups, is intricately tied to group theory.

In the modern approach a *geometry* or *geometric space* will consist of a set whose elements will be called *points*. There will be other notions defined on this set – *lines, incidence, congruence, betweenness, distance, planes and dimension*.

Note that not every one of these notions will be present in every geometry.

After the basic concepts there will be *axioms* describing the properties and relationships of these ideas.

The theory of this *geometry* will consist of the theorems proved about the basic concepts.

1.1.1 Geometric Notions

The basic elements of a geometry or geometric space are called *points*. The set of *lines* of a geometry will be a distinguished class of subsets of points. A point will be *incident* with a line if it is an element of that line.

A *geometric figure* is a subset of the set of points of the geometry. The study of the geometry will be about *congruence* of geometric figures where *congruence* is a specific equivalence relation on the class of geometric figures.

A *metric geometry* is one where there is a function which allows the measure of distance. *Distance* or *metric* is a real-valued function d defined on pairs of points of the geometry satisfying the ordinary distance properties:
(i) $d(x, y) \geq 0$ and $d(x, y) = 0$ if and only if $x = y$;
(ii) $d(x, y) = d(y, x)$;
(iii) $d(x, y) \leq d(x, z) + d(z, y)$.

If a geometry has a metric on it, it is a *metric geometry*. Otherwise it is a *non-metric geometry*.

The class of *planes* of a geometry is another class of subsets of points. If there is only one plane comprising the whole geometry, it is a *planar geometry*. The whole geometry is called *space*. Important in this regard is the concept of *dimension*. *Dimension* is a positive integer-valued function on certain subsets of the geometry describing a "size". If there is a dimension function, usually a line will be 1-dimensional and a plane 2-dimensional.

1.2 Overview of Euclid's Method and Approaches to Geometry

Euclid's Elements introduced the axiomatic method to the study of geometry and mathematics in general. Besides this there are several different approaches to the study of geometry.

The first is the *axiomatic approach* which was the method applied by Euclid. In the *axiomatic approach* the undefined notions are given, the axioms chosen and theorems proven in the spirit of Euclid. In general, an *axiom system* and the *axiomatic method* consists of:

(i) A set of undefined terms; all other terms are to be defined from these.

(ii) A set of axioms.

(iii) A set of theorems which are logical consequences of the axioms. The undefined terms and the axioms are subject to the interpretation of the reader.

Important to the axiomatic approach is the concept of *models*. An *interpretation* of an axiom system is giving a particular meaning to the undefined terms. If all axioms are true in an interpretation then it's a *model*. If there exists a bijection between the sets of interpretations of undefined terms which preserves each relationship between undefined terms then we get *isomorphic models*. In *isomorphic models* there is a bijection between them which preserves geometric structure, that is, the bijection takes congruent figures to congruent figures.

Models serve as testing grounds for potential theorems. If it is true in the geometry, it must be true in every model. Hence if we have an assertion about a geometry and we find a model where this assertion is false then this assertion cannot be a theorem in the geometry.

A set of axioms is *consistent* if there are no contradictory theorems. The existence of a *concrete model* establishes *absolute consistency*. The existence of an *abstract model* (an interpretation in another abstract system) establishes *relative consistency*.

An axiom is *independent* if it cannot be logically deduced from the other axioms in the system. An axiom system is *independent* if it consists of independent axioms. To determine the independence of an axiom, produce a model in which one axiom is incorrect and the others are correct.

An axiom system is *complete* if it is impossible to add an additional consistent and independent axiom without adding additional undefined terms.

Finally, a system is *categorical* if all models are isomorphic. Categorical implies completeness.

The second approach to geometry was introduced by F. Klein (1849–1925) in his *Erlanger Programm* in 1885. This is called the *transformation group approach*. In this method a geometry is defined on a set by those properties of the set which are *invariant* or *unchanged* under the action of some group of transformations of the set. This group of transformations is called the *group of congruence motions* of the geometry and in this approach knowing the geometry is equivalent to knowing the congruence group.

For example, in this approach planar Euclidean geometry would consist of those properties of the Euclidean plane, defined as real two-dimensional space equipped with the ordinary metric, that are invariant under a group called the *group of Euclidean motions*. This group consists of all the *isometries* or distance preserving mappings of the plane to itself. In this geometry two figures would be *congruent* if one can be mapped to the other by an isometry or *congruence motion*.

A final approach to geometry is by imposing a coordinate system. This is called *analytic geometry* and uses linear algebra.

In the *analytic geometry* approach, a coordinate system is placed on the geometry. It is then considered as part of a vector space over some field. Geometric notions are then interpreted in terms of the linear algebraic properties of the coordinates.

The analytic approach leads to the same theorems as the other approaches but often makes the proofs much simpler.

If there is no coordinate system, a geometry is known as a *synthetic geometry*. The following is a very elementary theorem of Euclidean geometry. We give two proofs – one synthetic and one analytical to show the difference.

Theorem 1.1. *If A and B are two points in the Euclidean plane and ℓ is the perpendicular bisector of the line segment $\overline{AB}$ then any point on ℓ is equidistant from A and B.*

In school geometry this theorem is often stated as: The perpendicular bisector of $\overline{AB}$ is the locus of points equidistant from A and B.

We give a synthetic proof as well as an analytic one.

Proof. (synthetic) Let ℓ be the perpendicular bisector of $\overline{AB}$. Recall that this means ℓ goes through the midpoint of $\overline{AB}$ and is perpendicular to line segment $\overline{AB}$. We then have the diagram with P any point on ℓ and C the midpoint of $\overline{AB}$, see Figure 1.2.

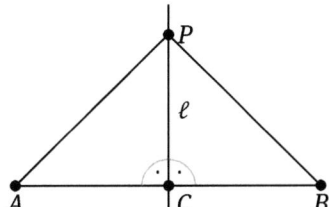

Figure 1.2: Any point P is equidistant from A and B.

Then $\|\overline{AC}\| = \|\overline{BC}\|$ by the definition of the midpoint. We denote the length of the line segment $\overline{AB}$ by $\|\overline{AB}\|$. We later also write $\|\overline{AB}\| = \|\overrightarrow{AB}\|$ if we consider the line segment $\overline{AB}$ as a vector $\overrightarrow{AB}$. Further, $\sphericalangle(\overline{CA}, \overline{CP}) = \sphericalangle(\overline{CB}, \overline{CP})$ for the angles between the respective line segments since all right angles are congruent. Finally, $\|\overline{CP}\| = \|\overline{CP}\|$. Hence the triangles ACP and BCP are congruent by the side–angle–side criterion or SAS criterion. Then $\|\overline{AP}\| = \|\overline{BP}\|$ using corresponding parts of congruent triangle being equal. □

Proof. (analytic) We impose the standard coordinate system on the Euclidean plane. Without loss of generality, we can place A at $(-1, 0)$ and B at $(1, 0)$. The perpendicular bisector ℓ of $\overline{AB}$ is then the y-axis and hence a general point P on ℓ has coordinates $(0, y)$. Then

$$\|\overline{AP}\| = \sqrt{(-1)^2 + y^2} = \sqrt{1 + y^2}$$

and

$$\|\overline{BP}\| = \sqrt{(1)^2 + y^2} = \sqrt{1 + y^2}.$$

Therefore $\|\overline{AP}\| = \|\overline{BP}\|$.   □

1.2.1 Incidence Geometries – Affine Geometries, Finite Geometries, Projective Geometries

Perhaps the simplest type of geometry is an *incidence geometry*. An *incidence geometry* is a geometry satisfying the following three incidence axioms:

Incidence Axiom 1. For every point P and point $Q \neq P$ there exists a unique line ℓ through P and Q.

Incidence Axiom 2. For every line ℓ there are at least two distinct points incident with ℓ.

Incidence Axiom 3. There exist three distinct points such that no line is incident with all three of them.

On a basic incidence geometry we impose some further conditions. An *affine geometry* is an incidence geometry which satisfies the Euclidean parallel postulate, that is, given a point P and a line ℓ not containing P, there exists a unique line through P parallel to ℓ. By parallel we mean that the lines ℓ and k are coplanar and $\ell \cap k = \emptyset$.

The parallel postulate is tied to solving linear equations. Hence given an n-dimensional vector space over a field A, we can always build an affine geometry on it defining lines in terms of linear equations. Hence A^n without any metric is called *affine n-space*.

The *elliptic parallel postulate* for an incidence geometry assumes that there are no parallel lines. A *projective geometry* is an incidence geometry which satisfies the elliptic parallel postulate and assumes that every line has at least three points on it.

Given an affine $(n + 1)$-space A^{n+1}, we can build a *projective n-space P^n* by considering points as 1-dimensional subspaces of it. We pursue this further in the exercises.

Classical projective geometry grew out of the perspective problem in art and is constructed by adding *ideal points* to Euclidean geometry.

A *finite incidence geometry* is an incidence geometry with finitely many points. In an affine geometry we have the following theorem.

Theorem 1.2. *In any affine geometry a line parallel to one of two intersecting lines must intersect the other. This implies that in an affine geometry parallelism is transitive.*

Being a finite affine geometry restricts the number of points that can be in it.

Theorem 1.3. *In a finite affine geometry the following is true:*
(a) *All lines contain the same number of points.*
(b) *If each line has n points then each point is on exactly n + 1 lines.*
(c) *If each line has n points then there exist exactly n^2 points and $n(n + 1)$ lines.*

Proof. Let ℓ be a line in a finite affine geometry. Suppose that $\ell = \{p_1, p_2, \ldots, p_m\}$ and $q \notin \ell$. Let ℓ_1 be the unique line through q which is parallel to ℓ. We claim that q is on exactly $m + 1$ lines. Each point $p_i \in \ell$ defines a unique line qp_i. We claim that the set of lines $\{\ell_1, qp_1, \ldots, qp_m\}$ is the complete set of lines containing q. Let ℓ_2 be any other line containing q. If ℓ_2 is parallel to ℓ then $\ell = \ell_2$ by the Euclidean axiom. If ℓ_2 is not parallel to ℓ then they intersect at some point p_i. But then $p_i, q \in \ell_2$ and hence $\ell_2 = qp_i$. The result then follows. If $\ell_1 = \{q_1, q_2, \ldots, q_t\}$ then q is on exactly $t + 1$ lines as above. Thus $m + 1 = t + 1$ and $m = t$. Therefore each line has the same number of points, and if this number is n, we have also proved that each point is on $n + 1$ lines. This proves (b).

For (c) there are n points and $n + 1$ lines, hence there are at least $n(n + 1)$ points. But here we have counted each point n times. It follows that the total number of points is $n(n + 1) - n = n^2$. □

Similar results follow for finite *projective geometries* where a projective geometry is an incidence geometry with no parallel lines.

Theorem 1.4. *In a finite projective geometry the following is true:*
(1) *All lines contain the same number of points.*
(2) *If each line has n points then each point is on exactly n lines.*
(3) *If each line has n points then there exist exactly $n^2 - n + 1$ points and $n^2 - n + 1$ lines.*

1.3 Euclidean Geometry

As we have remarked, mathematicians in the nineteenth century realized that there were many shortcomings in Euclid's treatment. In the latter part of the century D. Hilbert (1862–1943) gave a modern treatment in the spirit of Euclid and provided a complete set of axioms for Euclidean geometry. Undefined terms are *point, line, plane, incidence, between,* and *congruence.* Hilbert presented five groups of axioms: *axioms of incidence, axioms of order, axioms of congruence, axioms of parallels,* and *axioms of continuity.*

Axioms of Incidence. These are the three basic incidence axioms to make Euclidean geometry an incidence geometry.

Axioms of Order. Axioms to describe betweenness. We write $A - B - C$ to indicate that A, B, C are pairwise distinct collinear points and that B is between A and C.

(1) **Order Axiom 1.** If $A - B - C$ then $C - B - A$.

(2) **Order Axiom 2.** For each $A \neq C$ there exists a B such that $A - C - B$; this is the line extension axiom.

(3) **Order Axiom 3.** If A, B, C are collinear then not more than one is between the other two.

(4) **Order Axiom 4** (Pasch's Axiom after M. Pasch (1843–1930)). If A, B, C are non-collinear and ℓ is a line not incident with any of them then if ℓ passes through $\overline{AB}$ it will pass through $\overline{AC}$ or $\overline{BC}$. This indicates that if a line enters a triangle it must leave it.

(5) **Order Axiom 5.** Given $A \neq C$ there exists a B such that $A - B - C$; that is, lines are infinite.

Axioms of Congruence. Axioms to describe the congruence relation $\equiv$.

(1) **Congruence Axiom 1.** Given $\overline{AB}$ with $A \neq B$ and a point A', there exists a B' such that $\overline{AB} \equiv \overline{A'B'}$.

(2) **Congruence Axiom 2.** Congruence is an equivalence relation on line segments and angles.

(3) **Congruence Axiom 3.** If $\overline{AB} \cap \overline{BC} = B$ and $\overline{A'B'} \cap \overline{B'C'} = B'$ then if $\overline{AB} \equiv \overline{A'B'}, \overline{BC} \equiv \overline{B'C'}$ then $\overline{AC} \equiv \overline{A'C'}$.

(4) **Congruence Axiom 4.** In a plane, given an angle $\sphericalangle(\overline{BA}, \overline{BC})$ and a line $B'C'$, there exists exactly one line segment $\overline{B'A'}$ on each side of $\overline{B'C'}$ such that $\sphericalangle(\overline{B'A'}, \overline{B'C'}) = \sphericalangle(\overline{BA}, \overline{BC})$.

(5) **Congruence Axiom 5.** Side–angle–side, or SAS, criterion for the congruence of triangles.

Axioms of Continuity. Axioms to describe the completeness properties of lines, related to the completeness of the real numbers.

(1) **Continuity Axiom 1** (Archimedean Axiom). If $\overline{AB}$ and $\overline{CD}$ are line segments then there exists an integer n such that n copies of $\overline{CD}$ constructed contiguously from A along line AB will pass point B.

(2) **Continuity Axiom 2** (Line Completeness Axiom). Any extension of a set of points on a line with its order and congruence relation that would preserve all the axioms is impossible.

Axiom of Parallels. Axiom to describe the Euclidean parallel property – In a plane, given a line ℓ and a point $P \notin \ell$, there exists exactly one line ℓ' through P parallel to ℓ, that is, $\ell \cap \ell' = \emptyset$.

Remark 1.5. This is, in the context of Euclidean geometry, with just the other four axiom sets, equivalent to *Playfair's axiom*, named after F. Playfair (1748–1819). In a plane, given a line ℓ and a point $P \notin \ell$, there exists at most one line ℓ' through P parallel to ℓ.

1.3.1 Birkhoff's Axioms for Euclidean Geometry

In the 1930s G. Birkhoff (1911–1996) devised a shorter complete set of axioms for Euclidean geometry. He based his axioms, however, on algebra and on measurement. Birkhoff shifted the axioms from geometric axioms to algebraic axioms and results on the real number system.

The undefined terms in the Birkhoff system are *point*, *line*, *distance* and *angle*. He then has the following axiom system in addition to the completeness of the real number system. Again we write AB for a line and $\overline{AB}$ for a line segment.

(1) **Axiom 1** (Line measure–ruler axiom). The points on any line can be placed in one-to-one correspondence with the real numbers such that $d(A, B) = |B - A|$.

(2) **Axiom 2.** There exists a unique line on any two distinct points.

(3) **Axiom 3** (Protractor axiom). We call a straight line extending infinitely in a single direction from a point, a *half-line*. In a plane, a set of half-lines $\{\ell, m, n, \ldots\}$ through any point O can be put into one-to-one correspondence with the real numbers a modulo 2π so that if A and B are points, not equal to O, of ℓ and m respectively, the difference $a_m - a_\ell$ mod 2π of the numbers associated with the half-lines is $\sphericalangle(\overline{OA}, \overline{OB})$.

(4) **Axiom 4** (Axiom of Similarity). Given two triangles ABC and $A'B'C'$ and some constant $k > 0$ such that $d(A', B') = kd(A, B), d(A', C') = kd(A, C)$ and $\sphericalangle(\overline{A'B'}, \overline{A'C'}) = \pm\sphericalangle(\overline{AB}, \overline{AC})$, we have $d(B', C') = kd(B, C), \sphericalangle(\overline{B'C'}, \overline{B'A'}) = \pm\sphericalangle(\overline{BC}, \overline{BA})$ and $\sphericalangle(\overline{C'A'}, \overline{C'B'}) = \pm\sphericalangle(\overline{CA}, \overline{CB})$.

1.4 Neutral or Absolute Geometry

From Euclid's time until the discovery of non-Euclidean geometry there were many attempts to prove that the parallel axiom is not independent of the other axioms. One of the strongest such attempts was by G. G. Saccheri (1667–1733) in the 1700s. He was a Jesuit priest and began looking at what is now called neutral geometry and proved a collection of startling theorems. Although he said that many of these had to be false and therefore disproved Euclidean geometry, what he actually did was lay out the foundation of non-Euclidean and hyperbolic geometry.

Neutral geometry, which is also called *absolute geometry*, is the incidence geometry formed by assuming all the Euclidean axioms except the parallel postulate. A theorem that can be proved without recourse to the Euclidean parallel postulate or any of its equivalent formulations is a neutral geometric theorem, while a Euclidean result is one which requires the Euclidean parallel postulate. Note that any neutral theorem is also a Euclidean theorem.

We first show that in neutral geometry parallel lines do exist. We do this by proving the *alternate interior angles theorem* in one direction. We need certain preliminaries.

Theorem 1.6 (Exterior Angle Theorem). *The exterior angle of a triangle is greater than either of the nonadjacent interior angles.*

Note that each of our proofs is the same as the corresponding Euclidean proofs, however, we cannot use either the parallel postulate or any of its equivalent formulations.

Proof. Recall that we cannot use the Euclidean parallel postulate. However, it can be shown that assuming SAS, the angle–side–angle (ASA), side–angle–angle (SAA) and side–side–side (SSS) criteria are all valid congruence conditions. Let ABC be any triangle as in the diagram below. Let D be on the extension of side BC. We first show that the exterior angle $\sphericalangle(\overline{CA}, \overline{CB})$ is greater than angle $\sphericalangle(\overline{AB}, \overline{AC})$. Consider Figure 1.3.

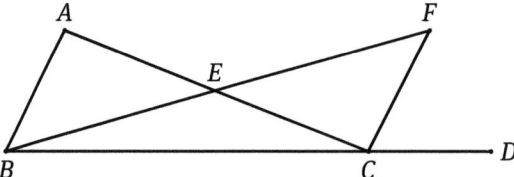

Figure 1.3: Diagram for the exterior angle theorem.

Let E be the midpoint of $\overline{AC}$ and let $\overline{BE}$ be extended its own length through E to F. Then $\overline{AE} = \overline{EC}, \overline{BE} = \overline{EF}$ and $\sphericalangle(\overline{EA}, \overline{EB}) = \sphericalangle(\overline{EC}, \overline{EF})$ since vertical angles are equal. Therefore $\triangle AEB \equiv \triangle CEF$ by SAS, and hence $\sphericalangle(\overline{AB}, \overline{AE}) = \sphericalangle(\overline{FC}, \overline{FE})$ by the corresponding parts of congruent triangles. Since $\sphericalangle(\overline{CA}, \overline{CD}) > \sphericalangle(\overline{CF}, \overline{CE})$, the whole being greater than one of its parts, it follows that $\sphericalangle(\overline{CA}, \overline{CD}) > \sphericalangle(\overline{AB}, \overline{AE})$.

We do an analogous construction to show that

$$\sphericalangle(\overline{CA}, \overline{CD}) > \sphericalangle(\overline{BA}, \overline{BC}). \qquad \square$$

We now prove in neutral geometry one direction of the alternate interior angle theorem. From this we can derive the existence of parallels.

Theorem 1.7 (The Alternate Interior Angle Theorem). *If the alternate interior angles formed by a transversal cutting two lines are congruent then the lines are parallel.*

We note that in school geometry this theorem is usually stated as an "if and only if". However, the other direction is equivalent to the Euclidean parallel postulate.

Proof. Recall that two lines in a plane are parallel if they do not meet. Let the line AB cross lines ℓ and m so that the alternate interior angles are equal. Suppose that ℓ and m meet at a point C as in Figure 1.4.

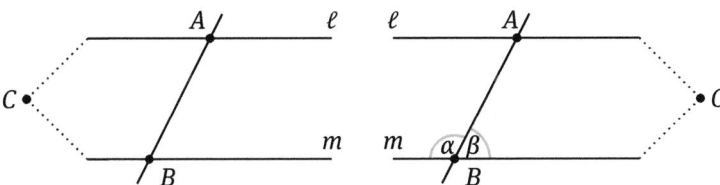

Figure 1.4: Alternate interior angles.

Then angle α is an exterior angle of the corresponding triangle ABC and therefore by the exterior angle theorem, angle α is greater than angle β, contradicting that they are equal. Therefore ℓ and m must be parallel. $\qquad\square$

Since all right angles are congruent the following corollary is immediate.

Corollary 1.8. *Two lines perpendicular to the same line are parallel.*

From this we obtain the existence of parallels.

Corollary 1.9. *Given a line ℓ and a point $P \notin \ell$, there exists a line ℓ_1 through P and parallel to ℓ. Therefore parallels exist in neutral geometry.*

Proof. From point P drop a perpendicular to ℓ with foot Q and at P erect a line m perpendicular to PQ. From the previous corollary m is parallel to ℓ, see Figure 1.5. $\qquad\square$

As we will see, the angle sum of a triangle will be crucial to distinguish Euclidean from non-Euclidean geometry. We prove one other neutral geometric theorem along these lines.

Theorem 1.10. *In neutral geometry the angle sum of a triangle must be less than or equal to 180°.*

Proof. From the exterior angle theorem (Theorem 1.6), the sum of any two angles of a triangle is less than 180°. Assume there exists a triangle ABC with angle sum $180° + p°$,

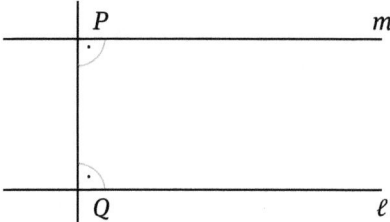

Figure 1.5: Parallels in neutral geometry.

$p > 0$. From exercise 13 we can find a triangle $A_1 B_1 C_1$ with the same angle sum as ABC but

$$\sphericalangle(\overline{A_1 B_1}, \overline{A_1 C_1}) \leq \frac{1}{2} \sphericalangle(\overline{AB}, \overline{AC}).$$

Now build a sequence of triangles $A_n B_n C_n$ with the same angle sum as ABC but with

$$\sphericalangle(\overline{A_n B_n}, \overline{A_n C_n}) \leq \frac{1}{2^n} \sphericalangle(\overline{AB}, \overline{AC}).$$

So,

$$\lim_{n \to \infty} \sphericalangle(\overline{A_n B_n}, \overline{A_n C_n}) = 0$$

which means that

$$\sphericalangle(\overline{B_n A_n}, \overline{B_n C_n}) + \sphericalangle(\overline{C_n A_n}, \overline{C_n B_n}) \geq 180°$$

for large n which contradicts the above. □

1.5 Euclidean and Hyperbolic Geometry

As we have seen in the previous section, given all the axioms of Euclidean geometry, there are only two possibilities for parallels. In a plane, given a point P and a line ℓ with $P \notin \ell$, either there is a unique parallel to ℓ through P or more than one parallel through P.

Theorem 1.11. *In neutral geometry, in a plane, if there exists a point P and a line ℓ so that there exists a unique parallel to ℓ through P then this property holds for any point P and any line ℓ with $P \notin \ell$. Further, in a plane, if there exists a point P and a line ℓ so that there are more than one parallel to ℓ through P then this property holds for any point P and any line ℓ with $P \notin \ell$.*

If two lines ℓ and m have a line that is perpendicular to both lines then the lines are parallel. Suppose that given a point P and a line ℓ there is a unique parallel. Then if a line is perpendicular to one of the two lines it must be perpendicular to the other by the uniqueness of parallels. It follows that if there is more than one parallel there must be a line which is parallel to one line with an angle less than 90° to the other line. Therefore there must be a minimal such angle. This is called the *angle of parallelism*. From this we can deduce the following. The proof we outline in the exercises.

Theorem 1.12. *In a neutral geometry, in a plane, given a point P and a line ℓ with P ∉ ℓ, if there exist more than one parallel to ℓ through P then there exist infinitely many parallels.*

From these results we see that in a neutral geometry, given a point P and a line $ℓ$ with $P ∉ ℓ$, there are exactly two possibilities: there is a unique parallel line to $ℓ$ through P or there are two or more parallels to $ℓ$ through P. Furthermore, this property is uniform throughout neutral geometry.

We now define two parallel postulates for a neutral geometry.

The Euclidean Parallel Postulate. Given a neutral geometry, the Euclidean parallel postulate, that we will abbreviate EPP, is that in a plane if P is a point and $ℓ$ a line with $P ∉ ℓ$ then there exists a unique parallel to $ℓ$ through P.

The Hyperbolic Parallel Postulate. Given a neutral geometry, the hyperbolic parallel postulate, that we will abbreviate HPP, is that in a plane if P is a point and $ℓ$ a line with $P ∉ ℓ$ then there exists at least 2 parallels to $ℓ$ through P.

A *non-Euclidean geometry* is any geometry which does not satisfy the Euclidean parallel postulate. A *hyperbolic geometry* is a neutral geometry together with the *hyperbolic parallel postulate*.

We thus have the following theorem.

Theorem 1.13. *A neutral geometry is either Euclidean geometry or hyperbolic geometry.*

Results that are true in Euclidean geometry but not in hyperbolic geometry are called *purely Euclidean results*, while results that are true in hyperbolic geometry but not in Euclidean geometry are called *purely hyperbolic results*.

Many of these types of results are tied to the angle sum of a triangle. The following theorem is crucial.

Theorem 1.14. *In Euclidean geometry the angle sum of any triangle is 180°, while in hyperbolic geometry the angle sum of any triangle is strictly less than 180°.*

1.5.1 Consistency of Hyperbolic Geometry

We have seen that given a neutral geometry it is either Euclidean or hyperbolic. How do we know that hyperbolic geometry is actually consistent, that is, perhaps there exist theorems that are both true and false in hyperbolic geometry.

This was handled by C. F. Gauss (1777–1855), N. I. Lobachevsky (1792–1856) and J. Bolyai (1802–1860), who are considered the discoverers of non-Euclidean geometry, by building a model of hyperbolic geometry within Euclidean geometry. Hence the consistency of Euclidean geometry implies the consistency of hyperbolic geometry. The particular planar models they constructed we will study in Chapter 8. Somewhat later E. Beltrami (1835–1900) and F. C. Klein (1849–1925) constructed a model of Euclidean geometry within hyperbolic geometry and thus the consistency of hyperbolic geometry implies the consistency of Euclidean geometry. We say that Euclidean geometry and hyperbolic geometry are *co-consistent*.

1.6 Elliptic Geometry

After the discovery of hyperbolic geometry B. Riemann (1826–1866) proposed a geometry which differed from both of these.

The Elliptic Parallel Postulate, or *Riemann Parallel Postulate*, which we abbreviate RPP, is that in the planar geometry there are no parallel lines.

There were historical reasons for studying this. When geometers studied projections of one plane onto another, they built a geometry that had no parallel lines. Every pair of lines either met in the plane or at a point at infinity. This was called *projective geometry*. Further, if we consider the surface of a sphere as a plane and lines as being great circles on a sphere then there are no parallel lines. This is called *spherical geometry*.

An *elliptic geometry* is a geometry which satisfies the *elliptic parallel postulate*.

We note that elliptic geometry cannot be a neutral geometry. Something must be lost from the neutral axioms. In *spherical geometry*, the geometry on a sphere, lines become great circles, but we either lose the betweenness axioms if we maintain axioms for an incidence geometry or we lose incidence geometry by allowing lines to intersect at two points – the antipodal points of great circles.

We will examine spherical geometry closely in Chapter 6. As we will see, whereas in hyperbolic geometry the angle sum of any triangle is less than 180°, the angle sum in any elliptic geometry is greater than 180°.

1.7 Differential Geometry

We close this chapter by mentioning briefly *differential geometry*. This is the study of geometry by using differential and integral calculus methods. Classically it was the study of curves and surfaces in Euclidean space. Differential geometry is a vast and important subject in modern mathematics.

We here consider curves in the real space $\mathbb{R}^3$, equipped with the canonical scalar product

$$\langle \vec{x}, \vec{y} \rangle = x_1 y_1 + x_2 y_2 + x_3 y_3$$

where

$$\vec{x} = \begin{pmatrix} x_1 \\ x_2 \\ x_3 \end{pmatrix} \quad \text{and} \quad \vec{y} = \begin{pmatrix} y_1 \\ y_2 \\ y_3 \end{pmatrix}.$$

More on the scalar product in general can be found in [12]. Curves are of course geometric objects. The area of differential geometry has wide-ranging applications in both other areas of mathematics and physics. The geometry of curves has been extended beyond real spaces to what are called manifolds. Roughly speaking a real n-manifold is a topological space that locally looks like a real n-space. We will not look at general manifolds but we will examine the basic differential geometry of curves in $\mathbb{R}^3$ to give a little insight into the richness of differential geometry and the way of working in this field, especially in that of curves in $\mathbb{R}^3$.

Curves in $\mathbb{R}^3$ and the Serret–Frenet apparatus

First, we have to introduce the *cross-product* of two vectors $\vec{u}$ and $\vec{w}$ from $\mathbb{R}^3$. The cross product is a bilinear map which assigns a new vector to $\vec{u}$ and $\vec{w}$,

$$\mathbb{R}^3 \times \mathbb{R}^3 \to \mathbb{R}^3, \quad (\vec{u}, \vec{w}) \mapsto \vec{u} \times \vec{w}.$$

This map is uniquely characterized by the following properties:
1. If $\varphi \in [0, \pi]$ is the angle between $\vec{u}$ and $\vec{w}$ then $\|\vec{u} \times \vec{w}\| = \|\vec{u}\| \|\vec{w}\| \sin(\varphi)$.
2. If $\|\vec{u} \times \vec{w}\| \neq 0$ then $\vec{u} \times \vec{w} \perp \vec{w}$ and further $\det(\vec{u}, \vec{w}, \vec{u} \times \vec{w}) > 0$.

We remark that we discuss determinants later in Chapter 2.

But anyway, for convenience we give here directly $\det(\vec{u}, \vec{w}, \vec{u} \times \vec{w})$ via the Sarrus principle. Let $\vec{u} = \begin{pmatrix} u_1 \\ u_2 \\ u_3 \end{pmatrix}$, $\vec{w} = \begin{pmatrix} w_1 \\ w_2 \\ w_3 \end{pmatrix}$ and $\vec{u} \times \vec{w} = \begin{pmatrix} v_1 \\ v_2 \\ v_3 \end{pmatrix}$. Then $\det(\vec{u}, \vec{w}, \vec{u} \times \vec{w}) = u_1 w_2 v_3 + u_2 w_3 v_1 + u_3 w_1 v_2 - u_1 w_3 v_2 - u_2 w_1 v_3 - u_3 w_2 v_1$.

Remark 1.15. Property 1. fixes the length of the vector $\vec{u} \times \vec{w}$ and property 2. its direction.

We get

$$\vec{u} \times \vec{w} = \|\vec{u} \times \vec{w}\| \sin(\varphi) \vec{m}$$

where $\vec{m}$ is a unit vector perpendicular to the plane containing $\vec{u}$ and $\vec{w}$ in the direction given by the right hand rule:

If $\vec{u} = \begin{pmatrix} u_1 \\ u_2 \\ u_3 \end{pmatrix}$ and $\vec{w} = \begin{pmatrix} w_1 \\ w_2 \\ w_3 \end{pmatrix}$ then $\vec{u} \times \vec{w} = \begin{pmatrix} u_2 w_3 - u_3 w_2 \\ u_3 w_1 - u_1 w_3 \\ u_1 w_2 - u_2 w_1 \end{pmatrix}$.

As already mentioned, curves in $\mathbb{R}^3$ are geometric objects. Traditionally, to study these objects, methods from calculus are employed. We now start with the concept of a curve in $\mathbb{R}^3$. Certainly, this includes the case of a curve in $\mathbb{R}^2$.

Definition 1.16.

1. A *parameterized curve* is a continuous map $c : [a, b] \to \mathbb{R}^3$ where $[a, b]$ is an interval in $\mathbb{R}$. The parameter $t \in [a, b]$ is called *time*. The *position vector* at the time t is $\overrightarrow{c(t)} = \left(\begin{smallmatrix} c_1(t) \\ c_2(t) \\ c_3(t) \end{smallmatrix} \right)$. We call the image of c the *path* of c.
2. A parameterized curve $c : [a, b] \to \mathbb{R}^3$ is differentiable if c is differentiable on the open interval (a, b).

Recall, if $\overrightarrow{c(t)} = \left(\begin{smallmatrix} c_1(t) \\ c_2(t) \\ c_3(t) \end{smallmatrix} \right)$ then $\overrightarrow{c'(t)} = \left(\begin{smallmatrix} c_1'(t) \\ c_2'(t) \\ c_3'(t) \end{smallmatrix} \right)$ stands for the derivative.

Definition 1.17.

1. A curve $c : [a, b] \to \mathbb{R}^3$ is n times continuously differentiable on (a, b) if it has continuous derivatives on (a, b) up to the order n.
2. A curve $c : [a, b] \to \mathbb{R}^3$ is *regular* if it is at least two times continuously differentiable on (a, b) with nonvanishing derivatives $\overrightarrow{c'(t)}$ and $\overrightarrow{c''(t)}$ for each time $t \in (a, b)$.

Agreement:

In what follows we only consider regular curves with the described properties. In the literature the definition is often a bit different than the one given here. One has to consider especially times with $\overrightarrow{c''(t)} = \overrightarrow{0}$. We want to avoid parts of the curve that are straight lines.

Definition 1.18. Let $c : [a, b] \to \mathbb{R}^3$ be a curve.

1. The *velocity vector* $\overrightarrow{v_c(t)}$ at t is the derivative $\overrightarrow{c'(t)}$ of the position vector at t, that is, $\overrightarrow{v_c(t)} = \overrightarrow{c'(t)}$. We call v_c the *velocity* of c.
2. The *speed* at time t is the length $\|\overrightarrow{v_c(t)}\|$ of the velocity vector $\overrightarrow{v_c(t)}$, and is a scalar quantity.
3. The *accelerative vector* $\overrightarrow{a_c(t)}$ at time t is the derivative of the velocity vector $\overrightarrow{v_c(t)}$ at time t, that is,

$$\overrightarrow{a_c(t)} = \overrightarrow{v_c'(t)} = \overrightarrow{c''(t)}.$$

We call a_c the acceleration of c.

Remark 1.19. It is useful to think of t as a time parameter and picture a curve as a particle moving along the path. Given a point P then t is the time it gets to point P.

Examples 1.20.
1. Let $c : [0, 2\pi] \to \mathbb{R}^3$,

$$\vec{c(t)} = \begin{pmatrix} r\cos(t) \\ r\sin(t) \\ 0 \end{pmatrix}, \quad r \in \mathbb{R}, r > 0.$$

Then

$$\vec{c'(t)} = \begin{pmatrix} -r\sin(t) \\ r\cos(t) \\ 0 \end{pmatrix} \quad \text{and} \quad \|\vec{c'(t)}\| = r \quad \text{(constant)}.$$

2. Let $c : [0, 2\pi] \to \mathbb{R}^3$,

$$\vec{c(t)} = \begin{pmatrix} r\cos(t) \\ r\sin(t) \\ ht \end{pmatrix}, \quad r, h \text{ constant}, r, h > 0.$$

Then

$$\vec{c'(t)} = \begin{pmatrix} -r\sin(t) \\ r\cos(t) \\ h \end{pmatrix} \quad \text{and} \quad \|\vec{c'(t)}\| = \sqrt{r^2 + h^2} \quad \text{(constant)}.$$

Definition 1.21. Let $c : [a, b] \to \mathbb{R}^3$ be a regular curve. Then *length of c* is defined by

$$\text{length of } c = \int_a^b \|\vec{c'(t)}\| dt,$$

that is, to get the length we integrate speed with respect to time.

Example 1.22. Let $\vec{c(t)} = \begin{pmatrix} r\cos(t) \\ r\sin(t) \\ 0 \end{pmatrix}, 0 \le t \le 2\pi$. Then

$$\text{length of } c = \int_0^{2\pi} \|\vec{c'(t)}\| dt = \int_0^{2\pi} r\, dt = 2\pi r$$

Definition 1.23. Let $c : [a, b] \to \mathbb{R}^3$ be a regular curve. Then c is a *constant speed curve* if $\|\vec{c'(t)}\| = \text{const} =: k$ for all times. It is a *unit speed curve* if $\|\vec{c'(t)}\| = 1$ for all times.

Reparameterizations

Given a regular curve $c : [a, b] \to \mathbb{R}^3$.

Traversing the same path at a different speed (and perhaps in the opposite directions) amounts what is called reparameterization.

Definition 1.24. Let $c : [a, b] \to \mathbb{R}^3$ be a regular curve. Let $h : [d, e] \to [a, b]$ be a continuous map such that $h : (d, e) \to (a, b)$ is a diffeomorphism between (d, e) and (a, b).

Then $\tilde{c} = c \circ h : [d, e] \to \mathbb{R}^3$ is a regular curve, called a reparameterization of c. We have $\overrightarrow{\tilde{c}(u)} = \overrightarrow{c(h(u))}$ where $t = h(u)$.

Example 1.25. Let $c : [0, 2\pi] \to \mathbb{R}^3$, $\overrightarrow{c(t)} = \left(\begin{smallmatrix} \cos(t) \\ \sin(t) \\ 0 \end{smallmatrix} \right)$. Take the reparameterization function $h : [0, \pi] \to [0, 2\pi]$, $t = h(u) = 2u$. Then

$$\overrightarrow{\tilde{c}(u)} = \overrightarrow{c(t)}, \quad \overrightarrow{\tilde{c}(u)} = \begin{pmatrix} \cos(2u) \\ \sin(2u) \\ 0 \end{pmatrix}.$$

Note, $\tilde{c}$ describes the same circle, but traversed twice as fast, so speed of $c = \|\overrightarrow{c'(t)}\| = 1$ and speed of $\tilde{c} = \|\overrightarrow{\tilde{c}'(u)}\| = 2$.

Remarks 1.26.
1. The curves c and $\tilde{c}$ describe the same path in $\mathbb{R}^3$, just traversed at different speed (and perhaps in opposite directions).
2. Compare velocities: Let $\overrightarrow{\tilde{c}(u)} = \overrightarrow{c(h(u))}$. By the chain rule, recall $t = h(u)$, we get

$$\frac{d\overrightarrow{\tilde{c}(u)}}{du} = \frac{d\overrightarrow{c(t)}}{dt} \cdot \frac{dt}{du} = \frac{d\overrightarrow{c(t)}}{du} h'.$$

If $h' > 0$ then we have an orientation preserving reparametrization, and if $h' < 0$ an orientation reversing one.

Lemma 1.27. *The length formula is independent of parameterization, that is, if $\tilde{c} : [d, e] \to \mathbb{R}^3$ is a reparameterization of $c : [a, b] \to \mathbb{R}^3$ then*

$$length\ of\ \tilde{c} = length\ of\ c.$$

Lemma 1.28. *Regular curves always admit a very important reparameterization. They can always be parameterized in term of arc length which we define now.*

Along a regular curve $c : [a, b] \to \mathbb{R}^3$ there is a distinguished parameter called arc length parameter or the natural parameter. Fix $t_0 \in [a, b]$. Define the following function (arc length function)

$$s = s(t), \quad t \in [a, b], \quad s(t) = \int_{t_0}^{t} \|\overrightarrow{c'(\tau)}\|\, d\tau.$$

Thus, if $t > t_0$, then $s(t) = length\ of\ c\ from\ t_0\ to\ t$, and if $t < t_0$, then $s(t) = -\ length\ of\ c\ from\ t_0\ to\ t$.

By the fundamental theorem of calculus $s = s(t)$ is strictly increasing and so can be solved for t in terms of s, $t = t(s)$ (reparameterization function). Then $\overline{c}(s) = \overrightarrow{c(t(s))}$ is the arc length reparameterization of c. Hence, we get the following.

Theorem 1.29. *A regular curve admits a reparameterization in terms of arc length.*

Example 1.30. Reparameterize the circle

$$\overrightarrow{c(t)} = \begin{pmatrix} r\cos(t) \\ r\sin(t) \\ 0 \end{pmatrix}, \quad t \in [0, 2\pi],$$

in terms of arc length parameter, that is, up to a trivial translation of parameters, $s = t$. Hence unit speed curves are already parameterized with respect to arc length (as measured from some point). Conversely, if c is a regular curve parameterized with respect to arc length s then c is unit speed, that is, $\|\overrightarrow{c'(s)}\| = 1$ for all s. Hence, the phrases "unit speed curve" and "curve parameterized with respect to arc length" are used interchangeably.

We may rewrite that in the following theorem.

Theorem 1.31. *Any regular curve $c : [a, b] \rightarrow \mathbb{R}^3$ has a unit speed parameterization. The parameterization is in terms of the arc length parameter.*

In the following we introduce the Serret–Frenet apparatus. It is named after J. Serret (1819–1885) and J. E. Frenet (1816–1902). Because of all the details of calculations we relax the notation and write c for the curve and often $\overrightarrow{c}$ for the position vector at a time t. The Serret–Frenet apparatus consists of three vector functions and two scalar functions which provide a complete characterization of a regular curve. Also for these and related vectors we mostly write just $\overrightarrow{x}$ instead of $\overrightarrow{x(t)}$ if there is the time parameter t for them.

We first introduce the Frenet vectors which provide a rectangular frame at every point on the curve that is called *moving trihedron*.

We will assume that we have a unit speed curve c parameterized by the natural parameter which is now called t.

The first vector is the unit *tangent vector* $\overrightarrow{t_c(t)}$ at the point $\overrightarrow{c(t)}$. This is the normalized velocity vector at time t, so that

$$\overrightarrow{t_c} = \frac{\overrightarrow{c'}}{\|\overrightarrow{c'}\|} = \frac{\overrightarrow{v_c}}{\|\overrightarrow{v_c}\|} \quad \text{(at time } t).$$

Since $\overrightarrow{t_c}$ is a unit vector, we have $\langle \overrightarrow{t_c}, \overrightarrow{t_c} \rangle = 1$ and hence the derivate of $\langle \overrightarrow{t_c}, \overrightarrow{t_c} \rangle$ is 0. Then we have by the product rule $\langle \overrightarrow{t_c}, \overrightarrow{t_c'} \rangle = 0$, and therefore $\overrightarrow{t_c'}$ is a perpendicular to $\overrightarrow{t_c}$. The vector $\overrightarrow{t_c'}$ is called the *curvature vector* (at time t).

The direction of $\overrightarrow{t'_c}$ tells us which way the curve is bending. Its magnitude $\|\overrightarrow{t'_c}\|$ tells us how much the curve is bending, $\|\overrightarrow{t'_c}\|$ is called the curvature (at time t). Recall that $\|\overrightarrow{t'_c}\| \neq 0$ at each time by agreement.

Definition 1.32. Let c be a unit speed curve. The *curvature* $\kappa = \kappa(t)$ of c is

$$\kappa = \kappa(t) = \|\overrightarrow{t'_c}\| = \|\overrightarrow{c''}\| \neq 0.$$

Conceptually, the definition of curvature is the right one.

Using the chain rule one can obtain a formula for computing curvature which does not require that the curve be parameterized with respect to arc length.

Let $c : [a, b] \rightarrow \mathbb{R}^3$ be a regular curve with time t. We reparameterize c with respect to the arc length s:

$$c : [d, e] \rightarrow \mathbb{R}^3, \quad s = s(t).$$

So by the chain rule

$$\frac{\overrightarrow{t_c(t)}}{dt} = \frac{\overrightarrow{dt_c(s)}}{ds} \cdot \frac{ds}{dt} = \frac{\overrightarrow{dt_c(s)}}{ds} \cdot \left\| \frac{\overrightarrow{dc(t)}}{dt} \right\|.$$

Hence,

$$\left\| \frac{\overrightarrow{dt_c(t)}}{dt} \right\| = \left\| \frac{\overrightarrow{dc(t)}}{dt} \right\| \cdot \underbrace{\left\| \frac{\overrightarrow{dt_c(s)}}{ds} \right\|}_{=\kappa}.$$

Lemma 1.33. *It holds*

$$\kappa = \frac{\left\| \frac{\overrightarrow{dt_c(t)}}{dt} \right\|}{\left\| \frac{\overrightarrow{dc(t)}}{dt} \right\|}.$$

We choose $\overrightarrow{n_c}$ to be a principal *unit normal vector* along t as

$$\overrightarrow{n_c} = \frac{\overrightarrow{t'_c}}{\|\overrightarrow{t'_c}\|} = \frac{\overrightarrow{t'_c}}{\kappa}.$$

To define the final Frenet vector we choose $\overrightarrow{b_c} = \overrightarrow{t_c} \times \overrightarrow{n_c}$. This is called the *binormal vector*. As already mentioned, at each time these three vectors are mutually perpendicular unit vectors which form a moving trihedron.

We now describe a set of relations between Frenet vectors and their derivatives, called the Serret–Frenet formulas.

Note, that the definition of $\overrightarrow{n_c}$ implies $\overrightarrow{t'_c} = \kappa \overrightarrow{n_c}$. The plane formed by $\overrightarrow{t_c}$ and $\overrightarrow{n_c}$ is called the *kissing plane* or *osculating plane* for c. We finally choose $\overrightarrow{b_c} = \overrightarrow{t_c} \times \overrightarrow{n_c}$ (at time t). This $\vec{b_c}$ is called the *binormal vector* (at t). As a unit vector the unit normal vector $\vec{n_c'}$ is orthogonal to $\vec{n_c}$, and so $\vec{n_c'}$ is in the $(\vec{t_c}, \vec{b_c})$-plane.

This implies $\vec{n_c'} = \alpha \vec{t_c} + \beta \vec{b_c}$ for some scalar functions α and β. Now the binormal vector is $\vec{b_c} = \vec{t_c} \times \vec{n_c}$.

Then $\|\vec{b_c}\| = 1$, so $\vec{b_c'}$ is orthogonal to $\vec{b_c}$. But

$$\vec{b_c'} = \vec{t_c'} \times \vec{n_c} + \vec{t_c} \times \vec{n_c'} = \vec{t_c} \times \vec{n_c'}$$

by the definition of $\vec{n_c}$ and $\vec{t_c'} \times \vec{t_c'} = \vec{0}$. Then

$$\vec{t_c} \times \vec{n_c'} = \vec{t_c} \times (\alpha \vec{t_c} + \beta \vec{b_c}) = -\tau \vec{n_c}$$

for a scalar function τ. This function $\tau = \tau(t)$ is called the *torsion*.

Finally, $\vec{n_c} = \vec{b_c} \times \vec{t_c}$ which gives

$$\vec{n_c'} = \vec{b_c} \times \vec{t_c'} + \vec{b_c'} \times \vec{t_c} = -\kappa \vec{t_c} + \tau \vec{b_c}.$$

Hence we have the following Theorem.

Theorem 1.34 (Serret–Frenet formulas for unit speed curves).

$$\vec{t_c'} = \kappa \vec{n_c},$$
$$\vec{n_c'} = -\kappa \vec{t_c} + \tau \vec{b_c},$$
$$\vec{b_c'} = -\tau \vec{n_c}.$$

Theoretically, any question about regular curves can be answered using the Serret–Frenet apparatus.

In the next subsection we continue to consider some important examples using the Serret–Frenet apparatus.

Remark 1.35. For a general speed curve we just get the Serret–Frenet formulas by multiplying by the speed (see Lemma 1.33).

Then we get the following corollary.

Corollary 1.36 (Serret–Frenet formulas for general speed curves).

$$\vec{t_c'} = \kappa \|\vec{c'}\| \vec{n_c}$$
$$\vec{n_c'} = -\kappa \|\vec{c'}\| \vec{t_c} + \tau \|\vec{c'}\| \vec{b_c}$$
$$\vec{b_c'} = -\tau \|\vec{c'}\| \vec{n_c}.$$

If we look at the Taylor series, named after B. Taylor (1685–1731), expansion for a unit speed curve we gain a better understanding of the curvature and the torsion. The first two terms give the best linear approximation and involve the unit tangent $\overrightarrow{t(0)}$. The first three terms give the best quadratic approximation.

This appears in the $(\overrightarrow{t_c}, \overrightarrow{n_c})$-plane and is controlled by the curvature. The torsion τ controls the twisting orthogonal to the kissing plane.

1.7.1 Some Special Curves

In this subsection we consider general speed curves. We use the Serret–Frenet apparatus to study special curves in $\mathbb{R}^3$. The first question we look at is to determine when a curve is planar, that is, lies completely in a plane. Since the torsion measures the twisting out of a plane the answer easily is that a curve will be planar if and only if its torsion is $\tau = 0$. Without loss of generality we may assume that the curve is of unit speed.

Theorem 1.37. *A unit speed curve c lies completely in a plane if and only if its torsion is $\tau = 0$.*

Proof. A plane E is determined by a point $P_0 \in E$ and a normal vector $\overrightarrow{v}$ orthogonal to E. Suppose that $c : [a, b] \to \mathbb{R}^3$ is a unit speed curve and planar.

Then there exists a fixed vector $\overrightarrow{q}$ such that $\langle (\overrightarrow{c(t)} - \overrightarrow{c(a)}), \overrightarrow{q} \rangle = 0$. This implies on taking derivatives that $\langle \overrightarrow{c'(t)}, \overrightarrow{q} \rangle = 0$, that is, $\langle \overrightarrow{t_c}, \overrightarrow{q} \rangle = 0$. Differentiating again gives $\langle \overrightarrow{c''(t)}, \overrightarrow{q} \rangle = 0$, that is, $\langle \overrightarrow{n_c}, \overrightarrow{q} \rangle = 0$.

Therefore, $\overrightarrow{q}$ is always orthogonal to $\overrightarrow{t_c}$ and $\overrightarrow{n_c}$, and hence in the direction of $\overrightarrow{b_c}$.

Since $\overrightarrow{q}$ is fixed this implies that $\overrightarrow{b_c}$ is constant, so $\overrightarrow{b_c'} = \overrightarrow{0}$. Since $\overrightarrow{b_c'} = -\tau \overrightarrow{n_c}$, it follows that $\tau = 0$.

Conversely, suppose $\tau = 0$. Then $\overrightarrow{b_c}$ is a constant. Let $g(t) = \langle (\overrightarrow{c(t)} - \overrightarrow{c(a)}), \overrightarrow{b_c(t)} \rangle$. Then $g'(t) = \langle \overrightarrow{t_c(t)}, \overrightarrow{b_c(t)} \rangle = 0$. Therefore $g(t)$ is a constant and since $g(a) = 0$ we have $g(t) = 0$ for all t and hence $\overrightarrow{c(t)}$ is planar with binormal vector $\overrightarrow{b_c}$ (in fact, it lies in the kissing plane). □

The next question we look at is when a curve is circular or part of a circle. Again we may assume that the curve is of unit speed.

Theorem 1.38. *A unit speed curve c is part of a circle of radius r if and only if $\tau = 0$ and $\kappa = \frac{1}{r}$.*

Proof. It is straightforward to see that $\tau = 0$ and $\kappa = constant \neq 0$. Now, suppose $\tau = 0$ and κ is constant. Then consider

$$\overrightarrow{g(t)} = \overrightarrow{c(t)} + \frac{1}{\kappa}\overrightarrow{n_c(t)}.$$

Then we get

$$\vec{g'} = \vec{c'} + \frac{1}{\kappa}\vec{n'}$$

$$= \vec{t_c} + \frac{1}{\kappa}\left(-\kappa\vec{t_c} + \tau\vec{b_c}\right)$$

$$= \vec{t_c} - \vec{t_c} = 0 \quad \text{at time } t.$$

Therefore $\overrightarrow{g(t)} = \vec{p}$ is a constant. Hence $\overrightarrow{c(t)} + \frac{1}{\kappa}\overrightarrow{n_c(t)} = \vec{p}$ and further $\overrightarrow{c(t)} - \vec{p} = -\frac{1}{\kappa}\overrightarrow{n_c(t)}$. Thus, $\|\overrightarrow{c(t)} - \vec{p}\| = \frac{1}{\kappa}$. Therefore c lies on a circle with center $\vec{p}$ and radius $\frac{1}{\kappa}$. □

A curve $c : [a, b] \to \mathbb{R}^3$ is a cylindrical helix if there exists a fixed vector $\vec{u}$ and a fixed angle α such that $\langle \vec{t_c}, \vec{u} \rangle = \cos(\alpha)$ at each point on the curve, that is, $\vec{u}$ makes the same angle with the tangent $\vec{t_c}$ at each point. We next completely characterize cylindrical helixes. Again we may assume that the curve is of unit speed.

Theorem 1.39. *A unit speed curve $c : [a, b] \to \mathbb{R}^3$ is a cylindrical helix if and only if the ratio $\frac{\kappa}{\tau}$ is constant.*

Proof. Certainly, we have $\tau \neq 0$ for a unit speed cylindrical helix.

Suppose c is a cylindrical helix. Then there exists an angle α and a fixed vector $\vec{u}$ with $\langle \vec{t_c}, \vec{u} \rangle = \cos(\alpha)$ for each time t. Hence, $\langle \kappa\vec{n_c}, \vec{u} \rangle = 0$ for each t, so $\vec{u}$ is a unit vector in the $(\vec{t_c}, \vec{b_c})$-plane. Therefore $\vec{u} = \cos(\theta)\vec{t_c} + \sin(\theta)\vec{b_c}$. Taking $\langle \vec{t_c}, \vec{u} \rangle = \cos(\alpha)$ implies $\theta = \alpha$ and $\vec{u} = \cos(\alpha)\overrightarrow{t_c(t)} + \sin(\alpha)\overrightarrow{b_c(t)}$.

Differentiating gives

$$0 = \cos(\alpha)\kappa\vec{n_c} - \sin(\alpha)\tau\vec{n_c}$$

$$= (\cos(\alpha)\kappa - \sin(\alpha)\tau)\vec{n_c}$$

and, hence,

$$\cos(\alpha)\kappa - \sin(\alpha)\tau = 0$$

and $\frac{\kappa}{\tau} = \cot(\alpha)$ is a constant.

Conversely, suppose $\frac{\kappa}{\tau}$ is a constant. Then there exists an angle with $\cot(\alpha) = \frac{\kappa}{\tau}$. Let

$$\vec{U} = \cos(\alpha)\vec{t_c} + \sin(\alpha)\vec{b_c}.$$

Then $\vec{U'} = \vec{0}$, so $\vec{U}$ is constant and $\langle \vec{t_c}, \vec{U} \rangle = \cos(\alpha)$ by the Serret–Frenet formulas. Therefore c is a cylindrical helix. □

In theory, the Serret–Frenet formulas can answer any question involving curves.

1.7.2 The Fundamental Existence and Uniqueness Theorem

In this subsection we prove that a regular curve is completely determined, except for the position in space, by its natural parameters, that is, its curvature and its torsion. Using the chain rule we may calculate curvature and torsion for an arbitrary parameterized regular curve (Lemma 1.33). Hence, the above statement holds for regular curves in general (see Theorem 1.29). Hence, we now assume that $c : [a,b] \to \mathbb{R}^3$ is a unit speed curve.

The *intrinsic equations* for c are the curvature and torsion functions. With respect to the positions we need isometries. We consider isometries in general in Chapter 2. For this subsection we just need the special case $\mathbb{R}^3$.

An *isometry* of $\mathbb{R}^3$ is a mapping $f : \mathbb{R}^3 \to \mathbb{R}^3$ with $\|\vec{x} - \vec{y}\| = \|\overrightarrow{f(x)} - \overrightarrow{f(y)}\|$ for all $\vec{x}, \vec{y} \in \mathbb{R}^3$, that is, f preserves distances. Two curves c_1 and c_2 are congruent if there exists an isometry f with $c_2 = f(c_1)$ for the paths of c_1 and c_2. We will discuss isometries in a more general setting in the next chapter.

Theorem 1.40 (Fundamental existence and uniqueness theorem for curves). *Let $\kappa(t)$, $\tau(t)$ be arbitrary continuous functions on $[a,b]$. Then except for the position in space there exists a unique unit speed curve with curvature $\kappa(t)$ and torsion $\tau(t)$ as functions of a natural parameter.*

Proof. We prove the uniqueness first. Suppose that $c : [a,b] \to \mathbb{R}^3$ is a unit speed curve with curvature $\kappa(t)$ and torsion $\tau(t)$ at time t. We show that any unit speed curve on $[a,b]$ with the same curvature and torsion is congruent to c.

Suppose that $c^* : [d,e] \to \mathbb{R}^3$ is another unit speed curve with the same curvature and torsion. Move c^* so that $\overrightarrow{c^*(d)} = \overrightarrow{c(a)}$. Next rotate $\overrightarrow{t_{c^*}}, \overrightarrow{n_{c^*}}$ and $\overrightarrow{b_{c^*}}$ so that they coincide at $t = a$ with $\overrightarrow{t_c}, \overrightarrow{n_c}$ and $\overrightarrow{b_c}$.

Let c^{**} be the new curve. We show that $\overrightarrow{c^{**}(t)} = \overrightarrow{c(t)}$. Now

$$\frac{d}{dt}\langle \overrightarrow{t_c}, \overrightarrow{t_{c^{**}}}\rangle = \langle \kappa\overrightarrow{n_c}, \overrightarrow{t_{c^{**}}}\rangle + \langle \kappa\overrightarrow{t_c}, \overrightarrow{n_{c^{**}}}\rangle,$$

$$\frac{d}{dt}\langle \overrightarrow{n_c}, \overrightarrow{n_{c^{**}}}\rangle = \langle -\kappa\overrightarrow{t_c} + \tau\overrightarrow{b_c}, \overrightarrow{n_{c^{**}}}\rangle + \langle \overrightarrow{n_c}, -\kappa\overrightarrow{t_{c^{**}}} + \tau\overrightarrow{b_{c^{**}}}\rangle,$$

$$\frac{d}{dt}\langle \overrightarrow{b_c}, \overrightarrow{b_{c^{**}}}\rangle = \langle -\tau\overrightarrow{n_c}, \overrightarrow{b_{c^{**}}}\rangle + \langle -\tau\overrightarrow{n_{c^{**}}}, \overrightarrow{b_c}\rangle.$$

These then sum to zero, so

$$\frac{d}{dt}\left(\langle \overrightarrow{t_c}, \overrightarrow{t_{c^{**}}}\rangle + \langle \overrightarrow{n_c}, \overrightarrow{n_{c^{**}}}\rangle + \langle \overrightarrow{b_c}, \overrightarrow{b_{c^{**}}}\rangle\right) = 0.$$

Hence

$$\langle \overrightarrow{t_c}, \overrightarrow{t_{c^{**}}}\rangle + \langle \overrightarrow{n_c}, \overrightarrow{n_{c^{**}}}\rangle + \langle \overrightarrow{b_c}, \overrightarrow{b_{c^{**}}}\rangle = m = \text{constant},$$

but $\overrightarrow{t_c(a)} = \overrightarrow{t_{c^{**}}(a)}$, $\overrightarrow{n_c(a)} = \overrightarrow{n_{c^{**}}(a)}$ and $\overrightarrow{b_c(a)} = \overrightarrow{b_{c^{**}}(a)}$, so $m = 3$. However $\|\langle \overrightarrow{t_c}, \overrightarrow{t_{c^{**}}} \rangle\| \leq 1$, $\|\langle \overrightarrow{n_c}, \overrightarrow{n_{c^{**}}} \rangle\| \leq 1$, $\|\langle \overrightarrow{b_c}, \overrightarrow{b_{c^{**}}} \rangle\| \leq 1$ and so $\langle \overrightarrow{t_c}, \overrightarrow{t_{c^{**}}} \rangle = 1$ and since $\overrightarrow{t_c}$ and $\overrightarrow{t_{c^{**}}}$ are unit vectors, it follows that $\overrightarrow{t_c} = \overrightarrow{t_{c^{**}}}$. Now

$$\overrightarrow{t_c} = \frac{d\overrightarrow{c(t)}}{dt} = \frac{d\overrightarrow{c^{**}(t)}}{dt} = \overrightarrow{t_{c^{**}}}.$$

Hence $\overrightarrow{c(t)} = \overrightarrow{c^{**}(t)} + \overrightarrow{k_1}$, $\overrightarrow{k_1}$ is a constant, but $\overrightarrow{c(a)} = \overrightarrow{c^{**}(a)}$, so $\overrightarrow{k_1} = \overrightarrow{0}$.

Given $\kappa(s)$ and $\tau(s)$, the existence of a unit speed curve with this curvature and torsion follows from the existence theorem for ordinary differential equations (see [26]).

(In fact, the Serret–Frenet equations give nine scalar differential equations. From the fundamental existence theorem for ordinary differential equations there exist solutions. Then given $\overrightarrow{t_c}$ we define $\overrightarrow{c(t)} = \int_a^t \overrightarrow{t_c(\sigma)} d\sigma$.) □

1.7.3 Computing Formulas for the Curvature, the Torsion and the Components of Acceleration

Using the velocity and the acceleration there are some straightforward formulas for a regular curve. First, let v_c be the velocity and a_c the acceleration. Then $\overrightarrow{t_c} = \frac{\overrightarrow{v_c}}{\|\overrightarrow{v_c}\|}$.

We now assume that the regular curve is in addition three times continuously differentiable. We differentiate the velocity v_c to get the acceleration a_c. We have then $\overrightarrow{v_c} = \|\overrightarrow{v_c}\|\overrightarrow{t_c}$ and further

$$\overrightarrow{a_c} = \|\overrightarrow{v_c}\|\overrightarrow{t_c'} + \|\overrightarrow{v_c}'\|\overrightarrow{t_c},$$

and hence

$$\overrightarrow{a_c} = \kappa\|\overrightarrow{v_c}\|^2\overrightarrow{n_c} + \overrightarrow{t_c'} \quad \text{at time } t.$$

This gives

$$\overrightarrow{v_c} \times \overrightarrow{a_c} = \|\overrightarrow{v_c}\|\overrightarrow{t_c} \times \left(\kappa\|\overrightarrow{v_c}\|^2\overrightarrow{n_c} + \overrightarrow{t_c'} \right) = \kappa\|\overrightarrow{v_c}\|^3\overrightarrow{b_c} \quad \text{at time } t.$$

Therefore $\overrightarrow{b_c}$ is in the direction of $\overrightarrow{v_c} \times \overrightarrow{a_c}$ at time t.

Since $\overrightarrow{b_c(t)}$ is a unit vector, we get then

$$\overrightarrow{b_c(t)} = \frac{\overrightarrow{v_c(t)} \times \overrightarrow{a_c(t)}}{\|\overrightarrow{v_c(t)} \times \overrightarrow{a_c(t)}\|}.$$

Further examination will lead us to how to compute the curvature and the torsion. We have

$$\|\overrightarrow{v_c} \times \overrightarrow{a_c}\| = \kappa\|\overrightarrow{v_c}\|^3,$$

that is,

$$\kappa = \frac{\|\vec{v_c} \times \vec{a_c}\|}{\|\vec{v_c}\|^3} \quad \text{at time } t.$$

To find the torsion, we consider $\overrightarrow{c'''}$, what we assume to exist in this subsection. We get

$$\overrightarrow{c'''} = \kappa\|\vec{v_c}\|^2\vec{n_c} + \left(\kappa\|\vec{v_c}\|^2\right)' \vec{n_c} + \|\vec{v_c}\|\vec{t_c} + \|\vec{v_c''}\|\vec{t_c}$$

$$= \alpha\vec{t_c} + \beta\vec{n_c} + \kappa\tau\|\vec{v_c}\|^3\vec{b_c},$$

hence

$$\left\langle \vec{v_c} \times \vec{a_c}, \overrightarrow{c'''} \right\rangle = \kappa^2\tau\|\vec{v_c}\|^6 \quad \text{at time } t,$$

and therefore

$$\tau = \frac{\left\langle \vec{v_c} \times \vec{a_c}, \overrightarrow{c'''} \right\rangle^6}{\kappa^2\|\vec{v_c}\|^6} = \frac{\left\langle \vec{v_c} \times \vec{a_c}, \overrightarrow{c'''} \right\rangle}{\|\vec{v_c} \times \vec{a_c}\|^2}.$$

Example 1.41. Consider the curve c with $\overrightarrow{c(t)} = \begin{pmatrix} 2 \\ t^2 \\ \frac{1}{3}t^3 \end{pmatrix}$ and compute the Serret–Frenet apparatus.

$$\vec{v_c} = \begin{pmatrix} 2 \\ 2t \\ t^2 \end{pmatrix}, \quad \vec{a_c} = \begin{pmatrix} 0 \\ 2 \\ 2t \end{pmatrix}, \quad \vec{t_c} = \frac{1}{t^2 + 2}\begin{pmatrix} 2 \\ 2t \\ t^2 \end{pmatrix},$$

$$\vec{v_c} \times \vec{a_c} = \begin{pmatrix} 2t^2 \\ -4t \\ 4 \end{pmatrix}, \quad \|\vec{v_c} \times \vec{a_c}\| = 2t^2 + 4, \quad \vec{b_c} = \frac{1}{2t^2 + 4}\begin{pmatrix} 2t^2 \\ -4t \\ 4 \end{pmatrix}$$

and hence

$$\kappa = \frac{2t^2 + 4}{(2t^2 + 2)^3}.$$

Now

$$\overrightarrow{c'''} = \begin{pmatrix} 0 \\ 0 \\ 2 \end{pmatrix}, \quad \tau = \frac{\left\langle \vec{v_c} \times \vec{a_c}, \overrightarrow{c'''} \right\rangle}{\|\vec{v_c} \times \vec{a_c}\|^2},$$

and hence

$$\tau = \frac{8}{(2t^2 + 4)^2}.$$

1.7.4 Integration of Planar Curves

We consider plane curves, so that $\tau = 0$. For such curves it is always possible to interpret the Serret–Frenet equations to obtain the parameterization of the curve. Let ϕ be the angle that is made by $\vec{t_c}$ with the x-axis. For planar curves the slope of this is the derivation (considering the plane as the (x, y)-plane). Then $\phi = \int \kappa(t)dt + y_1$, where $\frac{d\phi}{dt} = \kappa(t)$.

Example 1.42. We show that the equations $\kappa(t) = \frac{1}{t}$, $\tau = 0$, $t > 0$ are the intrinsic equations of a logarithmic spiral. We have $\frac{d\phi}{dt} = \kappa(t) = \frac{1}{t}$, then $\phi = \log(t) + y_1$, while this implies that $t = e^{\phi - y_1}$. This leads to $\kappa(t) = \frac{1}{t} = e^{-(\phi - y_1)}$.

Then by integrating the Serret–Frenet equations for planar curves ($\tau = 0$ implies planarity):

$$\overrightarrow{c(\phi)} = \int \frac{1}{\kappa(\phi)} \begin{pmatrix} \cos(\phi) \\ \sin(\phi) \end{pmatrix} d\phi = \int e^{\phi - y_1} \begin{pmatrix} \cos(\phi) \\ \sin(\phi) \end{pmatrix} d\phi$$

$$= \begin{pmatrix} \frac{1}{2} e^{\phi - y_1} (\cos(\phi) + \sin(\phi)) \\ \frac{1}{2} e^{\phi - y_1} (\sin(\phi) - \cos(\phi)) \end{pmatrix} + y_2.$$

Choose $y_1 = \frac{\pi}{4}$, $\vec{y_2} = \vec{0}$, $\phi = \Theta + \frac{1}{4}$.
Then

$$\overrightarrow{c(\phi)} = \frac{1}{\sqrt{2}} e^{\Theta} \begin{pmatrix} \cos(\Theta) \\ \sin(\Theta) \end{pmatrix}$$

which in polar coordinates is

$$r(\Theta) = \frac{1}{\sqrt{2}} e^{\Theta}$$

and hence a log spiral.

Exercises

1. Given an affine $(n + 1)$-space A^{n+1}, show that we obtain a projective n-space P^n by considering points as 1-dimensional subspaces of it.
2. Prove Theorem 1.2, that is, in any affine geometry a line parallel to one of two intersecting lines must intersect the other. This implies that in an affine geometry parallelism is transitive.
3. Prove the following theorem.

    Theorem. *In a finite projective geometry the following are true:*
    (a) *All lines contain the same number of points.*
    (b) *If each line has n points then each point is on exactly n lines.*

(c) *If each line has n points then there exists exactly $n^2 - n + 1$ points and $n^2 - n + 1$ lines.*

(*Hint*: Follow the proof for a finite affine geometry modified in that there are no parallel lines.)

4. Show that the minimal model for a planar affine geometry (the affine geometry with the smallest number of points) has 4 points and 6 lines.

5. Let K be a field. We define the affine plane $A_2(K)$ by taking as points the pairs $(x, y) \in K^2$ and as lines the sets

$$\ell_c = \{(x, y) \in K^2 \mid x = c\} \quad \text{with } c \in K$$

and

$$\ell_{m,b} = \{(x, y) \in K^2 \mid y = mx + b\} \quad \text{with } m, b \in K.$$

Show that $A_2(K)$ satisfies the axioms of a planar affine geometry. In particular, show that $A_2(\mathbb{Z}_2)$ represents the minimal model of a planar affine geometry.

6. Show that in a neutral geometry, given a point P and a line ℓ with $P \notin \ell$, if there is more than one parallel to ℓ through P, there must be infinitely many parallels.
(*Hint*: As discussed, if there is more than one parallel there is an angle of parallelism which is less than 90°. Show that any line which is interior to the angle of parallelism is also parallel to ℓ.)

7. In Euclidean geometry, given three lines l, g and k such that k intersects l and g, assume that one external angle is equal to the inner one opposite on the same side or that the sum of the inner angles on the same side is 180°. Show that l and g are parallel.

8. Complete the proof of Theorem 1.6. Show that

$$\sphericalangle(\overline{CA}, \overline{CD}) > \sphericalangle(\overline{BA}, \overline{BC}).$$

9. Show that in a neutral geometry the following are equivalent:
(a) The Euclidean Parallel Postulate.
(b) The angle sum of any triangle is 180°.

10. Given a finite set S of points in the Euclidean plane $\mathbb{R}^2$, either all points are on one line, or there exists one line which contains exactly two of the points (J. J. Sylvester (1814–1987)).
(*Hint*: Assume that not all points are on one line and that each line through two points of S contains a third point of S. Let L be the set of connecting lines which contain at least two points of S. Show that there cannot exist a line ℓ from L and a point P from S which is not on ℓ such that the distance from ℓ to P is minimal under all distances from such line–point distances.)

11. Each set of $n > 2$ points in the Euclidean plane which are not all in one line determines at least n distinct connecting lines (P. Erdős (1913–1996)).
    (*Hint*: Use Exercise 10 and induction on n.)
12. Show that, in the context of Euclidean geometry, with the other four axiom sets, the axiom of parallels is equivalent to the existence of non-congruent similar triangles, that is, non-congruent triangles for which the corresponding angles have the same measure.
13. In neutral geometry, given triangle ABC and angle $\sphericalangle(\overline{AB}, \overline{AC})$. Then there exists a triangle $A_1 B_1 C_1$ with the same angle sum as a triangle ABC and

$$\sphericalangle(\overline{A_1 B_1}, \overline{A_1 C_1}) \leq \frac{1}{2} \sphericalangle(\overline{AB}, \overline{AC}).$$

(*Hint*: Let E be the midpoint of $\overline{BC}$. Construct F such that $\overline{AE} \equiv \overline{EF}$. Show that $AEB \equiv CEF$.)
14. Prove Lemma 1.27.
    (*Hint*: Use the fact that $\|\overrightarrow{c'(t)}\| > 0$.)
15. Reparameterize the curve $c : [0, 2\pi] \to \mathbb{R}^3$, $\overrightarrow{c(t)} = \begin{pmatrix} r\cos(t) \\ r\sin(t) \\ ht \end{pmatrix}$, $r, h > 0$, in terms of arc length parameter.

2 Isometries in Euclidean Vector Spaces and their Classification in $\mathbb{R}^n$

2.1 Isometries and Klein's Erlangen Program

In the previous chapter we mentioned an approach to geometry introduced by F. Klein (1849–1925) in his *Erlanger Programm* in 1885. This is called the *transformation group approach*. In this method, a geometry is defined on a set by those properties of the set which are *invariant* or *unchanged* under the action of some group of transformations of the set. This group of transformations is called the *group of congruence motions* of the geometry and in this approach knowing the geometry is equivalent to knowing the congruence group.

For example, in this approach, planar Euclidean geometry would consist of those properties of the Euclidean plane, defined as a real two-dimensional space equipped with the ordinary metric, that are invariant under a group called the *group of Euclidean motions*. This group consists of all the *isometries* or distance preserving mappings of the plane to itself. In this geometry two figures would be *congruent* if one can be mapped to the other by an isometry or *congruence motion*. In this chapter we consider the transformation group approach to Euclidean geometry by studying the *Euclidean group of motions* [5]. This chapter can be considered as the bridge to the book [12]. In [12] we considered the algebraic part of Euclidean vector spaces. Here we start with their geometric part.

In order to form the Euclidean group, we start with Euclidean vector spaces V, which are real vector spaces equipped with a scalar product. We then consider *isometries* which are mappings $f : V \to V$ which preserve distance.

If, in addition, we define $F : V \to V$, $F(\vec{v}) := f(\vec{v}) - f(\vec{0})$, then F preserves angles and hence maps geometric figures to congruent geometric figures.

Recall that distance in a Euclidean vector space can be computed via the scalar product or inner product. We then define an isometry via scalar products.

Definition 2.1. Let V be a Euclidean vector space with scalar product $\langle \ , \ \rangle$. Let $f : V \to V$ be a mapping and $F : V \to V$, $F(\vec{v}) = f(\vec{v}) - f(\vec{0})$. We call f an *isometry* if

$$\langle F(\vec{v}), F(\vec{w}) \rangle = \langle \vec{v}, \vec{w} \rangle \quad \text{for all } \vec{v}, \vec{w} \in V.$$

An isometry $f : V \to V$ is called a *linear isometry* if in addition $f(\vec{0}) = \vec{0}$.

If f is an isometry then certainly F is a linear isometry.

Lemma 2.2. *Each isometry $f : V \to V$ is injective.*

Proof. Consider $\vec{v}, \vec{w} \in V$ with $f(\vec{v}) = f(\vec{w})$. Then also $F(\vec{v}) = F(\vec{w})$, and

$$0 = \langle F(\vec{v}) - F(\vec{w}), F(\vec{v}) - F(\vec{w}) \rangle$$

https://doi.org/10.1515/9783110740783-002

$$\begin{aligned}
&= \langle F(\vec{v}), F(\vec{v}) \rangle - 2\langle F(\vec{v}), F(\vec{w}) \rangle + \langle F(\vec{w}), F(\vec{w}) \rangle \\
&= \langle \vec{v}, \vec{v} \rangle - 2\langle \vec{v}, \vec{w} \rangle + \langle \vec{w}, \vec{w} \rangle \\
&= \langle \vec{v} - \vec{w}, \vec{v} - \vec{w} \rangle.
\end{aligned}$$

Hence, $\vec{v} = \vec{w}$. Therefore f is injective. $\qquad\square$

Remark 2.3. If V is infinite-dimensional, then an isometry $f : V \to V$ is not necessarily surjective. The reason is the following.

Let $\mathcal{B}$ be an infinite basis of V and $\vec{e} \in \mathcal{B}$. Then there is a bijection $\varphi : \mathcal{B} \to \mathcal{B} \setminus \{\vec{e}\}$. But we have the following.

Lemma 2.4. *If $G = \{f : V \to V \mid f$ an isometry and bijective$\}$ then G is a group with respect to the composition.*

Proof. We have $G \neq \emptyset$ because the identity id_V is in G.

Let $f \in G$ and $F : V \to V$, $F(\vec{v}) = f(\vec{v}) - f(\vec{0})$. F is a linear isometry and bijective. Hence, if $f \in G$ then also $F \in G$, and vice versa. Therefore it is enough to prove Lemma 2.4 for linear isometries. Let $f, g \in G$ be linear isometries. The composition of the two elements $f, g \in G$ is in G, because

$$\langle f \circ g(\vec{v}), f \circ g(\vec{w}) \rangle = \langle f(g(\vec{v})), f(g(\vec{w})) \rangle = \langle g(\vec{v}), g(\vec{w}) \rangle = \langle \vec{v}, \vec{w} \rangle,$$

and $f \circ g$ is bijective.

The associative rule holds in general for the composition of maps.

The identity id_V is the identity element of G.

Let $f \in G$ be a linear isometry. We finally have to show that the inverse mapping $f^{-1} : V \to V$ is an isometry, that is, we have to show that

$$\langle f^{-1}(\vec{v}), f^{-1}(\vec{w}) \rangle = \langle \vec{v}, \vec{w} \rangle.$$

Since f is bijective there exist $\vec{x}, \vec{y} \in V$ with $f(\vec{x}) = \vec{v}$, $f(\vec{y}) = \vec{w}$. This gives

$$\langle f^{-1} \circ f(\vec{x}), f^{-1} \circ f(\vec{y}) \rangle = \langle \vec{x}, \vec{y} \rangle = \langle f(\vec{x}), f(\vec{y}) \rangle,$$

which holds because f is a linear isometry.

Therefore

$$\langle f^{-1}(\vec{v}), f^{-1}(\vec{w}) \rangle = \langle \vec{v}, \vec{w} \rangle.$$

Hence, G is a group. $\qquad\square$

We next show how isometries preserve distance and angle.

Lemma 2.5. *Let V be an Euclidean vector space, $f : V \to V$ be an isometry on V and $F : V \to V$, $F(\vec{v}) = f(\vec{v}) - f(\vec{0})$. Then f preserves distance, and F preserves distance and angle. That is, if $\vec{u}, \vec{v} \in V$ then $\|\vec{u} - \vec{v}\| = \|f(\vec{u}) - f(\vec{v})\| = \|F(\vec{u}) - F(\vec{v})\|$ and $\sphericalangle(\vec{u}, \vec{v}) =$*

$\sphericalangle(F(\vec{u}), F(\vec{v}))$. *It follows that if M is any geometric figure in V then* $f(M) \equiv M$, *that is, f maps a geometric figure to a congruent geometric figure.*

Proof. Let $f : V \to V$ be an isometry and $\vec{v} \in V$. We may assume that f is a linear isometry. Then

$$\|\vec{v}\| = \sqrt{\langle \vec{v}, \vec{v} \rangle} = \sqrt{\langle f(\vec{v}), f(\vec{v}) \rangle} = \|f(\vec{v})\|.$$

Hence a linear isometry preserves norm and so directly preserves distance and angle, which are defined in terms of scalar product and norm. □

Lemma 2.6. *Let* $f : V \to V$ *be a linear isometry.*
(a) *If* $f(\vec{v}) = \lambda \vec{v}$ *for some* $\lambda \in \mathbb{R}$, $\vec{v} \neq \vec{0}$, *then* $|\lambda| = 1$.
(b) $\vec{v} \perp \vec{w} \Leftrightarrow f(\vec{v}) \perp f(\vec{w})$ *for all* $\vec{v}, \vec{w} \in V$.

Proof. Claim (b) holds by definition, but follows also from Lemma 2.5. We now prove claim (a):

Suppose $f(\vec{v}) = \lambda \vec{v}$ with $\vec{v} \neq \vec{0}$. Since f is injective we have $\lambda \neq 0$. Then

$$\|\vec{v}\| = \|f(\vec{v})\| = \|\lambda \vec{v}\| = |\lambda| \|\vec{v}\|,$$

hence $|\lambda| = 1$. □

Remark 2.7. Since isometries are not necessarily bijective, from now on we always assume that V is a finite-dimensional Euclidean vector space. In this case we have that isometries are not only injective, they are already bijective.

Theorem 2.8. *Let V be a finite-dimensional Euclidean vector space with scalar product* $\langle \, , \, \rangle$. *Then the following hold:*
(1) *Let* $f : V \to V$ *be an isometry with* $\vec{v}_0 = f(\vec{0})$. *We define* $\tau_{\vec{v}_0} : V \to V, \vec{v} \mapsto f(\vec{v}) + \vec{v}_0$. *Let* f' *be the map defined by* $f = \tau_{\vec{v}_0} \circ f'$, *that is,* $f' : V \to V, \vec{v} \mapsto f(\vec{v}) - \vec{v}_0$. *Then* f' *is a linear isometry.*
(2) *Let* $f : V \to V$ *be a linear isometry. Then f is a linear transformation of V.*
(3) *Let* $f : V \to V$ *be an isometry. Then f is bijective.*

Recall that a map $f : V \to V$ is a linear transformation if $f(\vec{v} + \vec{w}) = f(\vec{v}) + f(\vec{w})$, $f(r\vec{v}) = rf(\vec{v})$ for all $\vec{v}, \vec{w} \in V$ and $r \in \mathbb{R}$.

Proof. (1) Since $\vec{v}_0 = f(\vec{0})$ and f is an isometry, we get

$$\begin{aligned} \langle f'(\vec{v}), f'(\vec{w}) \rangle &= \langle f(\vec{v}) - \vec{v}_0, f(\vec{w}) - \vec{v}_0 \rangle \\ &= \langle f(\vec{v}) - f(\vec{0}), f(\vec{w}) - f(\vec{0}) \rangle \\ &= \langle \vec{v}, \vec{w} \rangle. \end{aligned}$$

Hence, f' is a linear isometry.

(2) Let $\mathcal{B} = \{\vec{b}_1, \vec{b}_2, \ldots, \vec{b}_n\}$ be an orthonormal basis of V and $\vec{b}_i' = f(\vec{b}_i)$ for $i = 1, 2, \ldots, n$. Then $\mathcal{B}' = \{\vec{b}_1', \vec{b}_2', \ldots, \vec{b}_n'\}$ also is an orthonormal basis of V because certainly

$$\langle \vec{b}_i', \vec{b}_j' \rangle = \langle f(\vec{b}_i), f(\vec{b}_j) \rangle = \langle \vec{b}_i, \vec{b}_j \rangle = \delta_{ij},$$

and by [12, Lemma 14.24], $\vec{b}_1', \vec{b}_2', \ldots, \vec{b}_n'$ are linearly independent and therefore form a basis of V. Hence, if $\vec{v} \in V$, we have a linear combination

$$\vec{v} = \sum_{i=1}^n x_i \vec{b}_i, \quad x_i \in \mathbb{R} \quad \text{and} \quad f(\vec{v}) = \sum_{j=1}^n \lambda_j \vec{b}_j', \quad \lambda_j \in \mathbb{R}.$$

We calculate the λ_j. From

$$\langle f(\vec{v}), \vec{b}_i' \rangle = \left\langle \sum_{j=1}^n \lambda_j \vec{b}_j', \vec{b}_i' \right\rangle$$

$$= \sum_{j=1}^n \lambda_j \langle \vec{b}_j', \vec{b}_i' \rangle$$

$$= \lambda_i$$

we get

$$\lambda_i = \langle f(\vec{v}), f(\vec{b}_i) \rangle = \langle \vec{v}, \vec{b}_i \rangle = x_i, \quad i = 1, 2, \ldots, n,$$

which is independent of f.

For each i the map

$$\vec{v} \mapsto \langle \vec{v}, \vec{b}_i \rangle \vec{b}_i'$$

is a linear transformation, and hence f is a linear transformation as a sum of linear transformations.

(3) In any case f is injective by Lemma 2.2. Consider $\vec{v}_0 = f(\vec{0})$, $\tau_{\vec{v}_0} : V \to V, \vec{v} \mapsto \vec{v} + \vec{v}_0$ and f' defined by $f = \tau_{\vec{v}_0} \circ f'$, that is, $f' : V \to V, \vec{v} \mapsto f(\vec{v}) - \vec{v}_0$.

By (1) and (2) then f' is an injective linear transformation. Since V is finite-dimensional, f' (as a linear transformation) is in fact bijective (recall if $\mathcal{B} = \{\vec{b}_1, \vec{b}_2, \ldots, \vec{b}_n\}$ is a basis of V then also $\mathcal{B}' = \{f'(\vec{b}_1), f'(\vec{b}_2), \ldots, f'(\vec{b}_n)\}$ is a basis of V because f' is an injective linear transformation). Therefore also f is bijective. $\qquad\square$

Remark 2.9. We call a map $\tau_{\vec{v}_0} : V \to V, \vec{v} \mapsto \vec{v} + \vec{v}_0$, a *translation by* $\vec{v}_0$.

Corollary 2.10. *Let $f : V \to V$ be an isometry, where V is as in Theorem 2.8. Then there exists a translation $\tau : V \to V$ and a linear isometry $f' : V \to V$ such that $f = \tau \circ f'$.*

Before proceeding to describe the various different types of isometries, we recall the importance that triangles play in the congruence theory of planar Euclidean geometry (see Chapter 1). The next theorem explains why in terms of isometries.

Theorem 2.11. *Let $\mathbb{R}^2$ be the two-dimensional real vector space equipped with the canonical scalar product. Then an isometry $f : \mathbb{R}^2 \to \mathbb{R}^2$ is completely determined by its action on three non-collinear points, that is, on a triangle.*

Proof. Let $f : \mathbb{R}^2 \to \mathbb{R}^2$ be an isometry and A, B, and C three non-collinear points. First we show that if $f(A) = A, f(B) = B, f(C) = C$ then $f(P) = P$ for any point. Suppose $P \neq f(P)$. Then, since f preserves distance, the distance from A to P is the same as the distance from $f(A) = A$ to $f(P)$. Therefore A is equidistant from P and $f(P)$ and hence on the perpendicular bisector of $\overline{Pf(P)}$. The same is true for B and C and hence all three are on the perpendicular bisector of $\overline{Pf(P)}$, contradicting that they are non-collinear. Hence $P = f(P)$. Now suppose that g is another isometry with $f(A) = g(A), f(B) = g(B)$, $f(C) = g(C)$. As explained, isometries have inverses and therefore

$$f \circ g^{-1}(A) = A, \quad f \circ g^{-1}(B) = B, \quad f \circ g^{-1}(C) = C.$$

It follows from the first argument that $f \circ g^{-1}(P) = P$ for any point P and therefore $f(P) = g(P)$ for any point P. $\qquad\square$

Remark 2.12. In what follows we need some facts about linear transformations and matrices which we describe now.

Let $\mathcal{B} = \{\vec{b}_1, \vec{b}_2, \ldots, \vec{b}_n\}$ be a basis of $\mathbb{R}^n$. Let $f : \mathbb{R}^n \to \mathbb{R}^n$ be a linear transformation, that is, we have $f(\vec{v} + \vec{w}) = f(\vec{v}) + f(\vec{w}), f(r\vec{v}) = rf(\vec{v})$ for all $\vec{v}, \vec{w} \in \mathbb{R}^n, r \in \mathbb{R}$. Since $\mathcal{B}$ is a basis we have representation

$$f(\vec{b}_1) = \sum_{i=1}^{n} a_{i1}\vec{b}_i,$$

$$f(\vec{b}_2) = \sum_{i=1}^{n} a_{i2}\vec{b}_i,$$

$$\vdots$$

$$f(\vec{b}_n) = \sum_{i=1}^{n} a_{in}\vec{b}_i$$

with uniquely determined $a_{ij} \in \mathbb{R}$. This means, that f is, with respect to the fixed basis, uniquely characterized by the real numbers a_{ij}, and these we may pool as a matrix

$$A_{\mathcal{B}}(f) = A = \begin{pmatrix} a_{11} & a_{12} & \cdots & a_{1n} \\ a_{21} & a_{22} & \cdots & a_{2n} \\ \vdots & \vdots & \ddots & \vdots \\ a_{n1} & a_{n2} & \cdots & a_{nn} \end{pmatrix} \in M(n \times n, \mathbb{R}).$$

We often write just $A = (a_{ij})$ and denote the special matrix

$$\begin{pmatrix} 1 & 0 & \cdots & \cdots & 0 \\ 0 & 1 & \ddots & & \vdots \\ \vdots & \ddots & \ddots & \ddots & \vdots \\ \vdots & & \ddots & 1 & 0 \\ 0 & \cdots & \cdots & 0 & 1 \end{pmatrix},$$

the nth unit matrix, by E_n. If $f, g : \mathbb{R}^n \to \mathbb{R}^n$ are linear transformations, and if, with a fixed basis $\mathcal{B}$, f corresponds to $A \in M(n \times n, \mathbb{R})$ and g corresponds to $B \in M(n \times n, \mathbb{R})$, then the matrix

$$AB = \begin{pmatrix} a_{11} & a_{12} & \cdots & a_{1n} \\ a_{21} & a_{22} & \cdots & a_{2n} \\ \vdots & \vdots & \ddots & \vdots \\ a_{n1} & a_{n2} & \cdots & a_{nn} \end{pmatrix} \begin{pmatrix} b_{11} & b_{12} & \cdots & b_{1n} \\ b_{21} & b_{22} & \cdots & b_{2n} \\ \vdots & \vdots & \ddots & \vdots \\ b_{n1} & b_{n2} & \cdots & b_{nn} \end{pmatrix} = \begin{pmatrix} c_{11} & c_{12} & \cdots & c_{1n} \\ c_{21} & c_{22} & \cdots & c_{2n} \\ \vdots & \vdots & \ddots & \vdots \\ c_{n1} & c_{n2} & \cdots & c_{nn} \end{pmatrix}$$

with

$$c_{ij} = \sum_{k=1}^{n} a_{ik} b_{kj}, \quad i = 1, 2, \ldots, n; \; j = 1, 2, \ldots, n,$$

corresponds to the linear transformation $f \circ g$.

The linear transformation $f : \mathbb{R}^n \to \mathbb{R}^n$ is *bijective* if and only if f is injective and surjective; moreover, the matrix A, which corresponds to f with respect to the fixed basis $\mathcal{B}$, is *invertible*, that is, there exists a $C \in M(n \times n, \mathbb{R})$ with $AC = CA = E_n$; C is called the *inverse matrix* of A, denoted by $C = A^{-1}$.

Now if $\vec{x} = \sum_{i=1}^{n} x_i \vec{b}_i$, with the basis $\mathcal{B} = \{\vec{b}_1, \vec{b}_2, \ldots, \vec{b}_n\}$, then we have a representation

$$\vec{y} = f(\vec{x}) = \sum_{i=1}^{n} y_i \vec{b}_i = f\left(\sum_{i=1}^{n} x_i \vec{b}_i \right)$$

$$= \sum_{i=1}^{n} x_i f(\vec{b}_i) = \sum_{i=1}^{n} x_i \left(\sum_{j=1}^{n} a_{ji} \vec{b}_j \right)$$

$$= \sum_{j=1}^{n} \left(\sum_{i=1}^{n} a_{ji} x_i \right) \vec{b}_j.$$

Hence, the coordinate vector

$$\vec{x}_B = \begin{pmatrix} x_1 \\ x_2 \\ \vdots \\ x_n \end{pmatrix}$$

is mapped onto the coordinate vector

$$\vec{y}_B = \begin{pmatrix} y_1 \\ y_2 \\ \vdots \\ y_n \end{pmatrix} = \begin{pmatrix} \sum_{i=1}^{n} a_{1i} x_i \\ \sum_{i=1}^{n} a_{2i} x_i \\ \vdots \\ \sum_{i=1}^{n} a_{ni} x_i \end{pmatrix} = A \cdot \vec{x}_B,$$

the matrix product of A with the column vector $\vec{x}_B$. In what follows we need some more properties and facts about matrices of $M(n \times n, \mathbb{R})$.

Definition 2.13.
(1) If

$$A = \begin{pmatrix} a_{11} & a_{12} & \cdots & a_{1n} \\ a_{21} & a_{22} & \cdots & a_{2n} \\ \vdots & \vdots & \ddots & \vdots \\ a_{n1} & a_{n2} & \cdots & a_{nn} \end{pmatrix}$$

then the matrix

$$\begin{pmatrix} a_{11} & a_{21} & \cdots & a_{n1} \\ a_{12} & a_{22} & \cdots & a_{n2} \\ \vdots & \vdots & \ddots & \vdots \\ a_{1n} & a_{2n} & \cdots & a_{nn} \end{pmatrix}$$

is called the *transpose of A* denoted by A^T.
We have the following facts for $A, B \in M(n \times n, \mathbb{R})$:
(1) $(A^T)^T = A$;
(2) $(AB)^T = B^T A^T$;
(3) If A is invertible, then $(A^T)^{-1} = (A^{-1})^T$.
Also, if

$$\vec{x} = \begin{pmatrix} x_1 \\ x_2 \\ \vdots \\ x_n \end{pmatrix}$$

is a column vector, then the transpose

$$\vec{x}^T = (x_1, x_2, \ldots, x_n)$$

is a row vector, and we have the property that $(A \cdot \vec{x}_B)^T = \vec{x}_B^T \cdot A^T$ is the product of the row vector $\vec{x}_B^T$ with the matrix A^T.

(2) Let $A \in M(n \times n, \mathbb{R})$. An important invariant of A is its determinant $\det(A)$. Let S_n be the symmetric group on n letters. We define the *sign of a permutation* $\sigma \in S_n$ by

$$\text{sign}(\sigma) = \begin{cases} -1, & \text{if } \sigma \in A_n, \\ +1, & \text{if } \sigma \notin A_n. \end{cases}$$

Here A_n is the alternating subgroup of S_n (see [12, Chapter 8]). If

$$A = \begin{pmatrix} a_{11} & a_{12} & \cdots & a_{1n} \\ a_{21} & a_{22} & \cdots & a_{2n} \\ \vdots & \vdots & \ddots & \vdots \\ a_{n1} & a_{n2} & \cdots & a_{nn} \end{pmatrix} \in M(n \times n, \mathbb{R})$$

then the *determinant* $\det(A)$ is defined by

$$\det(A) = \sum_{\sigma \in S_n} \text{sign}(\sigma) a_{1\sigma(1)} a_{2\sigma(2)} \cdots a_{n\sigma(n)}.$$

(a) In the special cases $n = 2$ and $n = 3$ we get:

$$\det\left(\begin{pmatrix} a_{11} & a_{12} \\ a_{21} & a_{22} \end{pmatrix}\right) = a_{11}a_{22} - a_{12}a_{21};$$

(b)

$$\det\left(\begin{pmatrix} a_{11} & a_{12} & a_{13} \\ a_{21} & a_{22} & a_{23} \\ a_{31} & a_{32} & a_{33} \end{pmatrix}\right)$$
$$= a_{11}a_{22}a_{33} + a_{12}a_{23}a_{31} + a_{13}a_{21}a_{32} - a_{31}a_{22}a_{13} - a_{11}a_{23}a_{32} - a_{12}a_{21}a_{33}.$$

We have the following facts. Let $A, B \in M(n \times n, \mathbb{R})$. Then these five statements hold:

(1) $\det(E_n) = 1$.
(2) $\det(AB) = \det(A) \cdot \det(B)$.
(3) $\det(A^T) = \det(A)$.
(4) $\det(A) \neq 0 \Leftrightarrow A$ is invertible.
(5) If A is invertible then $\det(A^{-1}) = \frac{1}{\det(A)}$.

The set of the invertible matrices forms a group under the matrix multiplication, denoted by $GL(n \times n, \mathbb{R})$, the *general linear group*. The subset of $GL(n \times n, \mathbb{R})$ of the matrices A with $\det(A) = 1$ forms a subgroup of $GL(n \times n, \mathbb{R})$, denoted by $SL(n \times n, \mathbb{R})$, the *special linear group*. Let $A \in GL(n \times n, \mathbb{R})$. There are several possibilities to calculate A^{-1}. The easiest way is probably with help of the Gauss–Jordan elimination which is an algorithm that can be used to determine whether a given matrix of $M(n \times n, \mathbb{R})$ is invertible and to find the inverse if it is invertible. We here just give the analytic solution. Let $A \in GL(n \times n, \mathbb{R})$. We define the *adjugate* $\mathrm{adj}(A)$ via the cofactors

$$\tilde{a}_{ij} = (-1)^{i+j} \det \left(\left(\begin{array}{ccccccc} a_{11} & \cdots & a_{1,j-1} & a_{1,j+1} & \cdots & a_{1n} \\ \vdots & & \vdots & \vdots & & \vdots \\ a_{i-1,1} & \cdots & a_{i-1,j-1} & a_{i-1,j+1} & \cdots & a_{i-1,n} \\ a_{i+1,1} & \cdots & a_{i+1,j-1} & a_{i+1,j+1} & \cdots & a_{i+1,n} \\ \vdots & & \vdots & \vdots & & \vdots \\ a_{n1} & \cdots & a_{n,j-1} & a_{n,j+1} & \cdots & a_{n,n} \end{array} \right) \right),$$

and

$$\mathrm{adj}(A) = \begin{pmatrix} \tilde{a}_{11} & \tilde{a}_{12} & \cdots & \tilde{a}_{1n} \\ \tilde{a}_{21} & \tilde{a}_{22} & \cdots & \tilde{a}_{2n} \\ \vdots & \vdots & \ddots & \vdots \\ \tilde{a}_{n1} & \tilde{a}_{n2} & \cdots & \tilde{a}_{nn} \end{pmatrix}.$$

Then $A^{-1} = \frac{1}{\det(A)} \mathrm{adj}(A)$.

As a reference we give [5].

In what follows we consider linear isometries $f : V \to V$, V a Euclidean vector space with scalar product $\langle \ , \ \rangle$ and $\dim(V) = n$. We know that f is bijective. Let $\mathcal{B} = \{\vec{b}_1, \vec{b}_2, \ldots, \vec{b}_n\}$ be now any orthonormal basis of V. Let

$$\vec{x} = x_1 \vec{b}_1 + x_2 \vec{b}_2 + \cdots + x_n \vec{b}_n \quad \text{and} \quad \vec{y} = y_1 \vec{b}_1 + y_2 \vec{b}_2 + \cdots + y_n \vec{b}_n.$$

Then

$$\langle \vec{x}, \vec{y} \rangle = x_1 y_1 + x_2 y_2 + \cdots + x_n y_n$$

because

$$\langle \vec{b}_i, \vec{b}_j \rangle = \delta_{ij} = \begin{cases} 1, & \text{if } i = j, \\ 0, & \text{if } i \neq j. \end{cases}$$

This we may write a bit differently. Let

$$\vec{y}_B = \begin{pmatrix} y_1 \\ y_2 \\ \vdots \\ y_n \end{pmatrix}$$

and

$$\vec{x}_B^T = (x_1, x_2, \ldots, x_n).$$

Then $\langle \vec{x}, \vec{y} \rangle = \vec{x}_B^T \cdot \vec{y}_B$, the matrix product of a row vector with a column vector.

Now, let $f : V \to V$ be a linear isometry. Let $A_B(f) = A$ be the matrix which corresponds to f with respect to B. Then

$$\langle f(\vec{x}), f(\vec{y}) \rangle = (f(\vec{x}))_B^T \cdot (f(\vec{y}))_B = (A\vec{x}_B)^T \cdot (A\vec{y}_B) = \vec{x}_B^T \cdot A^T A \vec{y}_B = \vec{x}_B^T \cdot \vec{y}_B = \langle \vec{x}, \vec{y} \rangle$$

because f is a linear isometry.

Remark 2.14. In the following the orthogonal matrices $A \in GL(n, \mathbb{R})$ play an important part. These are exactly those matrices which, for a fixed orthonormal basis, are allocated in a unique way to linear isometries.

Definition 2.15. A matrix $A \in GL(n, \mathbb{R})$ is called *orthogonal* if $A^{-1} = A^T$, where A^{-1} is the inverse and A^T the transposed matrix of A.

Theorem 2.16. *Let* $A \in GL(n, \mathbb{R})$ *be orthogonal. Then* $|\det(A)| = 1$.

Proof. From $A^T A = E_n$, E_n the identity matrix in $GL(n, \mathbb{R})$, we get

$$1 = \det(A^T A) = \det(A^T) \det(A) = (\det(A))^2$$

because $\det(A^T) = \det(A)$. It follows that $\det(A) = \pm 1$. $\square$

Remark 2.17. If $O(n) = \{A \in GL(n, \mathbb{R}) \mid A \text{ orthogonal}\}$ then $O(n)$ is a group with respect to matrix multiplication. If $A, B \in O(n)$ then

$$(AB)^{-1} = B^{-1} A^{-1} = B^T A^T = (AB)^T.$$

Theorem 2.18. *Consider an* $A \in M(n \times n, \mathbb{R})$, *and let* $\mathbb{R}^n$ *be equipped with the canonical scalar product. Then the following are equivalent:*
(1) *A is orthogonal.*
(2) *The rows of A form an orthonormal basis of* $\mathbb{R}^n$.
(3) *The columns of A form an orthogonal basis of* $\mathbb{R}^n$.

Proof. Transition from A to A^T shows the equivalence of (2) and (3).

(1) means $A^T A = E_n$, that is, $A^{-1} = A^T$.
(2) means $A A^T = E_n$, that is, $A^{-1} = A^T$.

This gives the equivalence of (1) and (2). □

Theorem 2.19. *Let V be a finite-dimensional Euclidean vector space with scalar product $\langle\ ,\ \rangle$. Let B be an orthonormal basis of V. Let $f : V \to V$ be a linear transformation and $A_B(f)$ the matrix allocated to f with respect to B. Then*

$$f \text{ is a linear isometry} \quad \Leftrightarrow \quad A_B(f) \text{ is orthogonal.}$$

Proof. Let $n = \dim(V)$ and $A := A_B(f) \in M(n \times n, \mathbb{R})$. Since B is an orthonormal basis we get $\langle \vec{v}, \vec{w} \rangle = \vec{x}_B^T \vec{y}_B$ for all $\vec{v}, \vec{w} \in V$, where $\vec{x}_B$ and $\vec{y}_B$ are the vectors of the coordinates with respect to B (written as rows) of $\vec{v}$ and $\vec{w}$, respectively.

If f is a linear isometry then

$$\vec{x}_B^T \vec{y}_B = (A\vec{x}_B)^T \cdot A\vec{y}_B = \vec{x}_B^T A^T A \vec{y}_B$$

for all rows $\vec{x}_B, \vec{y}_B \in \mathbb{R}^n$, and hence, $A^T A = E_n$, that is, A is orthogonal. The fact that $A^T A = E_n$ follows from the following observation:

If $C, D \in M(n \times n, K)$, K a field, and if $\vec{v}^T C \vec{w} = \vec{v}^T D \vec{w}$ for all $\vec{v}, \vec{w} \in \mathbb{R}^n$ (written as rows), then $C = D$.

Proof of the fact. Let $C = (c_{ij})$ and $D = (d_{ij})$. If we take the canonical basis

$$\vec{e}_k = (0, \ldots, 0, \underbrace{1}_{k\text{th place}}, 0, \ldots, 0)$$

of K^n then

$$c_{ij} = \vec{e}_i^T C \vec{e}_j = \vec{e}_i^T D \vec{e}_j = d_{ij}$$

for all $i, j = 1, 2, \ldots, n$. □

Now, let A be orthogonal. Then

$$\langle \vec{v}, \vec{w} \rangle = \vec{x}_B^T \vec{y}_B = \vec{x}_B^T A^T A \vec{y}_B = (A\vec{x}_B)^T A \vec{y}_B = \langle f(\vec{v}), f(\vec{w}) \rangle,$$

that is, f is a linear isometry.

Remark 2.20. Let V be a finite-dimensional Euclidean vector space with a scalar product $\langle\ ,\ \rangle$. Let B be a fixed orthogonal basis of V. Then we have essentially all linear isometries $f : V \to V$, if we have all orthogonal matrices $A \in O(n)$, where $n = \dim(V)$. Together with Corollary 2.10, we then have essentially all isometries by the translations and the linear isometries.

We can obtain better geometrical insight by a suitable choice of an orthonormal basis, which then gives for each linear isometry a suitable normal form for the respective orthogonal matrix.

Let $f : V \to V$ be a linear isometry. Let $\mathcal{B}_1$, $\mathcal{B}_2$ be two orthonormal bases of V and let $A_1 = A_{\mathcal{B}_1}(f)$, $A_2 = A_{\mathcal{B}_2}(f)$ be the matrices for f relative to $\mathcal{B}_1$ and $\mathcal{B}_2$, respectively. If $\mathcal{B}_1 = \{\vec{b}_1, \vec{b}_2, \ldots, \vec{b}_n\}$, $\mathcal{B}_2 = \{\vec{b}_1', \vec{b}_2', \ldots, \vec{b}_n'\}$ and

$$\vec{b}_j' = \sum_{i=1}^{n} x_{ij}\vec{b}_i, \quad C = (x_{ij}),$$

then as is well known

$$A_2 = C^{-1}A_1 C.$$

The point is to find for f an orthonormal basis $\mathcal{B}$ such that $A_{\mathcal{B}}(f)$ is an orthogonal matrix which describes f geometrically in a suitable manner.

By Theorem 2.19 and its proof, it is enough to assume that $V = \mathbb{R}^n$ equipped with canonical scalar product

$$\langle \vec{x}, \vec{y} \rangle = x_1 y_1 + x_2 y_2 + \cdots + x_n y_n = \vec{x}^T \vec{y},$$

where

$$\vec{x} = \begin{pmatrix} x_1 \\ x_2 \\ \vdots \\ x_n \end{pmatrix}, \quad \vec{y} = \begin{pmatrix} y_1 \\ y_2 \\ \vdots \\ y_n \end{pmatrix} \in \mathbb{R}^n$$

(written as columns).

This we assume from now on for the following parts in this chapter. We proceed in steps and consider first the most interesting cases $n = 2$ and $n = 3$. Finally, we give a complete description of the linear isometries for general n.

We remark that the $\mathbb{R}^n$ is equipped with an Euclidean geometry with respect to the canonical scalar product.

2.2 The Isometries of the Euclidean Plane $\mathbb{R}^2$

In the space $\mathbb{R}^2$ and $\mathbb{R}^3$ isometries are also often called *moves* or *congruence motions*. For consistency we always speak about isometries. We first describe the orthogonal matrices of $O(2)$. Let $A = \begin{pmatrix} a & b \\ c & d \end{pmatrix} \in O(2)$. Then $\det(A) = \pm 1$. Let first $\det(A) = 1$. Then

$$A^{-1} = \begin{pmatrix} d & -b \\ -c & a \end{pmatrix} = A^T = \begin{pmatrix} a & c \\ b & d \end{pmatrix},$$

and therefore

$$1 = ad - bc = a^2 + b^2.$$

Hence, there exists an $\alpha \in [0, 2\pi)$ with $a = \cos(\alpha)$, $c = \sin(\alpha)$, and we get

$$A = \begin{pmatrix} \cos(\alpha) & -\sin(\alpha) \\ \sin(\alpha) & \cos(\alpha) \end{pmatrix} =: D_\alpha.$$

The linear isometry f which corresponds to D_α (with respect to $\{\vec{e}_1, \vec{e}_2\}$) is a rotation (counterclockwise) with angle α and center $\vec{0} = (0, 0)$, see Figure 2.1.

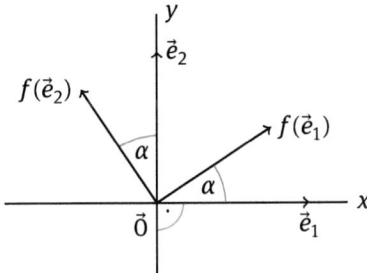

Figure 2.1: Linear isometry f corresponding to D_α.

$$f(\vec{e}_1) = \cos(\alpha)\vec{e}_1 + \sin(\alpha)\vec{e}_2,$$
$$f(\vec{e}_2) = -\sin(\alpha)\vec{e}_1 + \cos(\alpha)\vec{e}_2.$$

Now, let $\det(A) = -1$. Then

$$A^{-1} = \begin{pmatrix} -d & b \\ c & -a \end{pmatrix} = \begin{pmatrix} a & c \\ b & d \end{pmatrix} = A^T.$$

Hence, $-a = d$ and $b = c$, and therefore $\det(A) = -1 = -a^2 - c^2$ and

$$A = \begin{pmatrix} \cos(\alpha) & \sin(\alpha) \\ \sin(\alpha) & -\cos(\alpha) \end{pmatrix} =: S_\alpha$$

for some $\alpha \in [0, 2\pi)$.

The linear isometry which corresponds to S_α with respect to $\{\vec{e}_1, \vec{e}_2\}$ is a reflection at the line ℓ through the center $\vec{0}$ and which has the angle $\frac{\alpha}{2}$ to the x-axis, see Figure 2.2. We call α the *reflection angle*.

$$f(\vec{e}_1) = \cos(\alpha)\vec{e}_1 + \sin(\alpha)\vec{e}_2,$$
$$f(\vec{e}_2) = \sin(\alpha)\vec{e}_1 - \cos(\alpha)\vec{e}_2.$$

This can be seen as follows (see Figures 2.3 and 2.4).

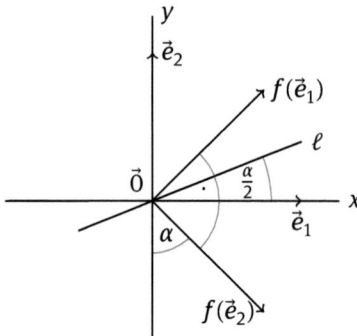

Figure 2.2: A reflection at the line ℓ through the center $\vec{0}$.

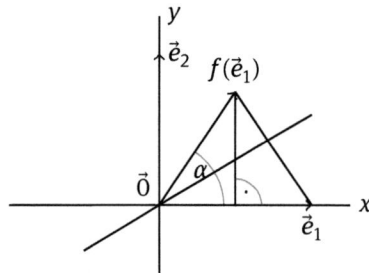

Figure 2.3: Calculation for $f(\vec{e}_1)$.

$$f(\vec{e}_1) = \cos(\alpha)\vec{e}_1 + \sin(\alpha)\vec{e}_2. \qquad (2.1)$$

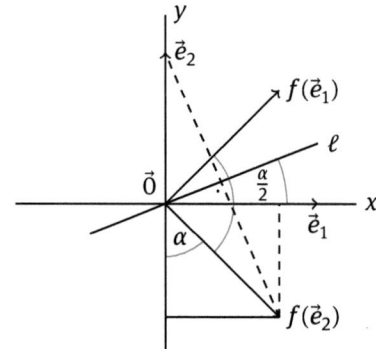

Figure 2.4: Calculation for $f(\vec{e}_2)$.

$$f(\vec{e}_2) = \cos\left(\frac{\pi}{2} - \alpha\right)\vec{e}_1 - \cos(\alpha)\vec{e}_2$$
$$= \sin(\alpha)\vec{e}_1 - \cos(\alpha)\vec{e}_2.$$

Hence we have the following result.

Theorem 2.21. *A linear isometry of the plane $\mathbb{R}^2$ is either a rotation or a reflection at a line ℓ through $\vec{0}$, that is, at a one-dimensional subspace.*

Remark 2.22. In the case of a reflection at a line ℓ through $\vec{0}$, we may choose an orthonormal basis. Such a basis we get from a given one just by a rotation, where one vector of the basis is on the line ℓ, see Figure 2.5.

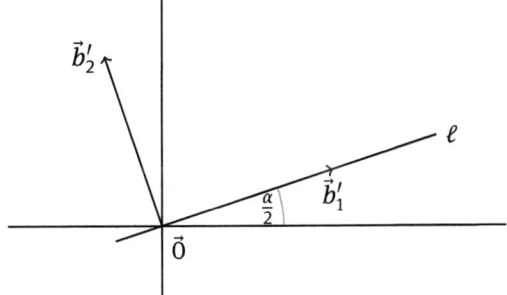

Figure 2.5: Orthonormal basis with one vector on a line ℓ.

For this orthonormal basis $\{\vec{b}_1', \vec{b}_2'\}$, the corresponding orthogonal matrix for the reflection at the line ℓ is just $\left(\begin{smallmatrix} 1 & 0 \\ 0 & -1 \end{smallmatrix}\right)$. We keep this in mind for later use.

Now we consider arbitrary isometries of the plane $\mathbb{R}^2$.

We have to add to the linear isometries for each vector $\vec{v} \in \mathbb{R}^2$ the translation $\tau_{\vec{v}} : \mathbb{R}^2 \to \mathbb{R}^2, \vec{w} \mapsto \vec{w} + \vec{v}$. Let $f : \mathbb{R}^2 \to \mathbb{R}^2$ be an isometry with $f(\vec{0}) = \vec{v}$. Let $\tau_{\vec{v}} : \mathbb{R}^2 \to \mathbb{R}^2$, $\vec{w} \mapsto \vec{w} + \vec{v}$; $\tau_{\vec{v}}$ be a translation and $f = \tau_{\vec{v}} \circ f'$ where f' is a linear isometry. We just write $f' = D_\alpha$ and $f' = S_\alpha$ if the corresponding matrix (with respect to $\{\vec{e}_1, \vec{e}_2\}$) is D_α and S_α, respectively.

First, let $f' = D_\alpha$ for some $\alpha \in [0, 2\pi)$. We begin with the remark that $\tau_{\vec{v}} \circ D_\alpha \circ \tau_{-\vec{v}}$ is a rotation with center $(v_1, v_2) = \vec{v}^T$, where $\vec{v} = \left(\begin{smallmatrix} v_1 \\ v_2 \end{smallmatrix}\right) \in \mathbb{R}^2$. Here we consider $\vec{v}^T$ as the point with the plane coordinates v_1 and v_2.

If $\alpha = 0$, then $f = \tau_{\vec{v}}$ is a translation.

Now, let $\alpha \neq 0$, that is, D_α is a nontrivial rotation. We show that there exists a vector $\vec{u} = \left(\begin{smallmatrix} u_1 \\ u_2 \end{smallmatrix}\right) \in \mathbb{R}^2$ such that

$$\tau_{\vec{u}} \circ D_\alpha \circ \tau_{-\vec{u}} = \tau_{\vec{v}} \circ D_\alpha,$$

that is,

$$\tau_{\vec{u} - D_{\alpha(\vec{u})}} = \tau_{\vec{v}}.$$

For this, we have to show that for each $\vec{v} = \left(\begin{smallmatrix} v_1 \\ v_2 \end{smallmatrix}\right)$ there exists a $\vec{u} = \left(\begin{smallmatrix} u_1 \\ u_2 \end{smallmatrix}\right)$ such that $\vec{u} - D_\alpha(\vec{u}) = \vec{v}$. This is equivalent with the solution of the linear system of equations:

$$(\cos(\alpha) - 1)u_1 - (\sin(\alpha))u_2 + v_1 = 0,$$
$$(\sin(\alpha))u_1 + (\cos(\alpha) - 1)u_2 + v_2 = 0.$$

This system always has a solution because

$$(\cos(\alpha) - 1)^2 + \sin^2(\alpha) > 0$$

if $\alpha \in (0, 2\pi)$ or, in general, if α is not a multiple of 2π. Hence, $f = \tau_{\vec{v}} \circ D_\alpha$, $\alpha \in (0, 2\pi)$, is a rotation with rotation center $\vec{u} \in \mathbb{R}^2$.

Now, let $f' = S_\alpha$ for some $\alpha \in [0, 2\pi)$. Then S_α is a reflection at a line g_0 through the center $\vec{0}$.

Choose a normed direction vector $\vec{u}$ for g_0, that is, $g_0 = \mathbb{R}\vec{u}$ with $\|\vec{u}\| = 1$. Add a vector $\vec{w}$ to $\vec{u}$ such that $\{\vec{u}, \vec{w}\}$ is an orthonormal basis for $\mathbb{R}^2$. Then there exist $\lambda, \mu \in \mathbb{R}$ with $\vec{v} = \lambda\vec{u} + \mu\vec{w}$.

Therefore we get

$$f = \tau_{\vec{v}} \circ S_\alpha = \tau_{\lambda\vec{u} + \mu\vec{w}} \circ S_\alpha = \tau_{\lambda\vec{u}} \circ (\tau_{\mu\vec{w}} \circ S_\alpha).$$

Since $\vec{w}$ is orthogonal to $\vec{u}$, we have that $\tau_{\mu\vec{w}} \circ S_\alpha$ is a reflection S at the line $\tau_{\frac{1}{2}\mu\vec{w}}(g_0) =: g$.

If $\lambda = 0$, then f is just a reflection at g, g is the line parallel to g_0 and $\tau_{\mu\vec{w}}(g_0)$ and halfway between them.

If $\lambda \neq 0$, then f is a *glide reflection*, that is, an isometry consisting of a reflection S at the line $g = \tau_{\frac{1}{2}\mu\vec{u}}(g_0)$ followed by a nontrivial translation $\tau_{\lambda\vec{u}}$, where $\vec{u}$ is parallel to a direction vector for g_0.

This we can also see as follows.

We have

$$f^2 = \tau_{\vec{v}} \circ S_\alpha \circ \tau_{\vec{v}} \circ S_\alpha = \tau_{\vec{v}} \circ \tau_{\vec{v}_1} \circ S_\alpha^2 = \tau_{\vec{v} + \vec{v}_1},$$

for some $\vec{v}_1 \in \mathbb{R}^2$.

If $\tau_{\vec{v} + \vec{v}_1} = \mathrm{id}_{\mathbb{R}^2}$, then f is just a reflection at the line $g = \tau_{\frac{1}{2}\vec{v}}(g_0)$. Now let $\tau_{\vec{v} + \vec{v}_1} \neq \mathrm{id}_{\mathbb{R}^2}$. Then f^2 is a nontrivial translation, and f^2 fixes the two lines $\ell_1 = Pf^2(P)$ and $\ell_2 = f(P)f^3(P)$, $P \in \mathbb{R}^2$, and therefore ℓ_1 and ℓ_2 are parallel (but not necessarily distinct). Moreover, f interchanges ℓ_1 and ℓ_2, and hence f leaves invariant the line ℓ parallel to ℓ_1 and ℓ_2 and halfway between them, see Figure 2.6.

Hence f is a reflection at ℓ followed by a translation $\tau_{\vec{w}}$, where $\vec{w}$ is parallel to a normed direction vector for ℓ, that is, f is a glide reflection, see Figure 2.7.

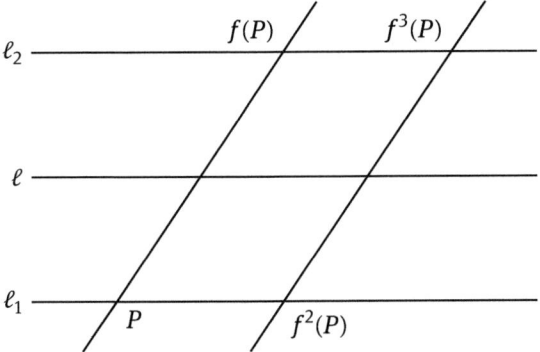

Figure 2.6: Invariant line ℓ.

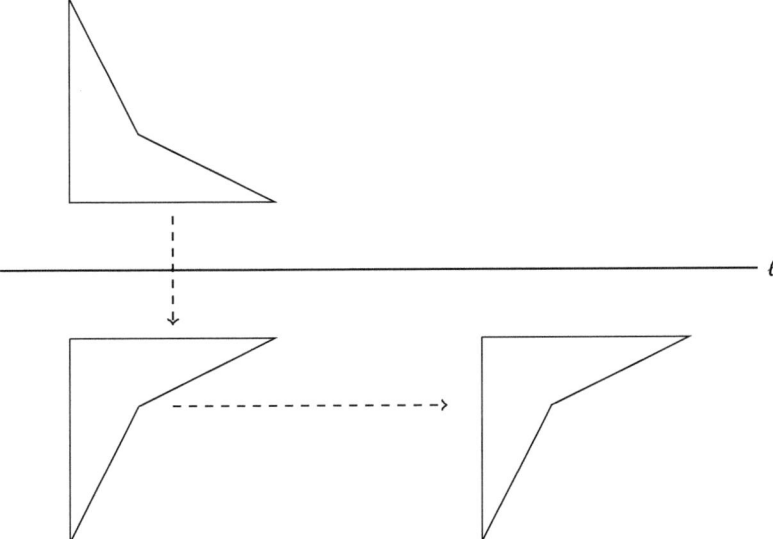

Figure 2.7: Glide reflection.

Theorem 2.23 (Classification of the isometries of the plane $\mathbb{R}^2$). *An isometry of the Euclidean plane $\mathbb{R}^2$ has one of the following forms:*

(1) *a rotation,*
(2) *a translation,*
(3) *a reflection,*
(4) *a glide reflection.*

Example 2.24.

1. Let $f : \mathbb{R}^2 \to \mathbb{R}^2$ be an isometry given by

$$\begin{pmatrix} x \\ y \end{pmatrix} \mapsto \begin{pmatrix} \frac{3}{5} & -\frac{4}{5} \\ \frac{4}{5} & \frac{3}{5} \end{pmatrix} \begin{pmatrix} x \\ y \end{pmatrix} + \begin{pmatrix} 1 \\ 1 \end{pmatrix}$$

(with respect to $\vec{e}_1, \vec{e}_2$).
Since

$$\begin{pmatrix} x \\ y \end{pmatrix} \mapsto \begin{pmatrix} \frac{3}{5} & -\frac{4}{5} \\ \frac{4}{5} & \frac{3}{5} \end{pmatrix} \begin{pmatrix} x \\ y \end{pmatrix}$$

is a rotation with center $\vec{0}$, then f is a rotation with a rotation center $\vec{u} = \begin{pmatrix} u_1 \\ u_2 \end{pmatrix}$. We calculate $\vec{u}$ by solving the system

$$\left(\frac{3}{5} - 1 \right) u_1 - \frac{4}{5} u_2 = -1,$$

$$\frac{4}{5} u_1 + \left(\frac{3}{5} - 1 \right) u_2 = -1.$$

This system has the unique solution $u_1 = -\frac{1}{2}$, $u_2 = \frac{3}{2}$. Hence the rotation center of f is $\vec{u}^T = (-\frac{1}{2}, \frac{3}{2})$.

2. Let $f : \mathbb{R}^2 \to \mathbb{R}^2$ be an isometry given by

$$\begin{pmatrix} x \\ y \end{pmatrix} \mapsto \begin{pmatrix} 0 & 1 \\ 1 & 0 \end{pmatrix} \begin{pmatrix} x \\ y \end{pmatrix} + \begin{pmatrix} 1 \\ 2 \end{pmatrix}$$

(with respect to $\vec{e}_1, \vec{e}_2$).
Certainly, f is a glide reflection. We calculate the reflection line g and the translation vector $\vec{x}$ parallel to ℓ.
The reflection line for the linear isometry

$$\begin{pmatrix} x \\ y \end{pmatrix} \mapsto \begin{pmatrix} 0 & 1 \\ 1 & 0 \end{pmatrix} \begin{pmatrix} x \\ y \end{pmatrix} = S \begin{pmatrix} x \\ y \end{pmatrix}$$

is $g_0 = \mathbb{R}\begin{pmatrix} 1 \\ 1 \end{pmatrix}$. As a direction vector for g_0 we may take $\begin{pmatrix} 1 \\ 1 \end{pmatrix}$. The vector $\vec{w} = \begin{pmatrix} 1 \\ -1 \end{pmatrix}$ is orthogonal to $\vec{u}$. We calculate $\lambda, \mu \in \mathbb{R}$ such that $\vec{v} = \begin{pmatrix} 1 \\ 2 \end{pmatrix} = \lambda \vec{u} + \mu \vec{w}$, that is, we have to solve the system

$$\lambda + \mu = 1,$$
$$\lambda - \mu = 2.$$

We get $\lambda = \frac{3}{2}$ and $\mu = -\frac{1}{2}$, that is, $\vec{v} = \frac{3}{2} \vec{u} - \frac{1}{2} \vec{w}$.
Since $\vec{w}$ is orthogonal to $\vec{u}$, we get that $\tau_{-\frac{1}{2}\vec{w}} \circ S$ is a reflection at the line $g = \mathbb{R}\begin{pmatrix} 1 \\ 1 \end{pmatrix} - \frac{1}{4}\begin{pmatrix} 1 \\ -1 \end{pmatrix}$ or $y = x + \frac{1}{2}$. The translation vector $\vec{x}$ parallel to g then is $\vec{x} = \lambda \vec{u} = \frac{3}{2}\begin{pmatrix} 1 \\ 1 \end{pmatrix}$.

Remark 2.25. If P is a polygon in the Euclidean plane $\mathbb{R}^2$ and $f : \mathbb{R}^2 \to \mathbb{R}^2$ an isometry, then P and $f(P)$ are of equal area.

From the definitions of length and angle we automatically get the congruence theorems for triangles in the Euclidean plane $\mathbb{R}^2$. Hereby two triangles are *congruent* if they are equal in the form and area.

Theorem 2.26 (Congruence Theorem 1, or SSS Criterion). *Two triangles, which coincide in the three side lengths, are congruent.*

Theorem 2.27 (Congruence Theorem 2, or ASA Criterion). *Two triangles, which coincide in two angles and the included side length, are congruent.*

Theorem 2.28 (Congruence Theorem 3, or SAS Criterion). *Two triangles, which coincide in two side lengths and the included angle, are congruent.*

Theorem 2.29 (Congruence Theorem 4, or SAA Criterion). *Two triangles, which coincide in one side length, one included angle and the excluded angle, are congruent.*

With these observations we may prove many results in the Euclidean Geometry of the plane $\mathbb{R}^2$ just by using isometries. We demonstrate this via two proofs of the Gougu Theorem (Theorem of Pythagoras).

1. Proof (China ca. 2000 BC). The four boundary triangles are all congruent (see Figure 2.8). Hence all have the area $\frac{1}{2}ab$.

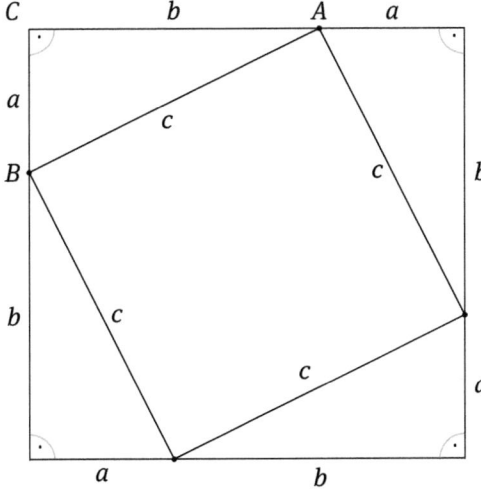

Figure 2.8: Classical proof of the Gougu Theorem.

For the area of the boundary square we get

$$(a + b)^2 = a^2 + 2ab + b^2 = 4 \cdot \frac{1}{2}ab + c^2,$$

hence

$$a^2 + b^2 = c^2. \qquad\qquad \square$$

2. Proof (James Garfield 1875).

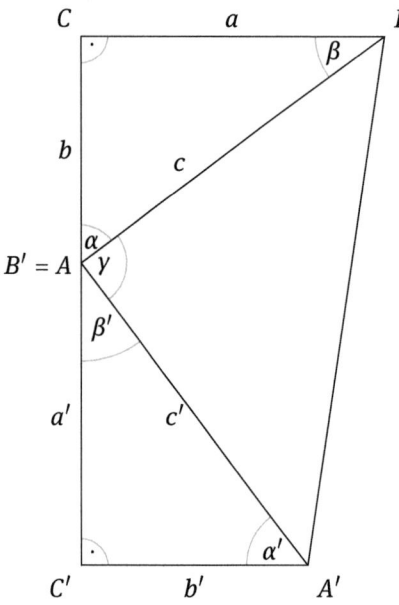

Figure 2.9: Garfield's proof of the Gougu Theorem.

Moving the triangle ABC along the line segment $\overline{AB}$ together with a rotation with center A and angle $90°$ gives the triangle $A'B'C'$. These two triangles are congruent (see Figure 2.9). Hence, $a = a'$, $b = b'$, $c = c'$. Further, $\alpha + \beta = 90°$, $\alpha' + \beta' = 90°$, and therefore $\gamma = 90°$. Therefore all three triangles are right triangles. The area F of the trapezoid is therefore

$$F = \frac{a+b}{2} \cdot \text{height} = \frac{1}{2}(a+b)(a+b).$$

We get

$$\frac{1}{2}(a+b)^2 = \frac{1}{2}ab + \frac{1}{2}ab + \frac{1}{2}c^2$$

using the three triangles. This gives

$$a^2 + b^2 = c^2. \qquad \square$$

Remark 2.30. We here used that the sum of the interior angle of a triangle in the Euclidean plane $\mathbb{R}^2$ is $180°$. This can be seen as follows

We have $\alpha' = \alpha$ and $\beta' = \beta$ by the alternate angle theorem (see Figure 2.10).
Hence we have $\alpha + \beta + \gamma = \alpha' + \beta' + \gamma = 180°$.

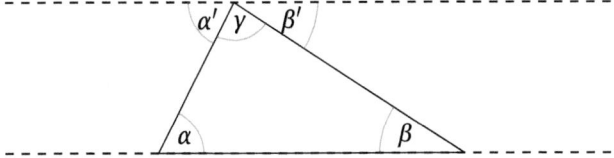

Figure 2.10: Sum of the interior angles of a triangle.

We finally want to give the elegant proof by A. Einstein (1879–1955, Nobel Prize in Theoretical Physics 1921) although it does not use isometries. We consider the following Figure 2.11 of right triangles.

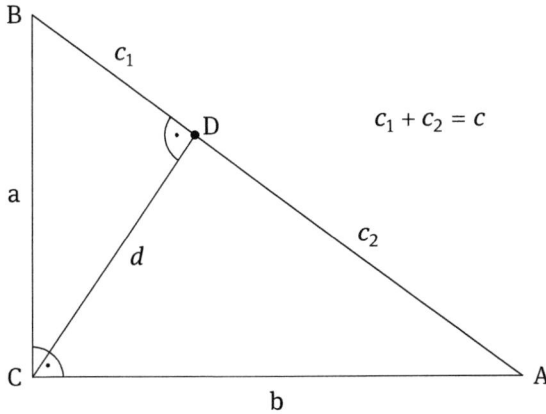

Figure 2.11: Right triangles.

The three right triangles in Figure 2.11 are similar, hence

$$\frac{ab}{c^2} = \frac{dc_1}{a^2} = \frac{dc_2}{b^2}.$$

Now $dc_1 + dc_2 = ab$. Hence

$$ab = \frac{ab}{c^2}(a^2 + b^2), \quad \text{that is,} \quad c^2 = a^2 + b^2.$$

We close this section with some number theoretical problems related to right triangles.

Let Δ be a right triangle with length c of the hypothenuse and a, b the two lengths of the legs. Then $a^2 + b^2 = c^2$.

If a, b, c are positive integers then we call the triple $(a, b; c)$ a Pythogarean triple. We consider the triples $(a, b; c)$ and $(b, a; c)$ as equal.

A Pythagorean triple is called primitive, if $gcd(a, b, c) = 1$. To classify all Pythagorean triples we may restrict ourselves to the primitive Pythagorean triples.

The following theorem from the book Arithmetica by Diophantus (300 BC) classifies all primitive Pythagorean triples.

Theorem 2.31 ([12]). *Let x, y be two relatively prime natural numbers with positive and odd difference $x - y$. Then $(x^2 - y^2, 2xy; x^2 + y^2)$ is a primitive Pythagorean triple, and further each primitive Pythagorean triple can be obtained in this manner.*

If only a and b are positive integers then $c = \sqrt{a^2 + b^2}$ which is not necessarily an integer.

The question arises when c^2 is an integer, that is, we ask, given a positive integer n, when is n a sum of two squares, that is, when is $n = a^2 + b^2$ for two positive integers a and b. The answer is given by P. Fermat (1601–1655).

Theorem 2.32 (see[12]). *Let $n \in \mathbb{N}, n \geq 2$. Then there exist $a, b \in \mathbb{Z}$ with $n = a^2 + b^2$ if and only if*

$$n = 2^\alpha p_1^{\beta_1} p_2^{\beta_2} \cdots p_k^{\beta_k} q_1^{\gamma_1} q_2^{\gamma_2} \cdots q_r^{\gamma_r}$$

with $\alpha \geq 0$, $p_i \equiv 1 \mod 4$ for $i = 1, 2, \ldots, k$ and $q_j \equiv 3 \mod 4$, γ_j even for $j = 1, 2, \ldots, r$.

The next number theoretical question related to right triangles is the following. When is the area $\frac{1}{2}ab$ a natural number if a, b, c are positive integers?

The following theorem gives a particular answer.

Theorem 2.33. *Let Δ be a right triangle with a, b, c positive integers, as above. Then the area of Δ is never the square of a natural number.*

Proof. Assume that the area of Δ is a square k^2 with $k \in \mathbb{N}$. Then we have the following equations

$$c^2 = a^2 + b^2 \quad \text{and} \quad ab = 2k^2.$$

If $gcd(a, b) = d \geq 2$ then necessarily $d|k$.

Hence, we may assume that $gcd(a, b) = 1$. Then $(a, b; c)$ is a primitive Pythagorean triple. From Theorem 2.31 we know that one of a, b is even, so we may assume that $a = 2r$ even. Then we have $rb = k^2$. Since $gcd(a, b) = 1$ we have $gcd(r, b) = 1$. From Euclid's Lemma (see Theorem 2.32 and exercise 2.3 in [12]) we see that r and b are both perfect squares, that is, $r = m^2$ and $b = n^2$ for some natural numbers m, n. Going back to a, b, we have $a = 2m^2$, $b = n^2$. Now the Pythagorean Equation becomes $4m^4 + n^4 = c^2$, that is, $(2m^2)^2 + (n^2)^2 = c^2$.

Now, $(2m^2, n^2, c)$ is a primitive Pythagorean triple. We now may use the method of infinite descent analogously as in the proof of Theorem 5.5 in [12]. This shows that the equation $4m^4 + n^4 = c^2$ has no solution in natural numbers. This gives a contradiction and proves Theorem 2.33. $\qquad\square$

The general question whether a natural number n is the area of some integral right triangle seems to be very difficult.

To get some information we should extend the questioning. Suppose we have a natural number n which is the area of some integral right triangle with side lengths a, b, c as above, and suppose that n has a square factor k^2, $k \in \mathbb{N}$, so $n = k^2 m$ with $m \in \mathbb{N}$. We may scale down the triangle by a factor of k to get a triangle with side lengths $\frac{a}{k}, \frac{b}{k}, \frac{c}{k}$. But these quotients may be not integers.

Example 2.34. Take the right triangle with side lengths $(8, 15, 17)$, its area is $60 = 2^2 \cdot 15$. The triangle with half side lengths $(4, \frac{15}{2}, \frac{17}{2})$, which are rational and not all integral, has area 15.

This leads to the following definition.

Definition 2.35. A natural number which is the area of a right triangle with rational sides is called a *congruent number*.

From Theorem 2.33 we know that 1 and 4 are not congruent numbers. It is easy to check that 2 and 3 are not congruent numbers. The smallest congruent number is 5. The area of the right triangle with rational sides $\frac{20}{3}, \frac{3}{2}, \frac{41}{6}$ is $5 = \frac{1}{2} \cdot \frac{20}{3} \cdot \frac{3}{2}$.

The problem of determining congruent numbers is related to study of rational solutions to certain cubic equations.

Theorem 2.36. *Let $n \in \mathbb{N}$. There is a one-to-one correspondence between the following sets:*

$$V = \left\{ (a, b, c) \in \mathbb{Q}^3 \mid a^2 + b^2 = c^2, \frac{1}{2} ab = n \right\}$$

and

$$W = \{ (x, y) \in \mathbb{Q}^2 \mid y^2 = x^3 - n^2 x, xy \neq 0 \}.$$

The correspondence is given by

$$f : V \to W,$$

$$(a, b, c) \mapsto \left(\frac{nb}{c - a}, \frac{2n^2}{c - a} \right)$$

and

$$g : W \to V,$$

$$(x, y) \mapsto \left(\frac{x^2 - n^2}{y}, \frac{2nx}{y}, \frac{x^2 + n^2}{y} \right).$$

The proof is straightforward, and we leave it to the reader.

Corollary 2.37. *A natural number n is a congruent number if and only if the equation $y^2 = x^3 - n^2 x$ has some rational solutions (x, y) with $xy \neq 0$.*

Example 2.38. The equation $y^2 = x^3 - 900x$ has a solution $(x, y) = (\frac{169}{4}, \frac{1547}{8})$. From Theorem 2.36 we obtain a right triangle with sides

$$(a, b, c) = \left(\frac{119}{26}, \frac{1560}{119}, \frac{42961}{3094} \right)$$

and area 30.

2.3 The Isometries of the Euclidean Space $\mathbb{R}^3$

In general, in an analogous manner we can describe all isometries of $\mathbb{R}^3$ if we can determine all linear isometries of $\mathbb{R}^3$.

We need some preliminary material because there are some additional situations.

Definition 2.39. Let $f : \mathbb{R}^3 \rightarrow \mathbb{R}^3$ be a linear transformation. An element $\lambda \in \mathbb{R}$ is called an *eigenvalue* of f if there exists a vector $\vec{v} \in \mathbb{R}^3$, $\vec{v} \neq \vec{0}$, such that $f(\vec{v}) = \lambda \vec{v}$.

The vector $\vec{v}$ with $f(\vec{v}) = \lambda \vec{v}$, $\vec{v} \neq \vec{0}$, is then called an *eigenvector* of f for the given value λ.

Let $\mathcal{B} = \{\vec{v}_1, \vec{v}_2, \vec{v}_3\}$ be a basis of $\mathbb{R}^3$ and $\vec{v} = x_1 \vec{v}_1 + x_2 \vec{v}_2 + x_3 \vec{v}_3$. Let $A \in M(3 \times 3, \mathbb{R})$ be the matrix which is allocated to f with respect to $\mathcal{B}$. Then $f(\vec{v}) = \lambda \vec{v}$ is equivalent to

$$A \begin{pmatrix} x_1 \\ x_2 \\ x_3 \end{pmatrix} = \lambda \begin{pmatrix} x_1 \\ x_2 \\ x_3 \end{pmatrix},$$

that is,

$$(A - \lambda E_3) \begin{pmatrix} x_1 \\ x_2 \\ x_3 \end{pmatrix} = \begin{pmatrix} 0 \\ 0 \\ 0 \end{pmatrix}.$$

This system of linear equations has a nontrivial solution (x_1, x_2, x_3) if and only if $\det(A - \lambda E_3) = 0$.

The polynomial $\chi_A(x) = \det(A - x E_3)$ is called the *characteristic polynomial* of A. The eigenvalues of f are the real zeros of $\chi_A(x)$. In general, if we extend the concept to the general space $\mathbb{R}^n$, $\chi_A(x)$ may have no real zeroes. But if $A \in M(3 \times 3, \mathbb{R})$, then $\chi_A(x)$ is a polynomial of degree 3, and hence $\chi_A(x)$ has a real zero. This gives the following.

If $\lambda \in \mathbb{R}$ is a zero of $\chi_A(x)$, then there exists a $\vec{v} \in \mathbb{R}^3$, $\vec{v} \neq \vec{0}$, with $f(\vec{v}) = \lambda \vec{v}$.

This we now want to apply for linear isometries of $\mathbb{R}^3$.

Theorem 2.40. *Let $f : \mathbb{R}^3 \rightarrow \mathbb{R}^3$ be a linear isometry. Then there exists a $\vec{v}_1 \in \mathbb{R}^3$ with $\|\vec{v}_1\| = 1$ and $f(\vec{v}_1) = \pm\vec{v}_1$. If we choose $\vec{v}_2$ and $\vec{v}_3$ so that $\{\vec{v}_1, \vec{v}_2, \vec{v}_3\}$ is an orthonormal basis of $\mathbb{R}^3$, then the matrix A which belongs to f with respect to this basis $\{\vec{v}_1, \vec{v}_2, \vec{v}_3\}$ is of the block form*

$$\begin{pmatrix} \pm 1 & 0 & 0 \\ 0 & a_{22} & a_{23} \\ 0 & a_{32} & a_{33} \end{pmatrix}$$

where

$$\begin{pmatrix} a_{22} & a_{23} \\ a_{32} & a_{33} \end{pmatrix}$$

is a matrix which corresponds to a linear isometry of the plane (more precisely, the plane spanned by the two orthonormal vectors $\vec{v}_2$ and $\vec{v}_3$, the $\vec{v}_2\vec{v}_3$-plane).

Proof. We know by the above remarks that there are a $\lambda \in \mathbb{R}$ and a $\vec{v}_1 \in \mathbb{R}^3$, $\vec{v}_1 \neq \vec{0}$, with $f(\vec{v}_1) = \lambda\vec{v}_1$. Since f is bijective, we have $\lambda \neq 0$. By Lemma 2.6 we get $\lambda = \pm 1$. By normalization we may assume $\|\vec{v}_1\| = 1$.

Starting with $\vec{v}_1$ we apply the Gram–Schmidt orthonormalization procedure to get an orthonormal basis $\{\vec{v}_1, \vec{v}_2, \vec{v}_3\}$ of $\mathbb{R}^3$. The matrix for f with respect to this basis has the block form

$$\begin{pmatrix} \pm 1 & 0 & 0 \\ 0 & a_{22} & a_{23} \\ 0 & a_{32} & a_{33} \end{pmatrix}.$$

This can be seen as follows. Since $f(\vec{v}_1) = \pm\vec{v}_1$, the first row has the form

$$\begin{pmatrix} \pm 1 \\ 0 \\ 0 \end{pmatrix}.$$

For the second and third rows we remark the following:
If $\vec{w} = r_2\vec{v}_2 + r_3\vec{v}_3$, then $\langle \vec{w}, \vec{v}_1 \rangle = 0$, and since f is an isometry, we get

$$\langle f(\vec{w}), \vec{v}_1 \rangle = \langle f(\vec{w}), \pm\vec{v}_1 \rangle = \langle f(\vec{w}), f(\vec{v}_1) \rangle = \langle \vec{w}, \vec{v}_1 \rangle = 0.$$

This gives the desired form of the matrix A. $\qquad\square$

We may use this to classify the linear isometries $f : \mathbb{R}^3 \rightarrow \mathbb{R}^3$ geometrically. We may choose an orthonormal basis $\mathcal{B} = \{\vec{v}_1, \vec{v}_2, \vec{v}_3\}$ of $\mathbb{R}^3$ such that the (orthogonal) matrix A of f with respect to the basis $\mathcal{B}$ has the form

$$A = \begin{pmatrix} \pm 1 & 0 & 0 \\ 0 & a_{22} & a_{23} \\ 0 & a_{32} & a_{33} \end{pmatrix},$$

where

$$\tilde{A} = \begin{pmatrix} a_{22} & a_{23} \\ a_{32} & a_{33} \end{pmatrix}$$

is the matrix of a linear isometry of the plane spanned by the orthonormal basis $\{\vec{v}_2, \vec{v}_3\}$. Now, there are four possibilities:

(1) $f(\vec{v}_1) = \vec{v}_1$ and $\tilde{A} = D_\alpha$ is the matrix of rotation in the plane spanned by $\vec{v}_2, \vec{v}_3$; f is called a *rotation around the axis given by $\vec{v}_1$*.

(2) $f(\vec{v}_1) = \vec{v}_1$ and $\tilde{A} = S_\alpha$ is the matrix of a reflection at a line ℓ in the plane spanned by $\vec{v}_2, \vec{v}_3$; f is called a *reflection at the plane through $\vec{v}_1$ and ℓ*.

(3) $f(\vec{v}_1) = -\vec{v}_1$ and $\tilde{A} = D_\alpha$ is the matrix of a rotation in the plane spanned by $\vec{v}_2, \vec{v}_3$; f is then a rotation around the axis given by $\vec{v}_1$ followed by a reflection at the plane spanned by $\vec{v}_2, \vec{v}_3$ which is orthogonal to $\vec{v}_1$; f is called a *rotation reflection*.

(4) $f(\vec{v}_1) = -\vec{v}_1$ and $\tilde{A}$ is the matrix of a reflection at a line ℓ spanned by $\vec{v}_2, \vec{v}_3$. As already mentioned, we may choose now $\vec{v}_2$ and $\vec{v}_3$ so that $f(\vec{v}_2) = -\vec{v}_2$ and $f(\vec{v}_3) = \vec{v}_3$, that is, $\tilde{A} = \begin{pmatrix} -1 & 0 \\ 0 & 1 \end{pmatrix}$. The matrix A for f (with respect to the chosen orthonormal basis $\vec{v}_1, \vec{v}_2, \vec{v}_3$) is then

$$A = \begin{pmatrix} -1 & 0 & 0 \\ 0 & -1 & 0 \\ 0 & 0 & 1 \end{pmatrix}.$$

Now f is a rotation with angle $180°$ around the axis given by $\vec{v}_3$.

Summarizing we get the following

Theorem 2.41. *Let $f : \mathbb{R}^3 \to \mathbb{R}^3$ be a linear isometry. Then f has one of the following forms:*

(1) *f is a rotation,*

(2) *f is a reflection,*

(3) *f is a rotation reflection.*

Using Theorem 2.41, we now may describe all isometries $f : \mathbb{R}^3 \to \mathbb{R}^3$.

Case 1. f has a fixed point, that is, there exists a $\vec{v} \in \mathbb{R}^3$ with $f(\vec{v}) = \vec{v}$. Then $\tilde{f} = \tau_{-\vec{v}} \circ f \circ \tau_{\vec{v}}$ is a linear isometry. Hence $\tilde{f}$, and therefore also $f = \tau_{\vec{v}} \circ \tilde{f} \circ \tau_{-\vec{v}}$, is of one of the three types in Theorem 2.41.

Case 2. f has no fixed point. As in the 2-dimensional case, we let $\vec{v} = f(\vec{0})$. Then $f = \tau_{\vec{v}} \circ \tilde{f}$, where $\tilde{f}$ is a linear isometry. If we argue analogously as for $\mathbb{R}^2$, we get the following two new possibilities (instead of one possibility of $\mathbb{R}^2$):

(1) $f = \tau_{\vec{v}} \circ D_{\ell,\alpha}$, where $D_{\ell,\alpha}$ is a rotation with angle α around an axis ℓ, and the line given by $\vec{v}$ is parallel to ℓ, that is, f is a *screw displacement* or *rotary translation*.

(2) $f = \tau_{\vec{v}} \circ S_M$, where S_M is a reflection at a plane M, and the line given by $\vec{v}$ is parallel to the plane M (that is, to all lines in M), that is, f is a glide (plane) reflection.

This can be seen as follows. Let first $\tilde{f}$ be a rotation around the axis $\vec{v}_1$ with rotation angle α. We have $f = \tau_{\vec{v}} \circ \tilde{f}$. We write $\vec{v} = \lambda_1 \vec{v}_1 + \lambda_2 \vec{v}_2 + \lambda_3 \vec{v}_3$. In the $\vec{v}_2 \vec{v}_3$-plane the isometry $\tau_{\lambda_2 \vec{v}_2 + \lambda_3 \vec{v}_3} \circ \tilde{f}$ is a rotation around some center $\vec{u}$. Hence in $\mathbb{R}^3$ the isometry $\tau_{\lambda_2 \vec{v}_2 + \lambda_3 \vec{v}_3} \circ \tilde{f}$ is around some axis ℓ through $\vec{u}$ and parallel to $\vec{v}_1$. If $\lambda_1 \neq 0$ then we get a rotary translation $\tau_{\lambda_1 \vec{v}_1} \circ (\tau_{\lambda_2 \vec{v}_2 + \lambda_3 \vec{v}_3} \circ \tilde{f})$.

The second possibility is that $\tilde{f}$ is a reflection at a plane spanned by $\vec{v}_1$ and a line g in the $\vec{v}_2 \vec{v}_3$-plane. Again, let $\vec{v} = \lambda_1 \vec{v}_1 + \lambda_2 \vec{v}_2 + \lambda_3 \vec{v}_3$. In the $\vec{v}_2 \vec{v}_3$-plane then $\tau_{\lambda_2 \vec{v}_2 + \lambda_3 \vec{v}_3 \circ \tilde{f}}$ is a glide reflection with a reflection line g' in the $\vec{v}_2 \vec{v}_3$-plane followed by a translation $\tau_{\vec{u}}$ with $\vec{u}$ parallel to g' in the $\vec{v}_2 \vec{v}_3$-plane.

Then $f = \tau_{\vec{v}} \circ \tilde{f}$ is a reflection at the plane spanned by $\vec{v}_1$ and g' followed by the translation $\tau_{\vec{u} + \lambda_1 \vec{v}_1}$, and $\vec{u} + \lambda_1 \vec{v}_1$ is parallel to the plane spanned by g' and $\vec{v}_1$. Hence, here f is a glide reflection.

Example 2.42.

(1) Let $f : \mathbb{R}^3 \to \mathbb{R}^3$ be given by

$$\begin{pmatrix} x \\ y \\ z \end{pmatrix} \mapsto \begin{pmatrix} 1 & 0 & 0 \\ 0 & -1 & 0 \\ 0 & 0 & -1 \end{pmatrix} \begin{pmatrix} x \\ y \\ z \end{pmatrix} + \begin{pmatrix} 2 \\ 1 \\ 1 \end{pmatrix}$$

(with respect to $\vec{e}_1, \vec{e}_2, \vec{e}_3$).

Certainly, f is a rotary translation. The rotation axis is given by

$$\begin{pmatrix} 1 & 0 & 0 \\ 0 & -1 & 0 \\ 0 & 0 & -1 \end{pmatrix} \begin{pmatrix} x \\ y \\ z \end{pmatrix} + \begin{pmatrix} 0 \\ 1 \\ 1 \end{pmatrix} = \begin{pmatrix} x \\ y \\ z \end{pmatrix},$$

that is, $y = z = \frac{1}{2}$ and x arbitrary. Hence, the rotation axis ℓ is

$$\mathbb{R} \begin{pmatrix} 1 \\ 0 \\ 0 \end{pmatrix} + \begin{pmatrix} 0 \\ \frac{1}{2} \\ \frac{1}{2} \end{pmatrix}.$$

The translation vector parallel to ℓ is then

$$\begin{pmatrix} 2 \\ 0 \\ 0 \end{pmatrix}.$$

(2) Let $f : \mathbb{R}^3 \to \mathbb{R}^3$ be given by

$$\begin{pmatrix} x \\ y \\ z \end{pmatrix} \mapsto \begin{pmatrix} 1 & 0 & 0 \\ 0 & 1 & 0 \\ 0 & 0 & -1 \end{pmatrix} \begin{pmatrix} x \\ y \\ z \end{pmatrix} + \begin{pmatrix} 2 \\ 1 \\ 1 \end{pmatrix}$$

(with respect to $\vec{e}_1, \vec{e}_2, \vec{e}_3$).

Since $\left(\begin{smallmatrix} 1 & 0 \\ 0 & -1 \end{smallmatrix}\right)\left(\begin{smallmatrix} y \\ z \end{smallmatrix}\right)$ is a reflection in the plane spanned by $\vec{e}_2$ and $\vec{e}_3$, we get that f is a glide reflection, and the reflection line g_0 in this plane is the y-axis, the line spanned by $\vec{e}_2$. Hence, the reflection line of

$$\begin{pmatrix} y \\ z \end{pmatrix} \mapsto \begin{pmatrix} 1 & 0 \\ 0 & -1 \end{pmatrix} \begin{pmatrix} y \\ z \end{pmatrix} + \begin{pmatrix} 1 \\ 1 \end{pmatrix}$$

is the line g' given by

$$\mathbb{R}\begin{pmatrix} 1 \\ 0 \end{pmatrix} + \begin{pmatrix} 0 \\ \frac{1}{2} \end{pmatrix}.$$

Therefore the reflection plane for f is the plane

$$\mathbb{R}\begin{pmatrix} 1 \\ 0 \\ 0 \end{pmatrix} + \mathbb{R}\begin{pmatrix} 0 \\ 1 \\ 0 \end{pmatrix} + \begin{pmatrix} 0 \\ 0 \\ \frac{1}{2} \end{pmatrix}$$

spanned by g' and $\vec{e}_1$, and the reflection is followed by the translation $\tau_{2\vec{e}_1 + \vec{e}_2}$.

(3) Let $f : \mathbb{R}^3 \to \mathbb{R}^3$ be given by

$$\begin{pmatrix} x \\ y \\ z \end{pmatrix} \mapsto \begin{pmatrix} -1 & 0 & 0 \\ 0 & 0 & -1 \\ 0 & 1 & 0 \end{pmatrix} \begin{pmatrix} x \\ y \\ z \end{pmatrix} + \begin{pmatrix} 2 \\ 1 \\ 2 \end{pmatrix}$$

(with respect to $\vec{e}_1, \vec{e}_2, \vec{e}_3$).

We get that f has to be a rotation reflection, more precisely, f is a rotation around an axis g parallel to the axis spanned by $\vec{e}_1$ (with rotation angle $\frac{\pi}{2}$) followed by a reflection at a plane M parallel to the plane spanned by $\vec{e}_2$ and $\vec{e}_3$. To get g and M

we have to find the fixed point

$$\vec{x} = \begin{pmatrix} x \\ y \\ z \end{pmatrix}$$

of f. From

$$f(\vec{x}) = \begin{pmatrix} -1 & 0 & 0 \\ 0 & 0 & -1 \\ 0 & 1 & 0 \end{pmatrix} \begin{pmatrix} x \\ y \\ z \end{pmatrix} + \begin{pmatrix} 2 \\ 1 \\ 2 \end{pmatrix} = \begin{pmatrix} x \\ y \\ z \end{pmatrix}$$

we get $x = 1$, $y = -\frac{1}{2}$ and $z = \frac{3}{2}$. Hence, g is the axis

$$\mathbb{R} \begin{pmatrix} 1 \\ 0 \\ 0 \end{pmatrix} + \begin{pmatrix} 0 \\ -\frac{1}{2} \\ \frac{3}{2} \end{pmatrix},$$

and M is the plane

$$\mathbb{R} \begin{pmatrix} 0 \\ 1 \\ 0 \end{pmatrix} + \mathbb{R} \begin{pmatrix} 0 \\ 0 \\ 1 \end{pmatrix} + \begin{pmatrix} 1 \\ 0 \\ 0 \end{pmatrix}.$$

(4) Let $f : \mathbb{R}^3 \to \mathbb{R}^3$ be given by

$$\begin{pmatrix} x \\ y \\ z \end{pmatrix} \mapsto \begin{pmatrix} -1 & 0 & 0 \\ 0 & 1 & 0 \\ 0 & 0 & -1 \end{pmatrix} \begin{pmatrix} x \\ y \\ z \end{pmatrix} + \begin{pmatrix} 0 \\ 0 \\ 1 \end{pmatrix}$$

(with respect to $\vec{e}_1, \vec{e}_2, \vec{e}_3$).
We get that f is a rotation with rotation axis

$$\mathbb{R} \begin{pmatrix} 0 \\ 1 \\ 0 \end{pmatrix} + \begin{pmatrix} 0 \\ 0 \\ \frac{1}{2} \end{pmatrix}$$

and rotation angle π.
Now let $g : \mathbb{R}^3 \to \mathbb{R}^3$ be given by

$$\begin{pmatrix} x \\ y \\ z \end{pmatrix} \mapsto \begin{pmatrix} 1 & 0 & 0 \\ 0 & -1 & 0 \\ 0 & 0 & -1 \end{pmatrix} \begin{pmatrix} x \\ y \\ z \end{pmatrix}$$

(with respect to $\vec{e}_1$, $\vec{e}_2$, $\vec{e}_3$), g is a rotation around the axis

$$\begin{pmatrix} 1 \\ 0 \\ 0 \end{pmatrix}.$$

Now $f \circ g : \mathbb{R}^3 \to \mathbb{R}^3$ is the isometry

$$\begin{pmatrix} x \\ y \\ z \end{pmatrix} \mapsto \begin{pmatrix} -1 & 0 & 0 \\ 0 & -1 & 0 \\ 0 & 0 & 1 \end{pmatrix} \begin{pmatrix} x \\ y \\ z \end{pmatrix} + \begin{pmatrix} 0 \\ 0 \\ 1 \end{pmatrix}.$$

Hence $f \circ g$ is a rotary translation with rotation axis

$$\mathbb{R} \begin{pmatrix} 0 \\ 0 \\ 1 \end{pmatrix}$$

and translation vector

$$\begin{pmatrix} 0 \\ 0 \\ 1 \end{pmatrix}.$$

(5) Let $f : \mathbb{R}^3 \to \mathbb{R}^3$ be given by

$$\begin{pmatrix} x \\ y \\ z \end{pmatrix} \mapsto \begin{pmatrix} -1 & 0 & 0 \\ 0 & -1 & 0 \\ 0 & 0 & 1 \end{pmatrix} \begin{pmatrix} x \\ y \\ z \end{pmatrix} + \begin{pmatrix} 0 \\ 2 \\ 0 \end{pmatrix}$$

(with respect to $\vec{e}_1$, $\vec{e}_2$, $\vec{e}_3$).
We get that f is a rotation with rotation axis

$$\mathbb{R} \begin{pmatrix} 0 \\ 0 \\ 1 \end{pmatrix} + \begin{pmatrix} 0 \\ 1 \\ 0 \end{pmatrix}$$

and rotation angle π. Now let $g : \mathbb{R}^3 \to \mathbb{R}^3$ be given by

$$\begin{pmatrix} x \\ y \\ z \end{pmatrix} \mapsto \begin{pmatrix} -1 & 0 & 0 \\ 0 & -1 & 0 \\ 0 & 0 & 1 \end{pmatrix} \begin{pmatrix} x \\ y \\ z \end{pmatrix}$$

(with respect to $\vec{e}_1$, $\vec{e}_2$, $\vec{e}_3$). Then $f \circ g : \mathbb{R}^3 \to \mathbb{R}^3$ is a translation $\tau_{\vec{v}_0} : \mathbb{R}^3 \to \mathbb{R}^3$, $\vec{v} \mapsto \vec{v} + \vec{v}_0$ with

$$\vec{v}_0 = \begin{pmatrix} 0 \\ 2 \\ 0 \end{pmatrix}.$$

2.4 The General Case $\mathbb{R}^n$ with $n \geq 2$

The arguments for cases $n = 2$ and $n = 3$ can be taken as the start for an inductive procedure.

We have to extend the preparations of the case $n = 3$.

Let $f : \mathbb{R}^n \to \mathbb{R}^n$ be a linear transformation. An element $\lambda \in \mathbb{R}$ is called an *eigenvalue* of f if there exists a vector $\vec{v} \in \mathbb{R}^n$, $\vec{v} \neq \vec{0}$, such that $f(\vec{v}) = \lambda\vec{v}$. The vector $\vec{v}$ is then again called an *eigenvector* of f for the eigenvalue λ. Let $\mathcal{B} = \{\vec{v}_1, \vec{v}_2, \ldots, \vec{v}_n\}$ be a basis of $\mathbb{R}^n$ and $\vec{v} = x_1\vec{v}_1 + x_2\vec{v}_2 + \cdots + x_n\vec{v}_n$. Let $A \in M(n \times n, \mathbb{R})$ be the matrix which is allocated to f with respect to the basis $\mathcal{B}$. Then $f(\vec{v}) = \lambda\vec{v}$ is equivalent to

$$A \begin{pmatrix} x_1 \\ x_2 \\ \vdots \\ x_n \end{pmatrix} = \lambda \begin{pmatrix} x_1 \\ x_2 \\ \vdots \\ x_n \end{pmatrix},$$

that is,

$$(A - \lambda E_n) \begin{pmatrix} x_1 \\ x_2 \\ \vdots \\ x_n \end{pmatrix} = \begin{pmatrix} 0 \\ 0 \\ \vdots \\ 0 \end{pmatrix}.$$

The system of linear equations has a nontrivial solution $(x_1, x_2, \ldots, x_n)$ if and only if $\det(A - \lambda E_n) = 0$. Again, the polynomial $\chi_A(x) = \det(A - x E_n)$ is called the characteristic polynomial of A.

We remark that if we change the basis $\mathcal{B}$ to a basis $\mathcal{B}'$, then the matrix A' for f with respect to $\mathcal{B}'$ is of the form $A' = D^{-1}AD$ for a $D \in \mathrm{GL}(n \times n, \mathbb{R})$, and therefore $\chi_A(x) = \chi_{A'}(x)$, that is, the characteristic polynomial is independent of the choice of the basis. The eigenvalues of f are the real zeros of $\chi_A(x)$. If $\lambda \in \mathbb{R}$ is a zero of $\chi_A(x)$ then there exists a $\vec{v} \in \mathbb{R}^n$, $\vec{v} \neq \vec{0}$, with $f(\vec{v}) = \lambda\vec{v}$.

Now let $f : \mathbb{R}^n \to \mathbb{R}^n$ be a linear isometry. Let $\mathcal{B} = \{\vec{b}_1, \vec{b}_2, \ldots, \vec{b}_n\}$ be an orthonormal basis for $\mathbb{R}^n$ and A be the matrix for f with respect to $\mathcal{B}$. By the fundamental theorem of algebra (see [12]), the characteristic polynomial $\chi_A(x)$ can be written as

$$\chi_A(x) = p_1(x)p_2(x)\cdots p_\ell(x)q_1(x)q_2(x)\cdots q_k(x),$$

$0 \leq \ell, k$ and $\ell + 2k = n$, and the $p_1(x), p_2(x), \ldots, p_\ell(x)$ are linear polynomials over $\mathbb{R}$, the $q_1(x), q_2(x), \ldots, q_k(x)$ are quadratic polynomials over $\mathbb{R}$ which have no zeroes in $\mathbb{R}$.

If $\ell \geq 1$ then there exists a real zero of $\chi_A(x)$, that is, a real eigenvalue, and therefore there exists a one-dimensional subspace W of $\mathbb{R}^n$ with $f(W) = W$.

Now let $\ell = 0$. We consider the quadratic polynomial $q_1(x)$ as an element of $\mathbb{C}[x]$. If $\alpha \in \mathbb{C} \setminus \mathbb{R}$ is a zero of $q_1(x)$ in $\mathbb{C}$, then also $\bar{\alpha}$ is a zero of $q_1(x)$ in $\mathbb{C}$. The system

$$A\begin{pmatrix} z_1 \\ z_2 \\ \vdots \\ z_n \end{pmatrix} = \alpha \begin{pmatrix} z_1 \\ z_2 \\ \vdots \\ z_n \end{pmatrix}$$

of equations has a nontrivial solution

$$\begin{pmatrix} z_1 \\ z_2 \\ \vdots \\ z_n \end{pmatrix} \in \mathbb{C}^n.$$

Then

$$\begin{pmatrix} \bar{z}_1 \\ \bar{z}_2 \\ \vdots \\ \bar{z}_n \end{pmatrix}$$

is a solution of the system of equations

$$A\begin{pmatrix} \bar{z}_1 \\ \bar{z}_2 \\ \vdots \\ \bar{z}_n \end{pmatrix} = \bar{\alpha} \begin{pmatrix} \bar{z}_1 \\ \bar{z}_2 \\ \vdots \\ \bar{z}_n \end{pmatrix}.$$

Let $z_j = x_j + iy_j, j = 1, 2, \ldots, n$ and $\alpha = a + ib$, then $\bar{z}_j = x_j - iy_j, j = 1, 2, \ldots, n$. We get

$$A\begin{pmatrix} x_1 \\ x_2 \\ \vdots \\ x_n \end{pmatrix} = \begin{pmatrix} ax_1 - by_1 \\ ax_2 - by_2 \\ \vdots \\ ax_n - by_n \end{pmatrix}$$

and

$$A \begin{pmatrix} y_1 \\ y_2 \\ \vdots \\ y_n \end{pmatrix} = \begin{pmatrix} bx_1 + ay_1 \\ bx_2 + ay_2 \\ \vdots \\ bx_n + ay_n \end{pmatrix}.$$

Let $\vec{v} = x_1\vec{b}_1 + x_2\vec{b}_2 + \cdots + x_n\vec{b}_n$ and $\vec{w} = y_1\vec{b}_1 + y_2\vec{b}_2 + \cdots + y_n\vec{b}_n$. Then the above considerations give

$$f(\vec{v}) \in \text{span}(\vec{v}, \vec{w}) \quad \text{and} \quad f(\vec{w}) \in \text{span}(\vec{v}, \vec{w}).$$

Here $\text{span}(\vec{v}, \vec{w})$ is the subspace of $\mathbb{R}^n$ generated by $\vec{v}$ and $\vec{w}$. So we got a 2-dimensional subspace $W \subset \mathbb{R}^n$ with $f(W) \subset W$. Since f is bijective, we get $f(W) = W$.

So, altogether there is a 1- or 2-dimensional subspace $W \subset \mathbb{R}^n$ with $f(W) = W$.

Let $W^\perp$ be the orthogonal complement of W, that is, $W^\perp = \{\vec{v} \in \mathbb{R}^n | \{\vec{v}\} \perp W\}$. Since f^{-1} is also an isometry, we get

$$\langle f(\vec{v}), \vec{w} \rangle = \langle (f^{-1} \circ f(\vec{v})), f^{-1}(\vec{w}) \rangle = \langle \vec{v}, f^{-1}(\vec{w}) \rangle = 0$$

for $\vec{w} \in W$ and $\vec{v} \in W^\perp$. Therefore $f(W^\perp) = W^\perp$, too. By induction hypothesis, we have a suitable orthonormal basis for $W^\perp$ which we may extend by a suitable orthonormal basis for W to a suitable orthonormal basis for $\mathbb{R}^n$ (see [5]). Altogether, maybe after a suitable renumbering of the basis elements, we get the following result.

Theorem 2.43. *Let $f : \mathbb{R}^n \to \mathbb{R}^n$, $n \geq 2$, be a linear isometry. Then there exist an orthonormal basis $\mathcal{B}$ of $\mathbb{R}^n$ such that the matrix allocated to f with respect to $\mathcal{B}$ has the form*

$$\begin{pmatrix} 1 & 0 & \cdots & \cdots & \cdots & \cdots & \cdots & \cdots & 0 \\ 0 & \ddots & \ddots & & & & & & \vdots \\ \vdots & \ddots & 1 & \ddots & & & & & \vdots \\ \vdots & & \ddots & -1 & \ddots & & & & \vdots \\ \vdots & & & \ddots & \ddots & \ddots & & & \vdots \\ \vdots & & & & \ddots & -1 & \ddots & & \vdots \\ \vdots & & & & & \ddots & A_1 & \ddots & \vdots \\ \vdots & & & & & & \ddots & \ddots & 0 \\ 0 & \cdots & \cdots & \cdots & \cdots & \cdots & \cdots & 0 & A_k \end{pmatrix}$$

where

$$A_i = \begin{pmatrix} \cos(\alpha_i) & -\sin(\alpha_i) \\ \sin(\alpha_i) & \cos(\alpha_i) \end{pmatrix}, \quad \alpha_i \in [0, 2\pi) \text{ for } i = 1, 2, \ldots, k.$$

Remark 2.44. Again, an arbitrary isometry $f : \mathbb{R}^n \to \mathbb{R}^n$ can be described as $f = \tau_{\vec{v}} \circ f'$ where $\tau_{\vec{v}} : \mathbb{R}^n \to \mathbb{R}^n$, $\vec{w} \mapsto \vec{w} + \vec{v}$ is a translation and $f' : \mathbb{R}^n \to \mathbb{R}^n$ is a linear isometry.

Exercises

1. (a) Let $f : \mathbb{R}^n \to \mathbb{R}^n$ be a linear mapping and $\{\vec{b}_1, \vec{b}_2, \ldots, \vec{b}_n\}$ an orthonormal basis of $\mathbb{R}^n$.
    (i) Show that f is an isometry if and only if $\langle \vec{b}_i, \vec{b}_j \rangle = \langle f(\vec{b}_i), f(\vec{b}_j) \rangle$ for all $i, j \in \{1, 2, \ldots, n\}$.
    (ii) Consider the following linear mappings and decide which of them are isometries:
    (A) $n = 2, f(\vec{b}_1) = 2\vec{b}_1 + \vec{b}_2, f(\vec{b}_2) = -\vec{b}_1$;
    (B) $n = 2, f(\vec{b}_1) = \frac{1}{2}\vec{b}_1 - \frac{1}{2}\sqrt{3}\vec{b}_2, f(\vec{b}_2) = \frac{1}{2}\sqrt{3}\vec{b}_1 + \frac{1}{2}\vec{b}_2$;
    (C) $n = 3, f(\vec{b}_1) = -\vec{b}_1, f(\vec{b}_2) = -\vec{b}_2, f(\vec{b}_3) = \vec{b}_3$;
    (D) $n = 3, f(\vec{b}_1) = \vec{b}_1, f(\vec{b}_2) = -\vec{b}_2, f(\vec{b}_3) = \vec{b}_1 + \vec{b}_3$.
    (b) Let $f : \mathbb{R}^n \to \mathbb{R}^n$ be a linear mapping. Show that

    $$f \text{ is injective} \quad \Leftrightarrow \quad f \text{ is surjective} \quad \Leftrightarrow \quad f \text{ is bijective.}$$

    (*Hint:* Use the fact that any two bases $\mathcal{B}_1, \mathcal{B}_2$ of $\mathbb{R}^n$ have exactly n elements and that n linearly independent vectors form a basis of $\mathbb{R}^n$.)

2. (a) Let $f : \mathbb{R}^2 \to \mathbb{R}^2$ be an isometry given by

    $$\begin{pmatrix} x \\ y \end{pmatrix} \mapsto \begin{pmatrix} \frac{3}{5} & -\frac{4}{5} \\ \frac{4}{5} & \frac{3}{5} \end{pmatrix} \begin{pmatrix} x \\ y \end{pmatrix} + \begin{pmatrix} 2 \\ 1 \end{pmatrix}.$$

    Is f a rotation? If yes, determine the center of the rotation.
    (b) Let $f : \mathbb{R}^2 \to \mathbb{R}^2$ be a glide reflection given by

    $$\begin{pmatrix} x \\ y \end{pmatrix} \mapsto \begin{pmatrix} 0 & 1 \\ 1 & 0 \end{pmatrix} \begin{pmatrix} x \\ y \end{pmatrix} + \begin{pmatrix} 1 \\ 2 \end{pmatrix}.$$

    Write f in the form $f = \tau_{\vec{x}} \circ S$, whereby S is a reflection at a line g and $\vec{x}$ is parallel to a direction vector of g. Find the reflection line g and the vector $\vec{x}$.
    (c) Consider an $f : \mathbb{R}^2 \to \mathbb{R}^2$ with

    $$\begin{pmatrix} x \\ y \end{pmatrix} \mapsto \begin{pmatrix} 0 & -1 \\ 1 & 0 \end{pmatrix} \begin{pmatrix} x \\ y \end{pmatrix} - \begin{pmatrix} 3 \\ 2 \end{pmatrix}.$$

Is f a rotation? If yes, find the center of the rotation.

(d) Consider an $f : \mathbb{R}^2 \to \mathbb{R}^2$ with

$$\begin{pmatrix} x \\ y \end{pmatrix} \mapsto \begin{pmatrix} \frac{1}{2}\sqrt{3} & \frac{1}{2} \\ \frac{1}{2} & -\frac{1}{2}\sqrt{3} \end{pmatrix} \begin{pmatrix} x \\ y \end{pmatrix} + \begin{pmatrix} 2 \\ -1 \end{pmatrix}.$$

Show that f is a glide reflection and determine the reflection line and translation vector.

(e) Let $f : \mathbb{R}^2 \to \mathbb{R}^2$ be an isometry with

$$\begin{pmatrix} x \\ y \end{pmatrix} \mapsto \begin{pmatrix} \frac{3}{5} & -\frac{4}{5} \\ \frac{4}{5} & \frac{3}{5} \end{pmatrix} \begin{pmatrix} x \\ y \end{pmatrix} + \begin{pmatrix} 1 \\ 1 \end{pmatrix}.$$

Is f a rotation? If yes, determine the rotation center.

(f) Let $f : \mathbb{R}^2 \to \mathbb{R}^2$ be a glide reflection with

$$\begin{pmatrix} x \\ y \end{pmatrix} \mapsto \begin{pmatrix} 0 & 1 \\ 1 & 0 \end{pmatrix} \begin{pmatrix} x \\ y \end{pmatrix} + \begin{pmatrix} 1 \\ 2 \end{pmatrix}.$$

Write f in the form $f = \tau_{\vec{x}} \circ S$, whereby S is a reflection at a line g and $\vec{x}$ is parallel to a direction vector of g. Determine the reflection line g and the vector $\vec{x}$.

3. Let $f_1, f_2 : \mathbb{R}^2 \to \mathbb{R}^2$ be linear isometries.

(a) Describe $f_1 \circ f_2$, if both f_1 and f_2 are rotations.

(b) Describe $f_1 \circ f_2$, if both f_1 and f_2 are reflections.

(c) Describe $f_1 \circ f_2$, if one of f_1 and f_2 is a rotation and the other one is a reflection.

(d) When does $f_1 \circ f_2 = f_2 \circ f_1$ hold?

4. (a) Consider an $f : \mathbb{R}^3 \to \mathbb{R}^3$ with

$$\begin{pmatrix} x \\ y \\ z \end{pmatrix} \mapsto \begin{pmatrix} -1 & 0 & 0 \\ 0 & 0 & -1 \\ 0 & 1 & 0 \end{pmatrix} \begin{pmatrix} x \\ y \\ z \end{pmatrix} + \begin{pmatrix} 2 \\ 1 \\ 2 \end{pmatrix}.$$

What kind of mapping is f? Determine the fixed point of f, that is,

$$\vec{x} = \begin{pmatrix} x \\ y \\ z \end{pmatrix}$$

with $f(\vec{x}) = \vec{x}$.

(b) Consider an $f : \mathbb{R}^3 \to \mathbb{R}^3$ with

$$\begin{pmatrix} x \\ y \\ z \end{pmatrix} \mapsto \begin{pmatrix} 1 & 0 & 0 \\ 0 & 0 & -1 \\ 0 & 1 & 0 \end{pmatrix} \begin{pmatrix} x \\ y \\ z \end{pmatrix} + \begin{pmatrix} 1 \\ 1 \\ 1 \end{pmatrix}.$$

Then f is a rotary translation. Find the rotation axis and the translation vector.

(c) Consider an $f : \mathbb{R}^3 \to \mathbb{R}^3$ with

$$\begin{pmatrix} x \\ y \\ z \end{pmatrix} \mapsto \begin{pmatrix} 1 & 0 & 0 \\ 0 & 1 & 0 \\ 0 & 0 & -1 \end{pmatrix} \begin{pmatrix} x \\ y \\ z \end{pmatrix} + \begin{pmatrix} 1 \\ 1 \\ 0 \end{pmatrix}.$$

Calculate the reflection plane.

5. (a) Consider an $f : \mathbb{R}^3 \to \mathbb{R}^3$ with

$$\begin{pmatrix} x \\ y \\ z \end{pmatrix} \mapsto \begin{pmatrix} -1 & 0 & 0 \\ 0 & -1 & 0 \\ 0 & 0 & 1 \end{pmatrix} \begin{pmatrix} x \\ y \\ z \end{pmatrix} + \begin{pmatrix} 0 \\ 2 \\ 0 \end{pmatrix}.$$

   (i) Show that f is a rotation and calculate the rotation axis.
   (ii) Let $g : \mathbb{R}^3 \to \mathbb{R}^3$ be a linear isometry with

$$\begin{pmatrix} x \\ y \\ z \end{pmatrix} \mapsto \begin{pmatrix} -1 & 0 & 0 \\ 0 & -1 & 0 \\ 0 & 0 & 1 \end{pmatrix} \begin{pmatrix} x \\ y \\ z \end{pmatrix}.$$

   Show that the product $f \circ g$ is a translation.

   (b) Consider an $f : \mathbb{R}^3 \to \mathbb{R}^3$ with

$$\begin{pmatrix} x \\ y \\ z \end{pmatrix} \mapsto \begin{pmatrix} -1 & 0 & 0 \\ 0 & 1 & 0 \\ 0 & 0 & -1 \end{pmatrix} \begin{pmatrix} x \\ y \\ z \end{pmatrix} + \begin{pmatrix} 0 \\ 0 \\ 1 \end{pmatrix}.$$

   (i) Show that f is a rotation and determine the rotation axis and rotation angle.
   (ii) Let $g : \mathbb{R}^3 \to \mathbb{R}^3$ be a linear isometry with

$$\begin{pmatrix} x \\ y \\ z \end{pmatrix} \mapsto \begin{pmatrix} 1 & 0 & 0 \\ 0 & -1 & 0 \\ 0 & 0 & -1 \end{pmatrix} \begin{pmatrix} x \\ y \\ z \end{pmatrix}.$$

   Then g is a rotation around the axis

$$\begin{pmatrix} 1 \\ 0 \\ 0 \end{pmatrix}.$$

   Show that the product $f \circ g$ is a rotary translation.

6. Let

$$A := \begin{pmatrix} -2 & 0 & 3 \\ 2 & 4 & 0 \\ 1 & 0 & 0 \end{pmatrix}.$$

(a) Determine the characteristic polynomial $\chi_A(x)$ and the eigenvalues λ_1, λ_2 and λ_3 of A.

(b) Calculate the corresponding eigenvectors for each eigenvalue.

7. Let

$$A = \begin{pmatrix} 1 & 0 & 0 \\ 0 & \frac{3}{5} & -\frac{4}{5} \\ 0 & \frac{4}{5} & \frac{3}{5} \end{pmatrix}$$

be given. Show that A has exactly one real eigenvalue λ. Calculate one eigenvector for λ.

8. Let $A \in M(n \times n, \mathbb{R})$ be a real $(n \times n)$-matrix. Then A is called *symmetric*, if $A = A^T$. Show the following:

(a) $\langle A\vec{x}, \vec{y} \rangle = \langle \vec{x}, A\vec{y} \rangle$ for all $\vec{x}, \vec{y} \in \mathbb{R}^n$.
(*Hint*: Use the equation $\langle A\vec{x}, \vec{y} \rangle = (A\vec{x})^T \vec{y} = \vec{x}^T (A^T \vec{y})$.)

(b) Let $n = 3$ and $A \in M(3 \times 3, \mathbb{R})$ be a symmetric matrix. Let λ and μ be two different eigenvalues of A and $\vec{v}$ and $\vec{w}$ be the eigenvalues for λ and μ, respectively. Show that $\vec{v}$ and $\vec{w}$ are perpendicular to each other, that is, $\langle \vec{v}, \vec{w} \rangle = 0$.

9. Let G be a group. A subgroup N of G is called a *normal subgroup* of G, if $gN = Ng$ for all $g \in G$, with $gN = \{gh \mid h \in N\}$ and $Ng = \{hg \mid h \in N\}$. Therefore, if N is a normal subgroup of G, then there is for each $g \in G$ and $h \in N$ an element $h' \in N$ with $gh = h'g$.
Let $O^+(n)$ be the set of all matrices A in $O(n)$ with $\det(A) = 1$.
Show that $O^+(n)$ is a normal subgroup of $O(n)$.

10. (a) Let

$$A = \begin{pmatrix} \frac{1}{\sqrt{2}} & -\frac{1}{\sqrt{2}} & 0 \\ \frac{1}{\sqrt{3}} & \frac{1}{\sqrt{3}} & \frac{1}{\sqrt{3}} \\ \frac{1}{\sqrt{6}} & \frac{1}{\sqrt{6}} & -\frac{2}{\sqrt{6}} \end{pmatrix}.$$

Determine if A is orthogonal.

(b) Let

$$A = a \begin{pmatrix} 6 & -8 \\ c & 6 \end{pmatrix}$$

with $a, c \in \mathbb{R}$. For which values a and c is the matrix A orthogonal?

(c) For which $k \in \mathbb{R}$ is the matrix

$$A = \begin{pmatrix} 1 & k \\ k & 1 \end{pmatrix}$$

orthogonal?

11. Let T be the set of translations in the group G of all isometries of $\mathbb{R}^n$. Show that:

(a) T is an Abelian group with respect to the concatenation.

(b) T is a normal subgroup of G (see Exercise 9 for a definition).

(*Hint*: If $f \in G$, then $f = \tau \circ f'$ with f' a linear isometry and a translation. Therefore, it is enough to show that, if f is a linear isometry and $\tau \in T$, then $f^{-1} \circ \tau \circ f \in T$.)

3 The Conic Sections in the Euclidean Plane

3.1 The Conic Sections

To the classical Greek geometers the simplest geometric curve was a line. After a line their most studied geometric curves were the *conic sections*. The work of the ancient Greek mathematicians on conic sections culminated around 200 BC, when Appollonius of Perga (ca. 265–190 BC) undertook a systematic study of their properties.

The aim of this chapter is to give a classification and a geometric description of the planar conic sections.

In a Euclidean plane there are four conic sections: a *circle*, an *ellipse*, a *parabola* and a *hyperbola,* with the circle being a special type of an ellipse, see Figures 3.1 and 3.2. They are curves of intersection of a plane and a *double circular cone* which we define below.

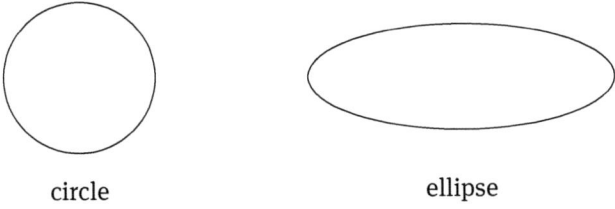

circle   ellipse

Figure 3.1: Conic sections: circles and ellipses.

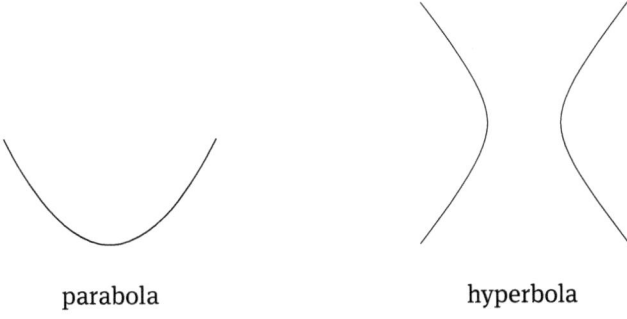

parabola   hyperbola

Figure 3.2: Conic sections: parabolas and hyperbolas.

A (double circular) cone is a pair of circular cones meeting at their common vertex. Formally, we give the following definition.

Definition 3.1.

(1) A (*double circular*) *cone* $K \subset \mathbb{R}^3$ is the rotation surface of a line g around an axis h which g cuts and which is not orthogonal to g, see Figure 3.3. We call S the peak of the cone.

https://doi.org/10.1515/9783110740783-003

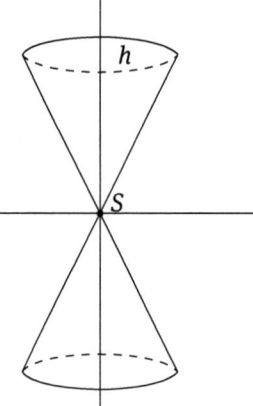

Figure 3.3: Double circular cone.

(2) A *conic section* (or simply a conic) is a curve obtained as the intersection of the surface of a double circular cone with a plane E.

Remark 3.2. For a geometric visualization we often assume that the axis h is the x_3-axis in $\mathbb{R}^3$ and the cone is symmetric with respect to the x_1x_2-plane. If we suppose in addition that the plane E does not contain the peak S of the cone, then our classification result will give the following:
(1) If the plane E is parallel to the x_1x_2-plane then the resulting conic section is a circle.
(2) If the plane E is not parallel to the x_1x_2-plane, not parallel to a line g' on the cone and not parallel to the x_3-axis h, then the resulting conic section is an ellipse.
(3) If the plane E is parallel to a line g' on the cone then the resulting conic section is a parabola.
(4) If the plane E is parallel to the x_3-axis h then the resulting conic section is a hyperbola.

We work in the Euclidean spaces $\mathbb{R}^2$ and $\mathbb{R}^3$ which we consider equipped with the canonical scalar product

$$\langle \vec{x}, \vec{y} \rangle \quad \text{for } \vec{x}, \vec{y} \in \mathbb{R}^n, n = 2 \text{ or } 3.$$

Our aim is to give a classification and a geometric description of the planar conic sections.

Remark 3.3. Before we start with this project, we need to develop the *Hessian Normal Form* for a plane in the Euclidean space $\mathbb{R}^3$.

A plane $E \subset \mathbb{R}^3$ is explained by a supporting vector $\vec{p} \in \mathbb{R}^3$ (considered as a point on E) and two linearly independent plane indication vectors $\vec{u}, \vec{v} \in \mathbb{R}^3$:

$$E = \{\vec{p} + s\vec{u} + t\vec{v} \mid s, t \in \mathbb{R}\}.$$

A vector $\vec{n} \in \mathbb{R}^3$, $\vec{n} \neq \vec{0}$, is a normal vector to E if $\langle \vec{n}, \vec{u} \rangle = \langle \vec{n}, \vec{v} \rangle = 0$, that is, $\vec{n}$ is orthogonal to $\vec{u}$ and $\vec{v}$ and, hence, to E, see Figure 3.4. Such an $\vec{n}$ always exists by the Gram–Schmidt orthonormalization procedure (see [12]).

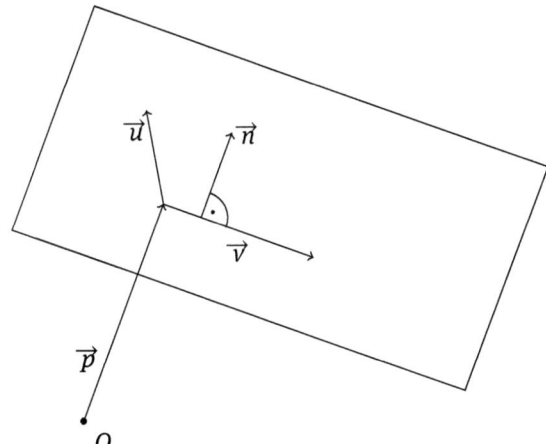

Figure 3.4: Hessian Normal Form.

Fact. $E = \{\vec{x} \in \mathbb{R}^3 \mid \langle \vec{x}, \vec{n} \rangle = d\}$, where $d = \langle \vec{p}, \vec{n} \rangle$, $\vec{p} \in E$ (considered again as a vector $\vec{p} = \overrightarrow{OP}$ from the center $O = \vec{0}$ to a point $P \in E$), and $\vec{n}$, $\vec{n} \neq \vec{0}$, is a normal vector to E.

Proof of the fact. We have to show that

$$E = \{\vec{x} = \vec{p} + s\vec{u} + t\vec{v} \mid s, t \in \mathbb{R}\} = \{\vec{x} \mid \langle \vec{x}, \vec{n} \rangle = d\}.$$

"$\subset$"

$$\langle \vec{x}, \vec{n} \rangle = \langle \vec{p} + s\vec{u} + t\vec{v}, \vec{n} \rangle = \langle \vec{p}, \vec{n} \rangle + \langle s\vec{u}, \vec{n} \rangle + \langle t\vec{v}, \vec{n} \rangle = \langle \vec{p}, \vec{n} \rangle = d.$$

"$\supset$"

$$\langle \vec{x}, \vec{n} \rangle = d = \langle \vec{p}, \vec{n} \rangle \Leftrightarrow \langle \vec{x} - \vec{p}, \vec{n} \rangle = \vec{0}.$$

Then $(\vec{x} - \vec{p}) \perp \vec{n}$, and hence there exist $s, t \in \mathbb{R}$ with $\vec{x} - \vec{p} = s\vec{u} + t\vec{v}$. $\qquad\qquad \square$

The form $E = \{\vec{x} \in \mathbb{R}^3 \mid \langle \vec{x}, \vec{n} \rangle = d\}$ where $d = \langle \vec{p}, \vec{n} \rangle$, $\vec{p} \in E$ and $\vec{n} \neq \vec{0}$, a normal vector to E, is called the *Hessian normal form* for the plane $E \subset \mathbb{R}^3$.

Lemma 3.4. *Consider a conic section, which is formed by the cone K and the plane E. If E contains the peak S of the cone, then $K \cap E$ is a single point, a line or a pair of lines.*

Proof. We may assume that the axis h is equal to the x_3-axis in $\mathbb{R}^3$ and the peak S is equal to the center $O = (0,0,0) =: 0$. Then the equation for the cone is given by

$$x_1^2 + x_2^2 = kx_3^2, \quad k > 0,$$

since the points on the cone, which have the same x_3-coordinate, are on a circle parallel to the x_1x_2-plane.

Further, we may assume that the plane E is orthogonal to the x_2x_3-plane. A normal vector $\vec{n}$ to E then is of the form $(0, n_2, n_3)^T$, and the Hessian normal form for E is $n_2x_2 + n_3x_3 = d$. The peak $S = O = 0$ of the cone is in E. Therefore $d = 0$.

If $n_3 \neq 0$ then $x_3 = \lambda x_2$ with $\lambda = -\frac{n_2}{n_3}$. Then $K \cap E$ is given by $x_1^2 + x_2^2 = kx_3^2$, and therefore $x_1^2 = (k\lambda^2 - 1)x_2^2$. Now, if $k\lambda^2 < 1$, then $K \cap E = \{0\}$. If $k\lambda^2 = 1$ then $K \cap E$ is the line $x_3 = \lambda x_2$, $x_1 = 0$.

Finally, if $k\lambda^2 > 1$ then the intersection $K \cap E$ is the pair of lines given by

$$x_3 = \lambda x_2, \quad x_1 = \pm\sqrt{(k\lambda^2 - 1)} \cdot x_2.$$

Now let $n_3 = 0$. Then the intersection $K \cap E$ is the pair of lines $x_2 = 0$, $x_1 = \pm\sqrt{k}x_3$. □

Remark 3.5. If $S \in E$, then we showed that $K \cap E$ is a point, that is, $K \cap E = \{S\}$, a line or a pair of lines.

We call these cases the *degenerate conic sections*.

From now on for the rest of this chapter, we do not consider the degenerate cases, that is, we only consider the case with $S \notin E$.

We call these cases *regular conic sections*, that is, those with $S \notin E$.

To get a general equation for the (regular) conic sections $K \cap E$, we choose a suitable system of coordinates. As the center $O = 0$ of the coordinate system we choose a point of $K \cap E$ which has the smallest distance from the peak S of the cone, and as x_1-axis the line OA through O and A, where A is the intersection point of the axis h of the cone with the plane E. In addition, the point A shall be on the positive ray of the x_1-axis.

Agreement. As already done in earlier chapters, we now write (x, y, z) instead of (x_1, x_2, x_3) for the points in $\mathbb{R}^3$. The points $X = (x, y, z)$ we also may consider as the vectors $\overrightarrow{OX}$. By CD we denote the line through the points C and D in $\mathbb{R}^3$, and by $\overline{CD}$ the line segment from C to D (this we may also interpret as the connection vector from C to D with starting point C).

We now choose as the y-axis the line in E through O which is orthogonal to the x-axis. As the z-axis we take the line through O which is orthogonal to E.

There are two cases to consider:

Case 1. The plane E cuts the cone K only in one half, see Figure 3.5.

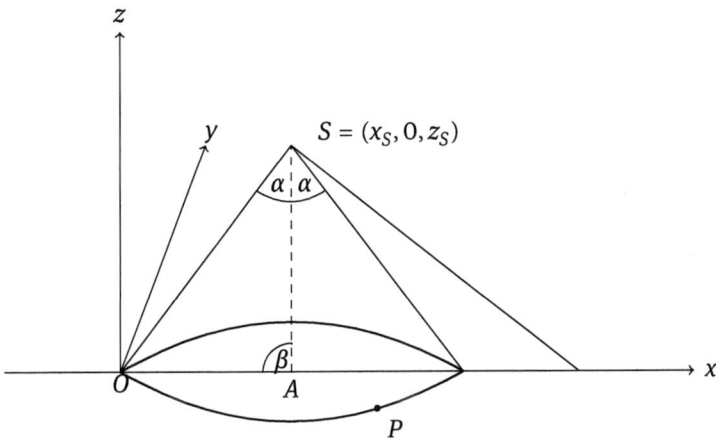

Figure 3.5: Case 1.

Case 2. The plane E cuts both halves of the cone K, see Figure 3.6.

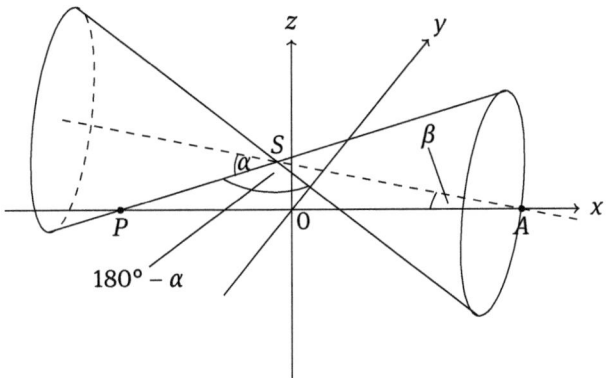

Figure 3.6: Case 2.

Since the plane E is identical to the xy-plane, we have $z = 0$ for each point P of the intersection $K \cap E$. Since the center O is chosen as the point of $K \cap E$ which has the smallest distance from S, the perpendicular from S onto the xy-plane is the line OA, that is, on the x-axis. Hence, $y_s = 0$.

Hence we know about the coordinates of the points S and P that $S = (x_s, 0, z_s)$ and $P = (x, y, 0)$, where $P \in K \cap E$ is arbitrary.

Since $P \in K \cap E$, we have either $\sphericalangle(\overrightarrow{SP}, \overrightarrow{SA}) = \alpha$ (Case 1) or $\sphericalangle(\overrightarrow{SP}, \overrightarrow{SA}) = 180° - \alpha$ (Case 2), where α is half of the opening angle of the cone. Hence,

$$\cos(\alpha) = \cos(\sphericalangle(\overrightarrow{SP}, \overrightarrow{SA})) = \frac{\langle \overrightarrow{SP}, \overrightarrow{SA} \rangle}{\|\overrightarrow{SP}\|\|\overrightarrow{SA}\|}$$

and

$$\cos(180° - \alpha) = -\cos(\alpha) = \cos(\sphericalangle(\overrightarrow{SP}, \overrightarrow{SA})) = \frac{\langle \overrightarrow{SP}, \overrightarrow{SA} \rangle}{\|\overrightarrow{SP}\|\|\overrightarrow{SA}\|}.$$

Let $\vec{n} = \frac{\overrightarrow{SA}}{\|\overrightarrow{SA}\|}$. Then $\langle \overrightarrow{SP}, \vec{n} \rangle = \|\overrightarrow{SP}\| \cdot \cos(\alpha)$, respectively $\langle \overrightarrow{SP}, \vec{n} \rangle = -\|\overrightarrow{SP}\| \cos(\alpha)$. In both cases we get

$$(\langle \overrightarrow{SP}, \vec{n} \rangle)^2 = \|\overrightarrow{SP}\|^2 \cos^2(\alpha).$$

Since also O is a point of the cone with $\sphericalangle(\overrightarrow{SO}, \overrightarrow{SA}) = \alpha$, we have

$$\langle \overrightarrow{SO}, \vec{n} \rangle = \|\overrightarrow{SO}\| \cos(\alpha).$$

Since $\langle \overrightarrow{SP}, \vec{n} \rangle = \langle \overrightarrow{SO}, \vec{n} \rangle + \langle \overrightarrow{OP}, \vec{n} \rangle$, it follows that

$$\|\overrightarrow{SP}\|^2 \cos^2(\alpha) = (\|\overrightarrow{SO}\| \cos(\alpha) + \langle \overrightarrow{OP}, \vec{n} \rangle)^2. \tag{3.1}$$

We now continue with the coordinates. Since $S = (x_s, 0, z_s)$ and $P = (x, y, 0)$, we get $\overrightarrow{SO} = (-x_s, 0, -z_s)^T$, $\overrightarrow{OP} = (x, y, 0)^T$ and $\overrightarrow{SP} = (x - x_s, y, -z_s)^T$.

Since the y-component of $\vec{n}$ must also be 0, we may represent $\vec{n}$ by $\vec{n} = (x_n, 0, z_n)$ with $x_n^2 + z_n^2 = 1$, because $\|\vec{n}\| = 1$. Hence, $\langle \overrightarrow{OP}, \vec{n} \rangle = x x_n$. If we plug this into (3.1), we get

$$(x^2 + x_s^2 - 2xx_s + y^2 + z_s^2) \cos^2(\alpha) = (x_s^2 + z_s^2) \cos^2(\alpha) + 2xx_n \cos(\alpha) \sqrt{x_s^2 + z_s^2} + x^2 x_n^2.$$

By solving this equation for y^2, we get

$$y^2 = 2\left(\frac{x_n}{\cos(\alpha)} \sqrt{x_s^2 + z_s^2} + x_s \right) x + \left(\frac{x_n^2}{\cos^2(\alpha)} - 1 \right) x^2. \tag{3.2}$$

We define $\epsilon := \frac{x_n}{\cos(\alpha)}$ and $p := \epsilon \sqrt{x_s^2 + z_s^2} + x_s$.

The quantity ϵ is called the *numerical eccentricity* and p the *parameter* of the regular conic section. Altogether we have the following.

Theorem 3.6 (Vertex equation of regular conic sections). *A regular conic section with numerical eccentricity ϵ and parameter p is described by the equation*

$$y^2 = 2px + (\epsilon^2 - 1)x^2. \tag{3.3}$$

Remark 3.7. In fact, $\epsilon = \frac{\cos(\beta)}{\cos(\alpha)}$ where β is the cutting angle between the plane E and the cone axis and 2α the opening angle of the cone (see Figures 3.5 and 3.6). This we see as follows.

From the choice of the coordinate system, β is the angle between $\vec{n}$ and the unit vector in the direction of the x-axis. Since $\|\vec{n}\| = 1$, we get $\cos(\beta) = \langle \vec{n}, (1,0,0)^T \rangle = x_n$ and hence, $\epsilon = \frac{\cos(\beta)}{\cos(\alpha)}$.

Since α is a sharp angle or a right angle, we have $\epsilon \geq 0$.

We now consider the vertex equation subject to ϵ. This is how we get a geometric description of the respective curve.

Case 1 ($\epsilon = 0$). Because of $\epsilon = \frac{\cos(\beta)}{\cos(\alpha)}$, we get in this case $\cos(\beta) = 0$, that is, $\beta = \frac{\pi}{2} = 90°$, the cone axis is orthogonal to the cutting plane, and the conic section is a circle.

If we consider the vertex equation (3.3) with $\epsilon = 0$, we get

$$y^2 = 2px - x^2 = p^2 - (x - p)^2 \quad \text{or} \quad (x - p)^2 + y^2 = p^2,$$

that is, the equation of a circle with center $M = (p, 0)$ and radius p.

Case 2 ($0 < \epsilon < 1$). In this case $\cos(\beta) < \cos(\alpha)$, and hence $\beta > \alpha$. With $k := -(\epsilon^2 - 1)$, which gives $k > 0$, the vertex equation (3.3) has the form

$$y^2 = 2px - kx^2 = \frac{p^2}{k} - \frac{p^2}{k} + 2px - kx^2 = \frac{p^2}{k} - k\left(x - \frac{p}{k}\right)^2,$$

that is,

$$y^2 + k\left(x - \frac{p}{k}\right)^2 = \frac{p^2}{k} \quad \text{or} \quad \frac{(x - \frac{p}{k})^2}{\frac{p^2}{k^2}} + \frac{y^2}{\frac{p^2}{k}} = 1.$$

This is by definition an ellipse in a position with one axis parallel to the x-axis and the other parallel to the y-axis.

We define $a := \frac{p}{k}$ and $b := \frac{p}{\sqrt{k}}$, and may assume that $a > 0$, $b > 0$. Then

$$\frac{(x - a)^2}{a^2} + \frac{y^2}{b^2} = 1,$$

and we get Figure 3.7.

The value a is called the *semi-major axis*, and b the *semi-minor axis* of the ellipse. By a translation we get the standard position defined by

$$\frac{x^2}{a^2} + \frac{y^2}{b^2} = 1$$

of the ellipse with center $(0, 0)$.

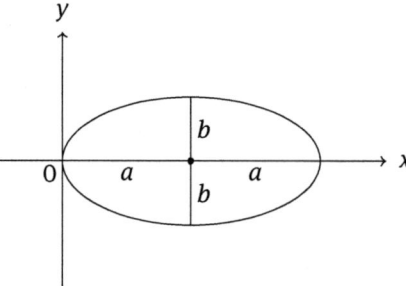

Figure 3.7: Ellipse.

Case 3 ($\epsilon = 1$). In this case we get $\alpha = \beta$, and the vertex equation (3.3) becomes

$$y^2 = 2px,$$

and this equation by definition describes a parabola.

Case 4 ($\epsilon > 1$). In this case $\beta < \alpha$. If $k := \epsilon^2 - 1$, then $k > 0$, and we get the vertex equation (3.3) in the form

$$y^2 = 2px + kx^2 = -\frac{p^2}{k} + \frac{p^2}{k} + 2px + kx^2 = -\frac{p^2}{k} + k\left(x + \frac{p}{k}\right)^2,$$

that is,

$$k\left(x + \frac{p}{k}\right)^2 - y^2 = \frac{p^2}{k} \quad \text{or} \quad \frac{(x + \frac{p}{k})^2}{\frac{p^2}{k^2}} - \frac{y^2}{\frac{p^2}{k}} = 1.$$

We define $a = \frac{p}{k}$ and $b = \frac{p}{\sqrt{k}}$, and may assume that $a > 0$, $b > 0$. Then

$$\frac{(x + a)^2}{a^2} - \frac{y^2}{b^2} = 1.$$

This is by definition a hyperbola where the principle axis is the x-axis, and we get the following Figure 3.8.

Again, a is called the semi-major axis and b the semi-minor axis. By a translation we get the standard position defined by

$$\frac{x^2}{a^2} - \frac{y^2}{b^2} = 1$$

of the hyperbola with center $(0,0)$.

Summary. In standard position we have the following equations:

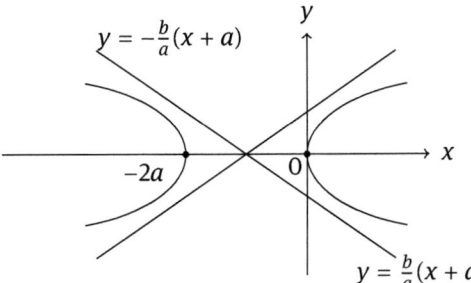

$$y = -\frac{b}{a}(x + a)$$

$$y = \frac{b}{a}(x + a)$$

Figure 3.8: Hyperbola.

(1) Ellipse:

$$\frac{x^2}{a^2} + \frac{y^2}{b^2} = 1$$

with $a, b > 0$. If $a = b$ we have a circle.

(2) Hyperbola:

$$\frac{x^2}{a^2} - \frac{y^2}{b^2} = 1,$$

with $a, b > 0$.

(3) Parabola:

$$y^2 = 2px$$

with $p \neq 0$.

Remark 3.8. We explained and described the conic sections by cuts of a plane E with a cone K. This legitimates the name conic sections for the circle, ellipse, hyperbola and parabola.

The conic sections can also be introduced, especially in school mathematics or in special courses, in a more natural and elementary manner.

We do this in the next three sections and show the correspondence with the previous definitions.

3.2 Ellipse

The ellipse is the locus of all points P in the Euclidean space $\mathbb{R}^2$ for which the sum of the distances $\|\overrightarrow{PF_1}\| + \|\overrightarrow{PF_2}\|$ from two fixed points F_1 and F_2, the *focal points*, is constant:

$$K_{\text{ell}} = \{P \in \mathbb{R}^2 \mid \|\overrightarrow{PF_1}\| + \|\overrightarrow{PF_2}\| = 2a\},$$

where $F_1, F_2 \in \mathbb{R}^2$ and $0 < \|\overrightarrow{F_1F_2}\| < 2a$, see Figure 3.9.

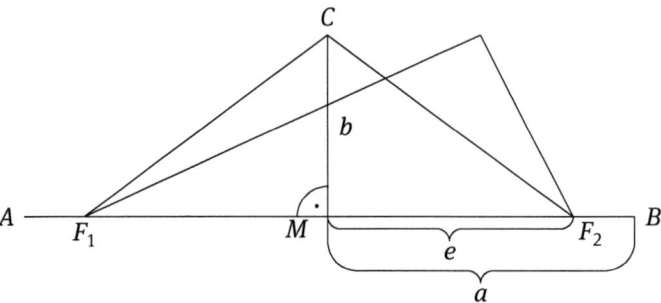

Figure 3.9: $A, B, C \in K_{\text{ell}}$ (A, B on the line F_1F_2).

We define the following:

$A, B, C \in K_{\text{ell}}$ with A, B on the line F_1F_2; M the center of the line segment $\overline{F_1F_2}$; $e := \|\overrightarrow{F_1M}\| = \|\overrightarrow{MF_2}\|$ the *linear eccentricity*; $a := \|\overrightarrow{MA}\| = \|\overrightarrow{MB}\|$ the *semi-major axis*; and $b := \|\overrightarrow{MC}\|$ the *semi-minor axis*.

From this we analytically derive the equation for K_{ell}. We take the coordinate system with $F_1 = (-e, 0)$, $F_2 = (e, 0)$, $e > 0$ in the xy-plane. For a point $P = (x, y)$ on the ellipse K_{ell}, we have

$$\|\overrightarrow{PF_1}\| + \|\overrightarrow{PF_2}\| = 2a,$$

that is,

$$\sqrt{(x+e)^2 + y^2} + \sqrt{(x-e)^2 + y^2} = 2a$$

from the Theorem of Pythagoras. By changing around and squaring, we get

$$(x+e)^2 + y^2 = 4a^2 - 4a\sqrt{(x-e)^2 + y^2} + (x-e)^2 + y^2,$$

that is,

$$ex - a^2 = -a\sqrt{(x-e)^2 + y^2}.$$

If we square again

$$e^2x^2 + a^4 - 2exa = a^2((x-e)^2 + y^2),$$
$$0 = a^2x^2 - e^2x^2 + a^2y^2 - a^4 + a^2e^2,$$
$$0 = x^2(a^2 - e^2) + a^2y^2 - a^2(a^2 - e^2).$$

Since $\|\overrightarrow{CF_1}\| = \|\overrightarrow{CF_2}\| = a$, we have

$$a^2 - e^2 = b^2$$

by the Theorem of Pythagoras, again. Therefore, $0 = x^2b^2 + a^2y^2 - a^2b^2$, and hence

$$\frac{x^2}{a^2} + \frac{y^2}{b^2} = 1,$$

the ellipse in standard position.

The characterization of an ellipse as the locus of points so that the sum of the distances to the focal points is constant leads to a method of drawing one using two drawing pins, a length of strings, and a pencil. In this method, pins are pushed into the paper at two points, which become the focal points of the ellipse. A string is tied at each end of the two pins and the tip of a pen is pulled to make the loop taut to form a triangle. The tip of the pen then traces an ellipse if it is moved while keeping the string taut. Using two pegs and a rope, gardeners use this procedure to outline an elliptical flower bed, thus it is called the *gardener's ellipse construction*.

We now determine the area I of the ellipse

$$\frac{x^2}{a^2} + \frac{y^2}{b^2} = 1$$

in standard position. We get

$$I = 4b \int_0^a \sqrt{1 - \left(\frac{x}{a}\right)^2} \, dx = 4ab \int_0^1 \sqrt{1 - x^2} \, dx = 4ab \int_0^{\frac{\pi}{2}} \cos^2(t) \, dt = \pi ab$$

by the rule of substitution and the fact that

$$\int_0^{\frac{\pi}{2}} \cos^2(t) \, dt = \int_0^{\frac{\pi}{2}} \left(\frac{1}{2}\cos(2t) + \frac{1}{2}\right) dt = \frac{\pi}{4}.$$

If the focal points of an ellipse are allowed to coincide then the ellipse is the locus of points equidistant from the single focal point and therefore a circle with center at this focus. It follows that a circle is a special type of an ellipse with only one focal point, and with equal major semi-axis a and minor semi-axis b. In the standard equation we then have $a = b$ and the circle in standard position is given by

$$x^2 + y^2 = r^2,$$

where r is the radius.

3.3 Hyperbola

The hyperbola is the locus of all points P in $\mathbb{R}^2$, for which the absolute value of the difference of the distance $|\|\overrightarrow{PF_1}\| - \|\overrightarrow{PF_2}\||$ from two fixed points F_1 and F_2, the *focal points*, is constant:

$$K_{\text{hyp}} = \{P \in \mathbb{R}^2 \mid \big| \|\overrightarrow{PF_1}\| - \|\overrightarrow{PF_2}\| \big| = 2a\},$$

where $F_1, F_2 \in \mathbb{R}^2$ and $0 < 2a < \|\overrightarrow{F_1F_2}\|$, see Figure 3.10.

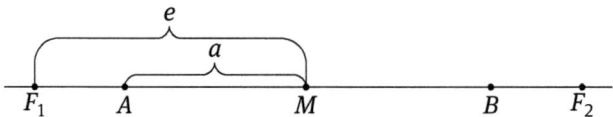

• P

Figure 3.10: $A, B, P \in K_{\text{hyp}}$ (A, B on the line F_1F_2).

Let $A, B \in K_{\text{hyp}}$. Again we define M to be the center of the line segment $\overline{F_1F_2}$, $e :=$ $\|\overrightarrow{MF_1}\| = \|\overrightarrow{MF_2}\|$ the linear eccentricity, $a := \|\overrightarrow{AM}\| = \|\overrightarrow{BM}\|$ the semi-major axis, $b :=$ $\sqrt{e^2 - a^2} > 0$ the semi-minor axis.

For the analytic determination of the equation for K_{hyp}, we take the coordinate system with $F_1 = (-e, 0)$ and $F_2 = (e, 0)$, $e > 0$. For a point $P = (x, y)$ on the hyperbola K_{hyp}, we have

$$\big| \|\overrightarrow{PF_1}\| - \|\overrightarrow{PF_2}\| \big| = 2a,$$

that is,

$$\left| \sqrt{(x + e)^2 + y^2} - \sqrt{(x - e)^2 + y^2} \right| = 2a,$$

or equivalently,

$$\pm\left(\sqrt{(x + e)^2 + y^2} - \sqrt{(x - e)^2 + y^2} \right) = 2a.$$

By changing around and squaring, we get

$$(x + e)^2 + y^2 = 4a^2 + 4a\sqrt{(x - e)^2 + y^2} + (x - e)^2 + y^2,$$

that is,

$$ex - a^2 = a\sqrt{(x - e)^2 + y^2}.$$

If we square again, we get

$$0 = x^2(a^2 - e^2) + a^2y^2 - a^2(a^2 - e^2).$$

Now, for the hyperbola, we define b by $b = \sqrt{e^2 - a^2}$, that is, $b^2 = e^2 - a^2$. Therefore $0 = -b^2x^2 + a^2y^2 + a^2b^2$, and finally, $\frac{x^2}{a^2} - \frac{y^2}{b^2} = 1$, the hyperbola in standard position.

3.4 Parabola

The parabola is the locus of all points P in $\mathbb{R}^2$, which have the same distance from a fixed point $F \in \mathbb{R}^2$, the *focal point*, and a fixed line ℓ in $\mathbb{R}^2$ with $F \notin \ell$, ℓ is called the *directrix*.

Remark 3.9. The distance of a point P (as a vector $\overrightarrow{OP}$ in $\mathbb{R}^2$) from a line ℓ is discussed in [12, Chapter 14].

Here it is enough to take for the distance of P and ℓ the equivalent version

$$d(P, \ell) = \inf_{Q \in \ell} \|\overrightarrow{PQ}\|.$$

This infimum is realized by the nearest point Q_ℓ from P on ℓ. It is the point at which the line segment from it to P is perpendicular to ℓ, that is,

$$d(P, \ell) = \|\overrightarrow{PQ_\ell}\|.$$

We have

$$K_{\text{par}} = \{P \in \mathbb{R}^2 \mid \|\overrightarrow{PF}\| = d(P, \ell)\},$$

where $F \in \mathbb{R}^2$, $\ell \subset \mathbb{R}^2$, $F \notin \ell$.

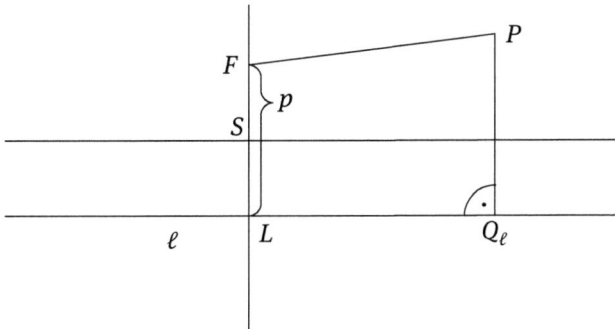

Figure 3.11: $P, S \in K_{\text{par}}$.

Parameter $p = \|\overrightarrow{PL}\| = d(F, \ell)$ is called the focal parameter; it is the distance from the focal point to the directrix; L is the nearest point from F on ℓ, the foot of the perpendicular from F to the directrix ℓ.

The center of the line segment $\overline{FL}$ is called the *vertex S.*

For the analytic determination of the equation for K_{par}, we take the coordinate system with $F = (0, \frac{p}{2})$, $S = (0,0) = O$ and $\ell : y = -\frac{p}{2}$, see Figure 3.11.

For a point $P = (x,y)$ on the parabola K_{par}, we have $\|\overrightarrow{PF}\| = d(P, \ell)$, that is,

$$\sqrt{x^2 + \left(y - \frac{p}{2}\right)^2} = \sqrt{\left(y + \frac{p}{2}\right)^2}.$$

By changing around and squaring, we get $x^2 = 2py$ as the equation for the parabola.

Reflection at the line $x = y$, that is, exchanging the axes, gives the parabola in standard position, that is, $y^2 = 2px$.

We remark that the reflection at the y-axis gives the parabola $y^2 = 2kx$ with $k = -p$.

3.5 The Principal Axis Transformation

We now consider a general equation of the form

$$a'x^2 + b'xy + c'y^2 + d'x + e'y + f' = 0$$

with $a', b', c', d', e', f' \in \mathbb{R}$. This equation describes a quadric over $\mathbb{R}$, that is, the set

$$\{(x,y) \in \mathbb{R}^2 \mid a'x^2 + b'xy + c'y^2 + d'x + e'y + f' = 0,\ a',b',c',d',e',f' \in \mathbb{R}\}.$$

If not all coefficients are 0, then the quadric may be transformed by isometries of $\mathbb{R}^2$ (see Chapter 2), so by translations, reflections and rotations, into

(a) the empty set ($\frac{x^2}{a^2} + \frac{y^2}{b^2} = -1$),

(b) one point ($\frac{x^2}{a^2} + \frac{y^2}{b^2} = 0$),

(c) a line ($ax + by + c = 0$),

(d) a pair of lines ($\frac{x^2}{a^2} - \frac{y^2}{b^2} = 0$),

(e) an ellipse ($\frac{x^2}{a^2} + \frac{y^2}{b^2} = 1$),

(f) a hyperbola ($\frac{x^2}{a^2} - \frac{y^2}{b^2} = 1$) or

(g) a parabola ($y^2 = 2px$).

We now consider a quadric over $\mathbb{R}$ as above such that not all coefficients are 0. We may transform the quadric by isometries of $\mathbb{R}^2$ into one of the forms (a) to (g). In the cases (b), (c) and (d) the quadric represents a degenerate conic section.

We now assume that the quadric represents a regular conic section. We may write the quadric in matrix form,

$$(x\ \ y) \begin{pmatrix} a' & \frac{b'}{2} \\ \frac{b'}{2} & c \end{pmatrix} \begin{pmatrix} x \\ y \end{pmatrix} + (d'\ \ e') \begin{pmatrix} x \\ y \end{pmatrix} + f' = 0$$

or

$$(x \quad y \quad 1) \begin{pmatrix} a' & \frac{b'}{2} & \frac{d'}{2} \\ \frac{b'}{2} & c' & \frac{e'}{2} \\ \frac{d'}{2} & \frac{e'}{2} & f' \end{pmatrix} \begin{pmatrix} x \\ y \\ 1 \end{pmatrix} = 0.$$

The regular conic sections described by the quadric can be classified in terms of the value $b'^2 - 4a'c'$, called the *discriminant* of the regular conic section.

Theorem 3.10. *A regular conic section is*
(1) *an ellipse if* $b'^2 - 4a'c' < 0$,
(2) *a circle if* $a' = c'$ *and* $b' = 0$,
(3) *a parabola if* $b'^2 - 4a'c' = 0$ *and*
(4) *a hyperbola if* $b'^2 - 4a'c' > 0$.

We leave the proof of Theorem 3.10 as exercise 7.

The discriminant $b'^2 - 4a'c'$ of the regular conic section and the quantify $a' + c'$ are invariant under arbitrary rotations and translations of the coordinate axes.

Applications

Regular conic sections are important in physics, optics and astronomy. Circles are applicable in uniform circular motions, and parabolas are applicable in kinematic problems. An object that is moving laterally with constant velocity traces a parabolic path subject to gravity. The shape of a planet around the sun is described by Kepler's law (named after J. Kepler, 1571–1630) as an ellipse with the sun as a focal point. More generally the orbits of two massive objects that interact according to Newton's law of universal gravitation (named after I. Newton, 1643–1727) are conic sections if their common center of mass is considered to be rest. If they are bound together, they will trace out ellipses; if they are moving apart, they will both follow parabolas or hyperbolas. Parabolas are used in optics. A parabola is a two-dimensional, mirror-symmetric curve. A mirror that has a cross-sectional parabolic shape has the property that a ray of light directly towards the mirror will be reflected towards the focal point of the parabola. A parabolic reflector is a mirror that uses this property to concentrate reflected light onto a single point. These kinds of mirrors are used in mircophone and satellite technology. More generally, the reflective properties of the conic sections are used in the design of searchlight, radiotelescopes and optical telescopes. A searchlight uses a parabolic mirror as a reflector. An optical telescope often uses a primary parabolic mirror to reflect light towards a secondary hyperbolic mirror which reflects it again to a focal point behind the first mirror.

Exercises

1. Let a plane $E \subset \mathbb{R}^3$ be given by the supporting vector $\vec{p} = (1, 1, -1)$ and the plane indication vectors $\vec{u} = (0, 1, 3)$ and $\vec{v} = (2, -1, 0)$. Find a normal vector to E and the Hessian normal form for E.

2. Let CS be a conic section in standard position, that is,
   - $CS = \{(x, y) \in \mathbb{R}^2 \mid \frac{x^2}{a^2} + \frac{y^2}{b^2} = 1\}$ if CS is an ellipse,
   - $CS = \{(x, y) \in \mathbb{R}^2 \mid \frac{x^2}{a^2} - \frac{y^2}{b^2} = 1\}$ if CS is a hyperbola and
   - $CS = \{(x, y) \in \mathbb{R}^2 \mid y^2 = 2px, p \neq 0\}$ if CS is a parabola.

   Let $P_0 = (x_0, y_0) \in CS$. Calculate the tangent $t = \{(x, y) \in \mathbb{R}^2 \mid y = mx + n\}$ of CS at P_0.

3. Let CS be an ellipse with focal points F_1, F_2, semi-major axis a and eccentricity ϵ, $0 < \epsilon < 1$.

   Show that there exists a line ℓ with the following property:
   The points $P \in CS$ are exactly those points in the plane with

   $$\frac{\|\overrightarrow{PF_1}\|}{\|P\ell\|} = \epsilon = \frac{\|\overrightarrow{F_1F_2}\|}{2a}$$

   where $\|P\ell\| = \inf\{\|\overrightarrow{PQ}\| \mid Q \in \ell\}$.

4. Let CS be an ellipse with focal points F_1, F_2, and let $P \in CS$. Show that the normal vector $\vec{n}$ to CS at P bisects the angle $\sphericalangle(\overrightarrow{PF_1}, \overrightarrow{PF_2})$.

5. Let CS be a parabola with focal point F, and let $P \in CS$. Let L be the foot of the perpendicular from P to the directrix ℓ.

   Show that the tangent line t to CS at P bisects the angle between the line segments $\overline{PF}$ and $\overline{PL}$.

6. Show the statement in Section 3.5 in detail.

7. Prove Theorem 3.10.

8. Describe the quadric defined by the equation
   (a) $x^2 + y^2 - xy - x - y = 0$,
   (b) $x^2 + y^2 - xy - x - y + 1 = 0$,
   (c) $x^2 + y^2 - 3xy = 0$,
   (d) $x^2 + y^2 - 3xy - 1 = 0$,
   (e) $xy - 1 = 0$.

9. Let CS be an ellipse in standard position

   $$\left\{ (x, y) \in \mathbb{R}^2 \mid \frac{x^2}{a^2} + \frac{y^2}{b^2} = 1 \right\}.$$

   Show that CS can be written parametrically as $(x, y) = (a\cos(\Theta), b\sin(\Theta))$.

10. Let CS be a hyperbola in standard position

$$\left\{(x,y) \in \mathbb{R}^2 \mid \frac{x^2}{a^2} - \frac{y^2}{b^2} = 1\right\}.$$

Show that CS can be written parametrically as $(x,y) = (\pm a \cosh(u), b \sinh(u))$.
We remind that the hyperbolic functions are defined as

$$\cosh(u) = \frac{1}{2}(e^u + e^{-u}) \quad \text{and} \quad \sinh(u) = \frac{1}{2}(e^u - e^{-u}).$$

11. Let CS be a parabola in standard position

$$\{(x,y) \in \mathbb{R}^2 \mid y^2 = 2px, p \neq 0\}.$$

Show that CS can be written parametrically as $(x,y) = (\frac{p}{2}t^2, pt)$.

4 Special Groups of Planar Isometries

In Chapter 2 we saw that the study of planar Euclidean geometry depended upon the knowledge of the *Euclidean group*, the group of all isometries of the Euclidean plane $\mathbb{R}^2$. In this chapter we describe certain special groups of planar isometries, that is, certain special subgroups of the Euclidean group. These special groups are tied in many cases to chemistry and physics, especially the structure of crystals.

In order to proceed, we need to explain some basic group-theoretical material. For details see, for instance, [8] or [13]. We first explain presentations of groups by generators and relations. Let G be a group and $X \subset G$. The elements of X are called a set of *generators* of G if every element of G is expressible as a finite product of their powers (including negative powers). Here X is almost always a finite subset of G, and then G is finitely generated. Each element $w \in G$ can be written in the form

$$w = x_1^{\alpha_1} x_2^{\alpha_2} \cdots x_k^{\alpha_k},$$

$x_i \in X$, $\alpha_i \in \mathbb{Z} \setminus \{0\}$, $k \geq 0$ with $i = 1, 2, \ldots, k$.

Recall that $k = 0$ means $w = 1$, the identity element in G. Now, let $\{x_1, x_2, \ldots, x_n\}$ be a generating set of G. There are always relations of the type $x_i x_i^{-1} = x_i^{-1} x_i = 1$. These we do not count, they are trivial. A set of relations $r_i = r_i(x_1, x_2, \ldots, x_n) = 1$, $i \in I$, satisfied by the generators $x_1, x_2, \ldots, x_n$ of G is called a defining set of relations (with respect to X) of G if every relation satisfied by the generators is an algebraic consequence of these particular relations r_i, $i \in I$. Here I is almost always finite, say $I = \{1, 2, \ldots, m\}$, and we call then G *finitely presented* if $|X| < \infty$ and $|I| < \infty$.

Let $X = \{x_1, x_2, \ldots, x_n\}$ be a set of generators of G, and $R = \{r_1, r_2, \ldots, r_m\}$ be a set of defining relations of G. Then we write

$$G = \langle X \mid r_1 = r_2 = \cdots = r_m = 1 \rangle \quad \text{or just} \quad G = \langle X \mid R \rangle$$

and call G presented by the set X of generators and the set R of defining relations. If $R = \emptyset$, we write $G = \langle X \mid \; \rangle$ instead of $G = \langle X \mid \emptyset \rangle$. We call $G = \langle X \mid R \rangle$ *commutative* or *Abelian*, after N. H. Abel (1802–1829), if $aba^{-1}b^{-1} = 1$ is a relation for all $a, b \in X$.

The group G is completely described by the set X and R. The presentation of G is very helpful, but certainly not unique. For instance, we may change the generating system and then the defining relations system, respectively.

Example 4.1. We consider some basic examples of Abelian groups:
(1) $\mathbb{Z} = \langle x \mid \; \rangle$.
(2) Let C_n be a cyclic group of order $n < \infty$. Then $C_n = \langle x \mid x^n = 1 \rangle$.
(3) Let $G = \mathbb{Z} \times \mathbb{Z}$ be the Cartesian product. Then G becomes a group via the multiplication $(m, n)(p, q) = (m + p, n + q)$. Let $a = (1, 0)$ and $b = (0, 1)$. Then

$$G = \langle a, b \mid aba^{-1}b^{-1} = 1 \rangle.$$

https://doi.org/10.1515/9783110740783-004

More generally, if H and K are groups and $G = H \times K$ their Cartesian product then G becomes a group via the multiplication $(h_1, k_1)(h_2, k_2) = (h_1 h_2, k_1 k_2)$. G is called the *direct product* of H and K.

Remark 4.2. If in a presentation some relation r_i is given as $r_i = uv^{-1} = 1$, then we also write in the presentation $u = v$ instead of $r_i = 1$. In this sense

$$G = \mathbb{Z} \times \mathbb{Z} = \langle a, b \mid ab = ba \rangle.$$

Occasionally, we need isomorphisms between groups.

Let G and H be groups. A map $f : G \to H$ is called a *homomorphism* if $f(gh) = f(g)f(h)$ for all $g, h \in G$, that is, f respects the multiplication in G. The homomorphism f is called an *isomorphism* if f is bijective; we call groups G and H isomorphic and write $G \cong H$. In case $H = G$, we also call f an automorphism of G.

We remark that certainly $f(G)$ is a subgroup of H. The kernel $\ker(f)$ is defined as $\ker(f) = \{g \in G \mid f(g) = 1 \text{ in } H\}$. Then $\ker(f)$ is a normal subgroup of G. Recall that a subgroup N of G is a normal subgroup of G if $gN = Ng$ for all $g \in G$, and we denote this by $N \lhd G$. Here gN and Ng are the cosets of N in G defined by $gN = \{gn \mid n \in N\}$ and $Ng = \{ng \mid n \in N\}$. Since $(gN)(hN) = ghNN$ and $NN = \{nn' \mid n, n' \in N\} = N$, we have that the set G/N is a group under the multiplication $(gN)(hN) = ghN$, the *factor group* or quotient of G by N. Now let $f : G \to H$ be again a homomorphism between groups. We mentioned that $\ker(f)$ is a normal subgroup of G. Then $G/\ker(f)$ is isomorphic to $f(G)$, written as $G/\ker(f) \cong f(G)$ (for more details, see [8] or [13]).

More generally, if U is a subgroup of a group G then the number (or cardinality) of left cosets gU is equal to the number of right cosets Ug of U in G, and this number is called the *index* $|G : U|$ of U in G.

Finally, we need the notation of a *semidirect product* of two groups. Consider a group G with identity element 1, a subgroup H, and a normal subgroup $N \lhd G$.

If $N \cap H = \{1\}$, and if G is generated by H and N, then G is called the (*inner*) *semidirect product* of N and H written $G = N \rtimes H$.

In this chapter we again consider the plane $\mathbb{R}^2$ as a two-dimensional vector space equipped with the scalar product (inner product)

$$\langle \vec{x}, \vec{y} \rangle = x_1 y_1 + x_2 y_2$$

for vectors $\vec{x} = (x_1, x_2)$, $\vec{y} = (y_1, y_2)$.

In Chapter 2 we saw that an isometry of $\mathbb{R}^2$ is a rotation, a translation, a reflection or a glide reflection.

If $\varphi : \mathbb{R}^2 \to \mathbb{R}^2$ is an isometry then $\varphi = \tau_{\vec{v}} \circ f$ with f a linear isometry and $\tau_{\vec{v}}$ a translation. We call φ *oriented* if $\det(A) = 1$ for the orthogonal matrix $A \in O(2)$ of f (with respect to an orthogonal basis of $\mathbb{R}^2$), otherwise *non-oriented*.

Hence, rotations and translations are oriented, reflections and glide reflections are non-oriented. Geometrically, an oriented isometry φ preserves the cyclic order of vertices around a triangle and a non-oriented isometry does not, see Figure 4.1.

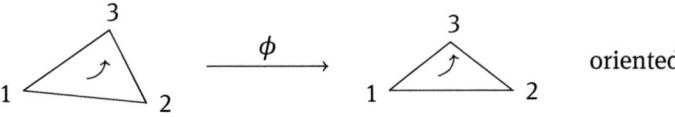

oriented

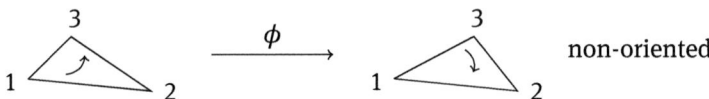

non-oriented

Figure 4.1: Oriented and non-oriented isometries.

If G is a group of isometries of $\mathbb{R}^2$ then the set of oriented isometries in G forms a subgroup G^+ of G of index $|G : G^+| \leq 2$, and hence G^+ is a normal subgroup of G; G^+ is called the oriented subgroup of G. If $\varphi = \tau_{\vec{v}} \circ f$, with f oriented, then $\det(A) = 1$ for the matrix A which belongs to f (with respect to an orthonormal basis of $\mathbb{R}^2$). The elements $A \in O(2)$ with $\det(A) = 1$ therefore form a subgroup $SO(2)$ of $O(2)$ of index 2, the *special orthogonal group*.

Before we consider special groups of isometries of $\mathbb{R}^2$, we make some group-theoretical remarks. We call an isometry of $\mathbb{R}^2$ now a *planar isometry* and denote by E the set of all planar isometries and by E^+ the set of the oriented elements of E. As seen more generally, E^+ is a normal subgroup of E of index 2. Let T be the set of all planar translations.

Let $\tau_1, \tau_2 \in T$. If $\tau_1 \neq \tau_2^{-1}$ then $\tau_1 \circ \tau_2$ is a nontrivial translation since $\tau_1 \circ \tau_2$ is not the identity 1 and has no fixed points. Recall that $P \in \mathbb{R}^2$ is a fixed point of the planar isometry φ if $\varphi(P) = P$.

Hence T forms a subgroup of E^+. In fact, T is Abelian because $\tau_1 \circ \tau_2 = \tau_2 \circ \tau_1$ for $\tau_1, \tau_2 \in T$. If $\alpha \in E$ and $\tau \in T, \tau \neq 1$, then $\alpha \circ \tau \circ \alpha^{-1}$ is a translation. This can be seen as follows. Let $P, Q \in \mathbb{R}^2$ with $\tau(P) = Q$, we have $Q \neq P$ because $\tau \neq 1$. Then $\alpha \circ \tau \circ \alpha^{-1}(\alpha(P)) = \alpha(Q)$, that is, $\alpha \circ \tau \circ \alpha^{-1}$ maps $\alpha(P)$ to $\alpha(Q)$. Since τ is a translation, $\alpha \circ \tau \circ \alpha^{-1}$ must also be a translation. Therefore T is a normal subgroup of E, and hence also of E^+.

As a conclusion, we get the chain $E \supset E^+ \supset T \supset \{1\}$ of normal subgroups. The factor groups have the structures

$$E/E^+ \cong \mathbb{Z}_2, \quad E/T \cong O(2), \quad E^+/T \cong SO(2) \quad \text{and} \quad T = T/\{1\} \cong \mathbb{R}^2,$$

where $\mathbb{R}^2$ means the additive group under vector addition.

The purpose of this chapter is to consider symmetry groups of figures in $\mathbb{R}^2$ and groups which act discontinuously on $\mathbb{R}^2$. Certainly there are many overlappings between these two concepts. The main reference for this chapter is the book [21].

Definition 4.3. A (*plane*) *figure F* is just a subset of $\mathbb{R}^2$. A symmetry of *F* is a planar isometry α with the property that $\alpha(F) = F$.

The set of all symmetries of *F* forms a group which we denote by Sym(*F*).

Example 4.4.
(1) If $F = \mathbb{R}^2$ then Sym(*F*) = *E*.
(2) Sym(*F*) = {1, γ} where γ is the reflection at *g*, and *F* is as in Figure 4.2.

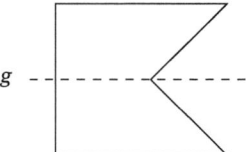

g

Figure 4.2: Reflection at *g*.

(3) Sym(*F*) = {1, γ, δ, $\gamma \circ \delta$} where γ is a reflection at *g*, δ is a reflection at *h*, and $\gamma \circ \delta$ is a rotation with center *M* and rotation angle π, and *F* is as in Figure 4.3. We have $\gamma^2 = \delta^2 = (\gamma \circ \delta)^2 = 1$, hence Sym(*F*) is the Klein four group (see Section 4.1).

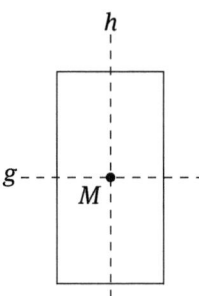

h

g

M

Figure 4.3: Reflections at *g* and *h*.

(4) Sym(*F*) = {1}, and *F* is as in Figure 4.4.

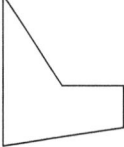

Figure 4.4: Only trivial symmetry.

Definition 4.5. A subgroup *G* of *E* is *discontinuous* if, for every point $P \in \mathbb{R}^2$, there is some disc *D* with center *P* that contains no image $\alpha(P)$, $\alpha \in G$, of *P* other than *P* itself.

We note that this condition is equivalent to the following:

If $P \in \mathbb{R}^2$ is any point and $G(P) = \{\alpha(P) \mid \alpha \in G\}$ the orbit of P, and if D is any disc in $\mathbb{R}^2$, then the intersection $G(P) \cap D$ consists of only finitely many points.

Remarks 4.6.

(1) If G is a discontinuous subgroup of E, then every subgroup H of G is also discontinuous, in particular $T_G = G \cap T$ is discontinuous.

(2) Let G be discontinuous and $\alpha \in G$ be a rotation around $P \in \mathbb{R}^2$. Assume that α has infinite order. Let D be a disk in $\mathbb{R}^2$ with center P and positive radius. Then, for each $Q \in D \setminus \{P\}$, all $\alpha^n(Q) \in D$, $n \in \mathbb{N}$, are in D and pairwise different, which gives a contradiction. Hence we get the following.

Theorem 4.7. *Let G be discontinuous and $\alpha \in G$ be a rotation. Then α has finite order.*

A *fixed point group* is a subgroup of the Euclidean group for which there exists a common fixed point for all elements.

Curiously, only few fixed point groups, that is, groups for which there exists a common fixed point for the elements, are found to be compatible with nontrivial translations to generate discontinuous groups, more precisely, frieze groups and wallpaper groups or planar crystallographic groups. In fact, lattice compatibility implies such a severe restriction that we have only 7 geometrically distinct frieze groups and 17 geometrically distinct wallpaper groups (see Sections 4.5 and 4.6).

Some art historians claim that all 17 wallpaper designs are present in the mosaic tiles found in the Alhambra in Granada, Spain, constructed in the 13th century.

Discontinuous groups play a fundamental role throughout this chapter. We start with the consideration of symmetry groups of regular polygons and regular tessellations of the plane $\mathbb{R}^2$.

4.1 Regular Polygons

A regular polygon in $\mathbb{R}^2$ is a polygon that is equiangular (all interior angles are equal in measure) and equilateral (all sides have the same length). If $n \in \mathbb{N}$, $n \geq 3$, we call a regular polygon a regular n-gon if it has n sides.

Let F be a regular n-gon, $n \geq 3$. The sum of the interior angles of a polygon is $(n-2)\pi$, hence, each interior angle is $\frac{(n-2)\pi}{n} = \pi - \frac{2\pi}{n}$.

All vertices of a regular n-gon, $n \geq 3$, lie on a common circle, the circumscribed circle. Together with the property of equal-length sides, this implies that every regular n-gon, $n \geq 3$, also has an inscribed circle that is tangent to every side at the midpoint.

If $n = 3$ we have an equilateral triangle and if $n = 4$ we have a square.

Now, let the figure F be a regular n-gon, $n \geq 3$, and $G = \text{Sym}(F)$.

Let O be the center of F. Then F has a rotational symmetry of order n, that is, G contains a rotation σ around O of order n. This rotation forms a cyclic group of order

n, that is, $\{1, \sigma, \sigma^2, \ldots, \sigma^{n-1}\} = \langle \sigma \mid \sigma^n = 1 \rangle$. There is no other rotation, because G does not contain a translation.

Figure F also has reflection symmetry in n axes that pass through the center O of F. If n is even then half of these axes pass through two opposite vertices, and the other half through the midpoints of opposite sides.

If n is odd all axes pass through a vertex and the midpoint of the opposite side. In any case, G contains n rotations and n reflections.

Since $G^+ = \langle \sigma \mid \sigma^n = 1 \rangle$ and $|G : G^+| = 2$ for the index, we get $|G| = 2n$. Let $\delta \in G$ be one fixed reflection. Then necessarily

$$G = \{1, \sigma, \sigma^2, \ldots, \sigma^{n-1}, \rho, \rho \circ \sigma, \rho \circ \sigma^2, \ldots, \rho \circ \sigma^{n-1}\}.$$

All $\rho \circ \sigma^k$, $0 \le k \le n - 1$, are reflections, that is,

$$(\rho \circ \sigma^k)^2 = 1, \quad \text{or equivalently,} \quad \rho \circ \sigma^k \circ \rho = \sigma^{-k}.$$

But, for every k, $\rho \circ \sigma^k \circ \rho = (\rho \circ \sigma \circ \rho)^k = \sigma^{-k}$. Hence, $(\rho \circ \sigma^k)^2 = 1$ is a consequence of the relation $(\rho \circ \sigma)^2 = 1$.

Hence, we have the following result.

Theorem 4.8. *Let F be a regular n-gon, $n \ge 3$, and $G = \mathrm{Sym}(F)$. Then*

$$G = \langle \sigma, \rho \mid \sigma^n = \rho^2 = (\rho \circ \sigma)^2 = 1 \rangle = \langle \rho_1, \rho_2 \mid \rho_1^2 = \rho_2^2 = (\rho_1 \circ \rho_2)^n = 1 \rangle,$$

where $\rho_1 = \rho$ and $\rho_2 = \rho \circ \sigma$.

A group with a presentation

$$\langle x, y \mid x^2 = y^2 = (xy)^n = 1 \rangle, \quad n \ge 3,$$

is called the *dihedral group D_n*.

Hence, up to isomorphism, $G = \mathrm{Sym}(F)$ is a dihedral group D_n, $n \ge 3$.

We extend the notation.

If $n = 2$ we define

$$D_2 = \langle x, y \mid x^2 = y^2 = (xy)^2 = 1 \rangle.$$

We also define the dihedral group D_2 to be the *Klein four group*. D_2 is Abelian but not cyclic.

If $n = 1$ we define $D_1 = \langle x \mid x^2 = 1 \rangle \cong \mathbb{Z}_2$. The group $D_\infty = \langle x, y \mid x^2 = y^2 = 1 \rangle$ is called the *infinite dihedral group*. We remark that $D_2 \cong S_3$, the symmetric group on $\{1, 2, 3\}$; S_3 is the smallest non-Abelian group.

4.2 Regular Tessellations of the Plane

A *tessellation* (*tiling*) of the plane $\mathbb{R}^2$ is a division of $\mathbb{R}^2$ into non-overlapping closed regions, which we shall always assumed to be bounded by finite polygons. A tessellation is regular if all the faces (tiles) into which it divides the plane are bounded by congruent regular n-gons, $n \geq 3$.

Let F be a regular tessellation by regular n-gons with $n \geq 3$. Since all the interior angles of a regular n-gon are equal, there must be the same number $m \geq 3$ of such n-gons at each vertex. Hence the interior angle of the n-gon must be $\frac{2\pi}{m}$. On the other hand, we know that the interior angle of the n-gon is $\pi - \frac{2\pi}{n}$. Hence, $\frac{2\pi}{m} = \pi - \frac{2\pi}{n}$ or $\frac{1}{m} + \frac{1}{n} = \frac{1}{2}$. There are exactly three solutions:

$$(n, m) = (3, 6), (4, 4), (6, 3).$$

This means that there are only three types under similarity, that is, maps $f : \mathbb{R}^2 \to \mathbb{R}^2$ with $\|f(P) - f(Q)\| = k\|P - Q\|$ for some real $k > 0$ and all $P, Q \in \mathbb{R}^2$ of regular tessellations F. Here P and Q just denote the local vectors $\overrightarrow{OP}$ and $\overrightarrow{OQ}$, respectively, and $\| \; \|$ denotes the length of a vector. For each of these, we determine its symmetry group $G = \mathrm{Sym}(F)$. To determine the structure of G, we could start with its translation subgroup T_G and extend T_G by adding rotations, reflections and glide reflections. However, at this stage we would like to introduce an important method by Poincaré, by which we obtain a presentation of G from more intrinsic geometrical considerations.

We confine the attention to the tessellation by squares, the most visualized case. The treatment of the other two cases, the tessellation by equilateral triangles and the tessellation by regular hexagons, follows exactly the same pattern. After having done the presentations using the method by Poincaré, we continue with the classification of the discontinuous subgroups of E starting with the translation subgroups.

Hence let F be a regular tessellation of the plane by squares. We may assume that we have unit squares, that is, squares with side length 1.

Theorem 4.9. *The symmetry group* $G = \mathrm{Sym}(F)$, F *being a regular tessellation of the plane by squares, has the presentation*

$$G = \langle \alpha, \beta, \gamma \mid \alpha^2 = \beta^2 = \gamma^2 = (\gamma \circ \beta)^4 = (\alpha \circ \gamma)^2 = (\beta \circ \alpha)^4 = 1 \rangle.$$

Proof. As already mentioned we may assume that we have a regular tessellation by unit squares. Let Q be a single square. It is incidental that we get G by the translations of the squares and the symmetry group $G_Q = \mathrm{Sym}(Q) \cong D_4$ of the single square Q, see Figure 4.5.

We have

$$\mathbb{R}^2 = \bigcup_{g \in G} g(\Delta)$$

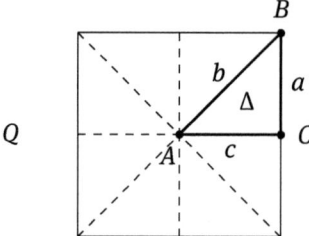

Figure 4.5: Fundamental region for G.

and $\overline{g(\mathring{\Delta})} \cap \overline{h(\mathring{\Delta})} = \emptyset$ if $g \neq h$, where $\overline{g(\mathring{\Delta})} = g(\Delta) \setminus \vartheta(g(\Delta))$ and $\vartheta(g(\Delta)) = g(a) \cup g(b) \cup g(c)$.

This, in fact, means that Δ is a *fundamental region for* G.

Let α, β, γ be the reflections at the sides a, b, c of Δ, respectively. Certainly $\alpha, \beta, \gamma \in G$.

Claim. $G = \langle \alpha, \beta, \gamma \rangle$.

Proof of the claim. Let $H = \langle \alpha, \beta, \gamma \rangle$, H is a subgroup of G. Assume that $H \neq G$. Let

$$U = \bigcup_{h \in H} h(\Delta).$$

By assumption we have $U \neq \mathbb{R}^2$. Thus some side s of some triangle $h(\Delta)$, $h \in H$, must lie on the boundary of U, separating $h(\Delta)$ from some $g(\Delta)$, $g \in G$, $g \notin H$.

Now s is the image $h(a)$, $h(b)$, $h(c)$ of a side a, b, c of Δ, and the reflection ρ in s is the corresponding element $h \circ \alpha \circ h^{-1}$, $h \circ \beta \circ h^{-1}$ or $h \circ \gamma \circ h^{-1}$. Since $h, \alpha, \beta, \gamma \in H$, we get that $\rho \in H$ and $g = \rho \circ h \in H$, which gives a contradiction. Therefore $H = G$, which proves the claim. □

We now seek a set of defining relations among α, β and γ. Since α, β and γ are reflections, we have the relations $\alpha^2 = \beta^2 = \gamma^2 = 1$. Further, $\gamma \circ \beta$ is a rotation at A through an angle twice the interior angle $\frac{2\pi}{8}$ of Δ at A, that is, through $\frac{2\pi}{4}$, therefore $(\gamma \circ \beta)^4 = 1$.

Similarly, $(\alpha \circ \gamma)^2 = 1$ and $(\beta \circ \alpha)^4 = 1$.

We show now that these six relations form a full set of defining relations.

Let D be the tessellation of $\mathbb{R}^2$ by the regions $g(\Delta)$, $g \in G$. Let O be the center of Δ, that is, the intersection of lines joining vertices to midpoints of opposite sides. We join $g(O)$ and $h(O)$ with an (directed) edge e if and only if $g(\Delta)$ and $h(\Delta)$ have a side s in common. This way we constructed a graph C with vertices being all centers $g(O)$ of triangles $g(\Delta)$ and edges as described above, see Figure 4.6. For a general discussion of graphs, see Chapter 5.

We assign to e the *label* $\lambda(e) = \alpha, \beta, \gamma$ if s is the image $g(a)$, $g(b)$, $g(c)$ of the side a, b, c of Δ, respectively.

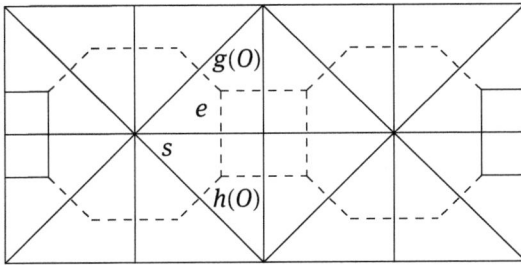

Figure 4.6: Dual graph C.

Let $p = e_1 e_2 \cdots e_n$, $n \geq 1$, be a path in C, that is, a sequence of edges such that e_{i+1} begins where e_i ends for $1 \leq i < n$. Let the path p begin at some $g_0(O)$ and have successive vertices $g_0(O), g_1(O), \ldots, g_n(O)$, see Figure 4.7.

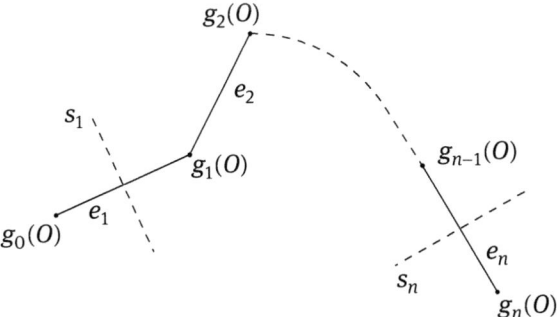

Figure 4.7: Path from $g_0(O)$ to $g_n(O)$.

Let $x_i = \lambda(e_i) = \alpha, \beta, \gamma$, $1 \leq i \leq n$, respectively.

The reflection ρ at the edge s_i between $g_i(\Delta)$ and $g_{i+1}(\Delta)$ is $\rho = g_i \circ x_{i+1} \circ g_i^{-1}$, hence $g_{i+1} = \rho \circ g_i = g_i \circ x_{i+1}$. It follows that $g_n = g_0 \circ x_1 \circ x_2 \circ \cdots \circ x_n$.

We define $\lambda(p) = x_1 \circ x_2 \circ \cdots \circ x_n$, and hence, $g_n = g_0 \circ \lambda(p)$.

Now, if p is a closed path, that is, $g_0(O) = g_n(O)$, then $g_0 = g_n$, and we have the relation $\lambda(p) = 1$.

On the other hand, if $w = x_1 \circ x_2 \circ \cdots \circ x_n = 1$ in G and $g_0 \in G$ arbitrary, then there exists a unique closed path p at $g_0(O)$ such that $\lambda(p) = 1 = x_1 \circ x_2 \circ \cdots \circ x_n$.

Now we have to show that each relation is a consequence of the described 6 relations. For this we consider the graph C and the closed paths in C more carefully.

The graph C divides the plane into regions Δ^* which are square or octagonal, and defines a tessellation D^* of $\mathbb{R}^2$ (the dual of D). Suppose that a path p runs between points P and Q along an arc on side of the boundary of a region Δ^*. Let p' be the path between P and Q obtained from p by replacing this arc by the other side of Δ^*.

We have $\lambda(p) = \lambda(p')$ because running around the boundary of Δ^* is one of $(\gamma \circ \beta)^{\pm 4}$, $(\alpha \circ \gamma)^{\pm 2}$, $(\beta \circ \alpha)^{\pm 4}$, see Figure 4.8.

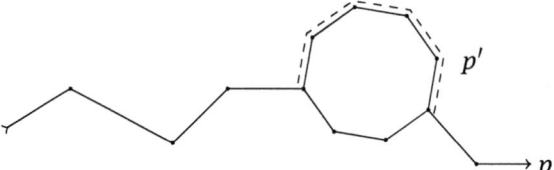

p'

p **Figure 4.8:** Replacing p by p'.

The relation $\alpha^2 = \beta^2 = \gamma^2 = 1$ corresponds to modifying p by deleting or inserting a spine, that is, an edge e_i followed by an edge e_{i+1}, that is, the same edge except traversed in the opposite direction, see Figure 4.9.

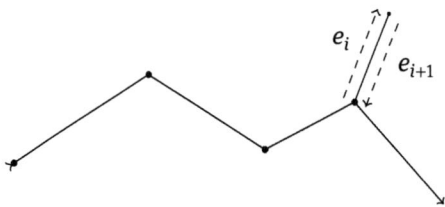

e_i e_{i+1}

p **Figure 4.9:** Cutting spines.

Showing that the given relations define G is therefore equivalent to showing that any closed path p in C can be reduced to the trivial path at some point (with $n = 0$ edges) by a succession of modifications of the two kinds from above. But this is clear. Inductively, by running around the other side of some region Δ^*, we can reduce the number of regions and spines enclosed by p, and finally reduce to the case that p is a simple loop. □

For the regular tessellation F of type $(n, m) = (3, 6)$, by equilateral triangular regions, six at each vertex, the argument differs from the case above only in that the interior angles of Δ are different.

We find a presentation

$$G = \langle \alpha, \beta, \gamma \mid \alpha^2 = \beta^2 = \gamma^2 = (\gamma \circ \beta)^3 = (\alpha \circ \gamma)^2 = (\beta \circ \alpha)^6 = 1 \rangle.$$

The remaining regular tessellation F of type $(6, 3)$ is dual to the one of type $(3, 6)$, and hence has the same group $G = \mathrm{Sym}(F)$. This can be seen also from the fact that a single fundamental region Δ serves for both cases, see Figure 4.10.

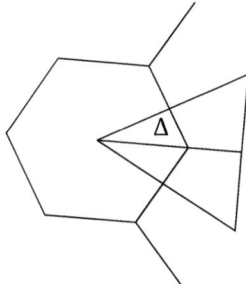

Figure 4.10: Dual types $(6, 3)$ and $(3, 6)$.

Remark 4.10. The triples $(2, 4, 4)$, $(2, 3, 6)$ and $(2, 6, 3)$ satisfy the equation

$$\frac{1}{p} + \frac{1}{q} + \frac{1}{r} = 1 \quad \text{where } p, q, r \in \mathbb{N} \setminus \{1\}.$$

For this equation, only one case remains, namely $(3, 3, 3)$. The corresponding group is

$$G = \langle \alpha, \beta, \gamma \mid \alpha^2 = \beta^2 = \gamma^2 = (\gamma \circ \beta)^3 = (\alpha \circ \beta)^3 = (\beta \circ \alpha)^3 = 1 \rangle.$$

This group is evidently a subgroup of index 2 of the full symmetry group of the regular tessellation of $\mathbb{R}^2$ by equilateral triangles.

Hence every group

$$G = \langle \alpha, \beta, \gamma \mid \alpha^2 = \beta^2 = \gamma^2 = (\gamma \circ \beta)^p = (\alpha \circ \gamma)^q = (\beta \circ \alpha)^r = 1 \rangle,$$

with $p, q, r \in \mathbb{N} \setminus \{1\}$ and $\frac{1}{p} + \frac{1}{q} + \frac{1}{r} = 1$, can be considered as a subgroup of E, generated by three reflections α, β, γ. The subgroup G^+ is generated by the rotations $x = \gamma \circ \beta$, $y = \beta \circ \alpha$ and $z = \alpha \circ \gamma$ and has a presentation

$$G^+ = \langle x, y, z \mid x^p = y^r = z^q = x \circ y \circ z = 1 \rangle = \langle x, y \mid x^p = y^r = (x \circ y)^q = 1 \rangle.$$

This can be easily seen if we go through the above proofs by deleting spines, that is, by deleting the elements α^2, β^2 and γ^2. In fact, for instance, we have just to consider $\Delta \cup \alpha(\Delta)$.

A group

$$\langle x, y \mid x^p = y^r = (x \circ y)^q = 1 \rangle, \quad 2 \le p, q, r,$$

is called a *triangle group*.

The connection with a triangle is given as follows:

Let $p, q, r \in \mathbb{N} \setminus \{1\}$ and $\frac{1}{p} + \frac{1}{q} + \frac{1}{r} = 1$. We consider in the plane a triangle Δ with interior angles $\frac{\pi}{p}, \frac{\pi}{q}, \frac{\pi}{r}$, see Figure 4.11.

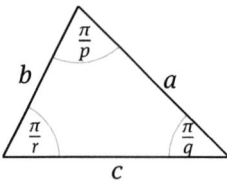

Figure 4.11: Triangle Δ with interior angles $\frac{\pi}{p}, \frac{\pi}{q}, \frac{\pi}{r}$.

Let G be the group generated by the reflections at the three sides a, b, c of Δ. Then G has a presentation

$$G = \langle \alpha, \beta, \gamma \mid \alpha^2 = \beta^2 = \gamma^2 = (\beta \circ \alpha)^p = (\alpha \circ \gamma)^q = (\gamma \circ \beta)^r = 1 \rangle.$$

This certainly works for the cases $(p, q, r) = (4, 2, 4)$ and $(3, 2, 6)$, up to the ordering in the triples, as we have seen from the symmetry groups of regular tessellations. It works also for the case $(p, q, r) = (3, 3, 3)$ if we consider an equilateral triangle, see Figure 4.12.

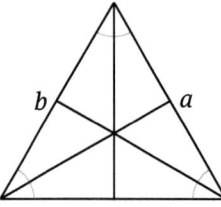

Figure 4.12: Equilateral triangle.

We now argue analogously as in the proof of Theorem 4.9.

4.3 Groups of Translations in the Plane $\mathbb{R}^2$

We now continue to consider discontinuous groups of isometries of the plane. We want to classify these via their translation subgroups.

Let G be a discontinuous group of isometries of the plane. Let $T_G = T \cap G$ be the translation subgroup.

In this section we describe all possible groups T_G.

To do this, let $G = T_G$ in this section.

A translation $\tau : \mathbb{R}^2 \to \mathbb{R}^2$ is given by $\vec{v} \mapsto \vec{v} + \vec{v}_0$ for some fixed vector $\vec{v}_0 \in \mathbb{R}^2$.

Theorem 4.11. *Let $G \subset T$ be a discontinuous group of translations of the plane. Then one of the following cases occur:*
(i) *$G = \{1\}$, that is, G contains only the identity map.*
(ii) *$G = \{\tau \mid \tau(\vec{v}) = \vec{v} + n\vec{v}_0, n \in \mathbb{Z}\}$ for some $\vec{v}_0 \neq \vec{0}$, that is, G is infinite cyclic. Such G is called a nontrivial simple-periodic group.*

(iii) $G = \{\tau \mid \tau(\vec{v}) = \vec{v} + m\vec{w}_1 + n\vec{w}_2,\ n, m \in \mathbb{Z}\}$ *for two linearly independent vectors* $\vec{w}_1, \vec{w}_2$. *Such group is called a double-periodic group.*

Proof. Suppose that $G \neq \{1\}$. If $\tau_1, \tau_2 \in G$ with $\tau_1(\vec{v}) = \vec{v} + \vec{w}_1$ and $\tau_2(\vec{v}) = \vec{v} + \vec{w}_2$ then also all translations τ with $\tau(\vec{v}) = \vec{v} + n\vec{w}_1 + m\vec{w}_2, n, m \in \mathbb{Z}$ are in G.

For a translation τ with $\tau(\vec{v}) = \vec{v} + \vec{c}$ and $\vec{v} \neq \vec{0}$ we call $\vec{c}$ a period for G.

We consider for G the set $M = \{\|\vec{c}\| \mid \vec{c}$ is a period for $G\}$ for periods for G.

Since G is discontinuous, the set M cannot have an accumulation point $\|\vec{c}\|$ (here $\|\vec{c}\|$ is the length of the vector $\vec{c} \in \mathbb{R}^2$).

Hence there exists a period $\vec{w}_1$ for G with minimal length $\|\vec{w}_1\|$. All transformations τ with $\tau(\vec{v}) = \vec{v} + m\vec{w}_1, m \in \mathbb{Z}$, belong to G.

Let $\tau \in G$ with $\tau(\vec{v}) = \vec{v} + r\vec{w}_1, r \in \mathbb{R}$. Then necessarily $r \in \mathbb{Z}$ because otherwise there exists a $m \in \mathbb{Z}$ with $0 < |r'| = |r - m| < 1$ and τ' with $\tau'(\vec{v}) = \vec{v} + r'\vec{w}_1$ belongs to G, contradicting the minimality of $\|\vec{w}_1\|$.

Suppose now that G contains a translation τ with $\tau(\vec{v}) = \vec{v} + \vec{b}$, and $\vec{b}$ is not of the form $\vec{b} = r\vec{w}_1, r \in \mathbb{R}$. Then $\vec{w}_1$ and $\vec{b}$ are linearly independent.

Again, since G is discontinuous, there is a period $\vec{w}_2$ of this kind, for which $\|\vec{w}_2\|$ is minimal. We have the following relations:

(1) $0 < \|\vec{w}_1\| \leq \|\vec{w}_2\|$,

(2) $\|\vec{a}\| \geq \|\vec{w}_1\|$ for all periods $\vec{a} \neq \vec{0}$ for G,

(3) $\|\vec{b}\| \geq \|\vec{w}_2\|$ for all periods $\vec{b}$ for G which are not of the form $\vec{b} = m\vec{w}_1$ for some $m \in \mathbb{Z}$.

Claim. Let $\vec{c}$ be an arbitrary period for G. Then necessarily

$$\vec{c} = n\vec{w}_1 + m\vec{w}_1 \quad \text{for some } n, m \in \mathbb{Z}.$$

Proof of the claim. First of all, $\vec{c} = r\vec{w}_1 + s\vec{w}_2$ for some $r, s \in \mathbb{R}$ because $\vec{w}_1$ and $\vec{w}_2$ form necessarily a basis of the vector space $\mathbb{R}^2$. We have

$$r = m + r_1, \quad -\frac{1}{2} < r_1 \leq \frac{1}{2},\ m \in \mathbb{Z},$$

$$s = n + s_1, \quad -\frac{1}{2} < s_1 \leq \frac{1}{2},\ n \in \mathbb{Z}.$$

Let $\vec{c}_1 = r_1\vec{w}_1 + s_1\vec{w}_2$. If $\vec{c}_1 \neq \vec{0}$ then $\vec{c}_1$ is a period for G.

If $s_1 = 0$, then also $r_1 = 0$ by the above arguments (r_1 has to be an integer). Analogously, if $r_1 = 0$ then also $s_1 = 0$ using the above relations.

Now, let $r_1 \neq 0 \neq s_1$. Then

$$\|\vec{c}_1\| < |r_1|\|\vec{w}_1\| + |s_1|\|\vec{w}_2\| \leq \frac{1}{2}\|\vec{w}_1\| + \frac{1}{2}\|\vec{w}_2\| \leq \|\vec{w}_2\|$$

because $\vec{w}_1, \vec{w}_2$ are linearly independent. This contradicts relation 3 from above.

Hence, we can not have $r_1 \neq 0 \neq s_1$. It follows $r_1 = s_1 = 0$, and therefore Theorem 4.11 is proven. $\qquad\square$

4.4 Groups of Isometries of the Plane with Trivial Translation Subgroup

Let G be a discontinuous group of isometries of the plane with $T_G = T \cap G = \{1\}$. By Theorem 4.7, we know that G does not contain rotations of infinite order.

Theorem 4.12. *G is a fixed point group, that is, there exists a $P \in \mathbb{R}^2$ with $f(P) = P$ for all $f \in G$.*

Proof. Since $T_G = T \cap G = \{1\}$, G also does not contain a glide reflection. This can be seen as follows. Assume that G contains a glide reflection ρ. Then we may express ρ as $\rho = \tau \circ f$ with $\tau \neq 1$ a translation and f a reflection such that $\tau \circ f = f \circ \tau$. Then

$$\rho^2 = (\tau \circ f) \circ (\tau \circ f) = \tau^2 \neq 1$$

is a nontrivial translation contradicting $T_G = \{1\}$. $\qquad\square$

Suppose now that G contains a nontrivial rotation σ. Without loss of generality, we may assume that σ has the center $\vec{0} = (0,0)$.

Let $\alpha \in G$. Then $\alpha \circ \sigma \circ \alpha^{-1}$ is a rotation with center $\alpha(\vec{0})$. Assume that $\alpha(\vec{0}) \neq \vec{0}$.

If α is oriented, then $\alpha \circ \sigma \circ \alpha^{-1} \circ \sigma^{-1}$ is a nontrivial translation giving a contradiction. If α is not oriented, then necessarily α is a reflection, and then $\alpha \circ \sigma \circ \alpha^{-1} \circ \sigma$ is a nontrivial translation giving a contradiction. Hence, $\alpha(\vec{0}) = \vec{0}$.

Therefore, if G contains a nontrivial rotation, then G has a fixed point. Now, suppose that G does not contain a nontrivial rotation. Now, since the product of two distinct reflections is a nontrivial translation or a nontrivial rotation, G can contain at most one reflection ρ.

Therefore, $G = \{1\}$ or $G = \{1, \rho\}$, ρ a reflection, and G fixes many points.

Theorem 4.13. *G is cyclic or a dihedral group D_n, $n \geq 2$.*

Proof. We know that $T_G = \{1\}$, and therefore G has a fixed point by Theorem 4.12. If G does not contain a nontrivial rotation, then $G = \{1\}$ or $G = \{1, \rho\}$, ρ a reflection, and G is cyclic.

Suppose that G contains a nontrivial rotation. Since G is discontinuous, G contains a rotation σ with smallest possible, positive angle θ for G. Especially, G does not contain a rotation with an angle θ_1 such that $n\theta < \theta_1 < (n+1)\theta$, $n \in \mathbb{Z}$. Hence, $G^+ = \langle \sigma \rangle$, and therefore $G = G^+$ or $G \cong D_n$ for some $n \in \mathbb{N}$. $\qquad\square$

Corollary 4.14. *Let H be a finite subgroup of E. Then H is cyclic or a dihedral group.*

Proof. Since H is finite, it is discontinuous with $T_H = \{1\}$. Now, the result follows from Theorem 4.13. $\qquad\square$

4.5 Frieze Groups

We now consider discontinuous subgroups G of E such that the translation subgroup $T_G = T \cap G$ is infinite cyclic, that is, $T_G = \langle \tau \mid \ \rangle$.

Such groups are the symmetry groups of certain infinite plane figures admitting as translation symmetries only powers of some translation τ along an axis (line) ℓ. Such figures are called *friezes*, and their symmetry groups are the *frieze groups*. In this section we enumerate the types of frieze groups and illustrate them by giving, for each type, a frieze whose symmetry group is of that type.

Let G be a frieze group, and let τ be a generator of T_G. Let $\tau : \mathbb{R}^2 \to \mathbb{R}^2$, $\vec{v} \mapsto \vec{v} + \vec{v}_0$. The vector $\vec{v}_0$ determines the direction of the translation. Let ℓ be the line through the origin $\vec{0}$ and $\vec{v}_0$ (as a point in $\mathbb{R}^2$). We call the line ℓ the *translation axis*; ℓ (and each parallel line) is mapped by the elements of T_G onto itself.

Case 1

G does not contain a nontrivial rotation. Then $G^+ = G \cap E^+ = T_G$. If $G = T_G = \langle \tau \mid \ \rangle$, then we have the first type.

Now let $G \neq T_G$. If $\alpha \in G$, $\alpha \notin T_G$, then $\alpha \circ \tau \circ \alpha^{-1}$ is a translation with translation axis $\alpha(\ell)$. Since $T_G = \langle \tau \rangle$ we must have $\alpha(\ell) = \ell$.

Since G does not contain a nontrivial rotation, we therefore must have $G = \langle \tau, \rho \rangle$, where ρ is a reflection or a glide reflection. This follows from the fact that $|G : G^+| = 2$. Let $\rho \in G$, $\rho \notin T_G$, a reflection or a glide reflection. Then necessarily $T_G = \langle \tau \rangle = \langle \rho \circ \tau \circ \rho^{-1} \rangle$, which gives $\rho \circ \tau \circ \rho^{-1} = \tau$ or $\rho \circ \tau \circ \rho^{-1} = \tau^{-1}$, which means that ρ has a reflection axis parallel or orthogonal to ℓ. We cannot have elements ρ of both kinds, for then their product would be a nontrivial rotation. If ρ is a reflection, it can be of either kind.

If ρ is a glide reflection, then ρ^2 is a nontrivial translation, hence $\rho^2 = \tau^h$ for some $h \neq 0$, and ρ has, without loss of generality, reflection axis ℓ. Now ρ commutes with τ, and $(\rho \circ \tau^k)^2 = \tau^{h+2k}$. Replacing ρ by $\rho \circ \tau^k$ for a suitable k, we may suppose that either $\rho^2 = 1$ and ρ is a reflection, or that $\rho^2 = \tau$ and ρ is a glide reflection. In the latter case F contains no reflection.

We have obtained four geometrically different types:

$$G_1 = \langle \tau \mid \ \rangle \cong \mathbb{Z},$$
$$G_1^1 = \langle \tau, \rho \mid \rho^2 = 1, \rho \circ \tau \circ \rho^{-1} = \tau \rangle \cong \mathbb{Z} \times \mathbb{Z}_2,$$
$$G_1^2 = \langle \tau, \rho \mid \rho^2 = 1, \rho \circ \tau \circ \rho^{-1} = \tau^{-1} \rangle \cong D_\infty,$$
$$G_1^3 = \langle \tau, \rho \mid \rho^2 = \tau, \rho \circ \tau \circ \rho^{-1} = \tau \rangle \cong \mathbb{Z}.$$

We emphasize that G_1 and G_1^3 are isomorphic as abstract groups, but are not geometrically equivalent since G_1 preserves orientation while G_1^3 does not.

Case 2

Group G contains a nontrivial rotation σ. Again, $\sigma \circ \tau \circ \sigma^{-1}$ must generate $\sigma T_G \sigma^{-1} = T_G$, and hence $\sigma \circ \tau \circ \sigma^{-1} = \tau$ or $\sigma \circ \tau \circ \sigma^{-1} = \tau^{-1}$. Since $\sigma \neq 1$, we cannot have $\sigma \circ \tau \circ \sigma^{-1} = \tau$. Therefore, $\sigma \circ \tau \circ \sigma^{-1} = \tau^{-1}$, and σ is a rotation of order 2. Without loss of generality, we may assume that the rotation center O of σ is on the axis ℓ. If σ' is any other rotation in G, then it also must have order 2, hence $\sigma' \circ \sigma$ is a translation, that is, $\sigma' \circ \sigma = \tau^h$ for some $h \in \mathbb{Z}$, and therefore $\sigma' = \tau^h \circ \sigma$.

Thus

$$G^+ = \langle \tau, \sigma \mid \sigma^2 = 1, \sigma \circ \tau \circ \sigma^{-1} = \tau^{-1} \rangle.$$

If $G = G^+$, then we have a first type.

Now let $G^+ \neq G$. Then $G = \langle \tau, \sigma, \rho \rangle$, where ρ is a reflection or a glide reflection because $|G : G^+| = 2$. As before, the axis of ρ must be either parallel or orthogonal to ℓ. If ρ has axis orthogonal to ℓ, then $\rho' = \sigma \circ \rho$ has axis parallel to ℓ. Thus, replacing ρ by ρ', if necessary, we may suppose that ρ has axis ℓ' parallel to ℓ. If $\ell' \neq \ell$, then $\rho(O) \neq O$, and the line segment $\overline{O\rho(O)}$ is orthogonal to ℓ. Now σ and $\rho \circ \sigma \circ \rho^{-1}$ are rotations of order 2 with rotation center O and $\rho(O)$, respectively. Then $\rho \circ \sigma \circ \rho^{-1} \circ \sigma$ is a translation in the direction of $\overrightarrow{O\rho(O)}$, meaning that the axis of $\rho \circ \sigma \circ \rho^{-1} \circ \sigma$ is orthogonal to ℓ. This contradicts $T_G = \langle \tau \mid \ \rangle$. Therefore $\ell' = \ell$, which means that ρ has reflection axis ℓ.

If ρ is a reflection, then $\rho_1 = \sigma \circ \rho$ is also a reflection with reflection axis orthogonal to ℓ through O. If ρ is a glide reflection, we may suppose as in case 1 that $\rho^2 = \tau$ or $(\rho \circ \tau^k)^2 = 1$ for some $k \in \mathbb{Z}$.

If $(\rho \circ \tau^k)^2 = 1$ then $\rho \circ \tau^k$ is a reflection.

Now let $\rho^2 = \tau$ and let m be the perpendicular bisector of the line segment $\overline{O\rho(O)}$, see Figure 4.13.

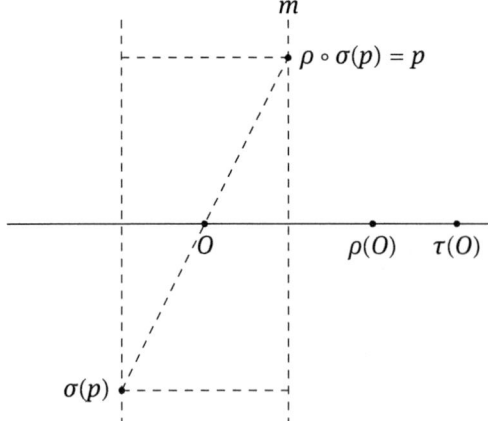

Figure 4.13: Bisector m.

Let $\rho_2 = \rho \circ \sigma$; ρ_2 fixes all points of m, that is, $\rho_2(Q) = Q$ for all $Q \in m$. Hence ρ_2 is a reflection with reflection axis m.

In this case G contains no reflection ρ_1 with axis orthogonal to ℓ at O, since $\rho_2 \circ \rho_1$ would then be a translation τ_0 carrying O to $\rho(O)$, hence $\tau_0^2 = \tau$ which contradicts $T_G = \langle \tau \mid \ \rangle$.

Hence altogether we have obtained three geometrically different types of group G containing a nontrivial rotation:

$$G_2 = \langle \tau, \sigma \mid \sigma^2 = 1, \sigma \circ \tau \circ \sigma^{-1} = \tau^{-1} \rangle \cong D_\infty,$$

$$G_2^1 = \langle \tau, \sigma, \rho \mid \sigma^2 = \rho^2 = 1, \sigma \circ \tau \circ \sigma^{-1} = \tau^{-1}, \rho \circ \tau \circ \rho^{-1} = \tau, \rho \circ \sigma \circ \rho^{-1} = \sigma \rangle$$

$$\cong D_\infty \times \mathbb{Z}_2,$$

$$G_2^2 = \langle \tau, \sigma, \rho \mid \sigma^2 = 1, \rho^2 = \tau, \sigma \circ \tau \circ \sigma^{-1} = \tau^{-1}, \rho \circ \tau \circ \rho^{-1} = \tau, \sigma \circ \rho \circ \sigma^{-1} = \rho^{-1} \rangle$$

$$= \langle \sigma, \rho \mid \sigma^2 = 1, \sigma \circ \rho \circ \sigma^{-1} = \rho^{-1} \rangle \cong D_\infty.$$

We remark that G_2 and G_2^2 are abstractly isomorphic, but geometrically different, since G_2 preserves orientation while G_2^2 does not. We summarize the result in the following Theorem.

Theorem 4.15. *There are exactly 7 geometrically different types of frieze groups with representation G_1, G_1^1, G_1^2, G_1^3, G_2, G_2^1, G_2^2 as given above.*

They fall algebraically into the four isomorphisms types $\mathbb{Z}$, D_∞, $\mathbb{Z} \times \mathbb{Z}_2$ and $D_\infty \times \mathbb{Z}_2$.

In the following Figure 4.14 we show friezes with symmetry groups of the seven types. In these figures, τ is a horizontal translation.

Broken lines indicate axes of reflection or glide reflections. Small circles mark centers of rotational symmetry.

4.6 Planar Crystallographic Groups

In this section we consider discontinuous groups G of E such that the translation subgroup $T_G = T \cap G$ is a double-periodic group. We have nontrivial translations

$$\tau_1 : \mathbb{R}^2 \to \mathbb{R}^2, \quad \vec{v} \mapsto \vec{v} + \vec{w}_1$$

and

$$\tau_2 : \mathbb{R}^2 \to \mathbb{R}^2, \quad \vec{v} \mapsto \vec{v} + \vec{w}_2,$$

such that

$$T_G = \{\tau \in T \mid \tau(\vec{v}) = \vec{v} + m\vec{w}_1 + n\vec{w}_2 \text{ for some } m, n \in \mathbb{Z}\}$$

where $\vec{w}_1 \neq \vec{0} \neq \vec{w}_2$ and $\vec{w}_1, \vec{w}_2$ are linearly independent.

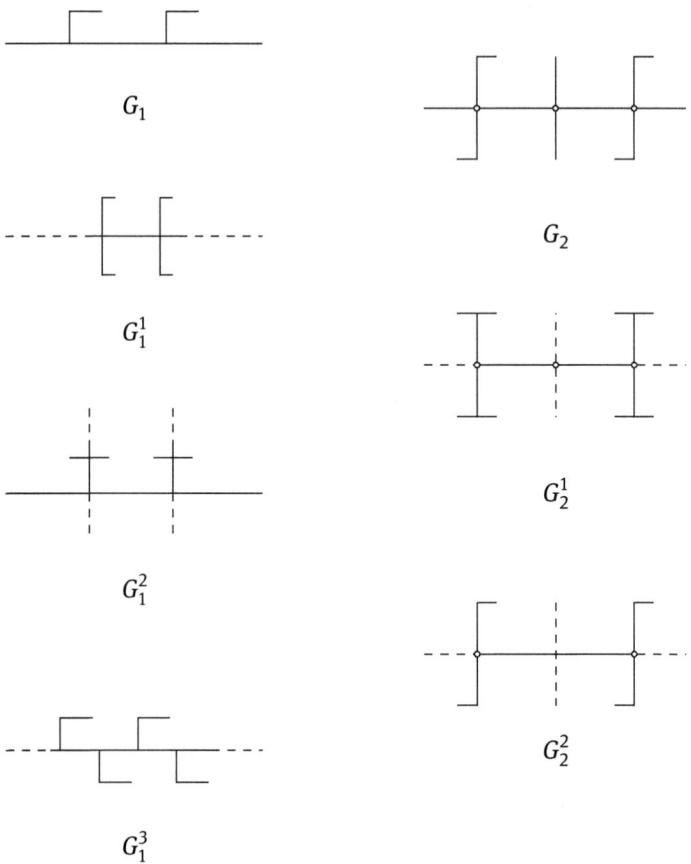

Figure 4.14: Friezes with symmetry groups of the seven types.

We may assume, without loss of generality, $\|\vec{w}_1\| \leq \|\vec{w}_2\|$, or more precisely, $\|\vec{w}_1\|$ is minimal among all $\|\vec{w}\|$ for nontrivial periods $\vec{w}$ of T_G, and $\|\vec{w}_2\|$ is minimal among all $\|\vec{w}\|$ for all nontrivial periods $\vec{w}$ which are not of the form $m\vec{w}_1$ for some $m \in \mathbb{Z}$ (see the proof of Theorem 4.11). In particular, we have

$$T_G = \langle \tau_1, \tau_2 \mid \tau_1 \circ \tau_2 = \tau_2 \circ \tau_1 \rangle \cong \mathbb{Z} \times \mathbb{Z}.$$

When a base point O has been chosen, the group T_G determines a lattice L, the orbit $L = \{\tau(O) \mid \tau \in T_G\}$ of O under T_G, and indeed the set $V(T_G)$ of vectors $\overrightarrow{O\tau(O)}$, $\tau \in T_G$, as a subset of $\mathbb{R}^2$, does not depend on the choice of O.

The first result is that a rotation contained in G can have only one of the orders 1, 2, 3, 4 or 6. This limitation is called the crystallographic restriction.

Theorem 4.16. *G^+ is generated by T_G together with a single rotation of order n where $n = 1, 2, 3, 4$ or 6, and is one of the following types (up to isomorphisms):*

$$G_1 = T_G = \langle \tau_1, \tau_2 \mid \tau_1 \circ \tau_2 = \tau_2 \circ \tau_1 \rangle,$$
$$G_2 = \langle \tau_1, \tau_2, \sigma \mid \tau_1 \circ \tau_2 = \tau_2 \circ \tau_1, \sigma^2 = 1, \sigma \circ \tau_1 \circ \sigma^{-1} = \tau_1^{-1}, \sigma \circ \tau_2 \circ \sigma^{-1} = \tau_2^{-1} \rangle,$$
$$G_3 = \langle \tau_1, \tau_2, \sigma \mid \tau_1 \circ \tau_2 = \tau_2 \circ \tau_1, \sigma^3 = 1, \sigma \circ \tau_1 \circ \sigma^{-1} = \tau_1^{-1} \circ \tau_2, \sigma \circ \tau_2 \circ \sigma^{-1} = \tau_1^{-1} \rangle,$$
$$G_4 = \langle \tau_1, \tau_2, \sigma \mid \tau_1 \circ \tau_2 = \tau_2 \circ \tau_1, \sigma^4 = 1, \sigma \circ \tau_1 \circ \sigma^{-1} = \tau_2, \sigma \circ \tau_2 \circ \sigma^{-1} = \tau_1^{-1} \rangle,$$
$$G_6 = \langle \tau_1, \tau_2, \sigma \mid \tau_1 \circ \tau_2 = \tau_2 \circ \tau_1, \sigma^6 = 1, \sigma \circ \tau_1 \circ \sigma^{-1} = \tau_2, \sigma \circ \tau_2 \circ \sigma^{-1} = \tau_1^{-1} \circ \tau_2 \rangle.$$

Proof. Suppose that G^+ contains a rotation σ of order $n \geq 2$, say with rotation center $O \in \mathbb{R}^2$. Then the n points $\tau_1(O), \sigma \circ \tau_1(O), \dots, \sigma^{n-1} \circ \tau_1(O)$ are evenly spaced around the circle at O of radius $r = \|\vec{w}_1\|$ (recall that $\tau_1(\vec{v}) = \vec{v} + \vec{w}_1$ for $\vec{v} \in \mathbb{R}^2$).

Assume that $n > 6$. The isometries τ_1, $\sigma \circ \tau_1 \circ \sigma^{-1}$ and $\sigma \circ \tau_1 \circ \sigma^{-1} \circ \tau_1^{-1}$ are all in T_G. Now $\sigma \circ \tau_1 \circ \sigma^{-1} \circ \tau_1^{-1}(\tau_1(O)) = \sigma \circ \tau_1(O)$, see Figure 4.15.

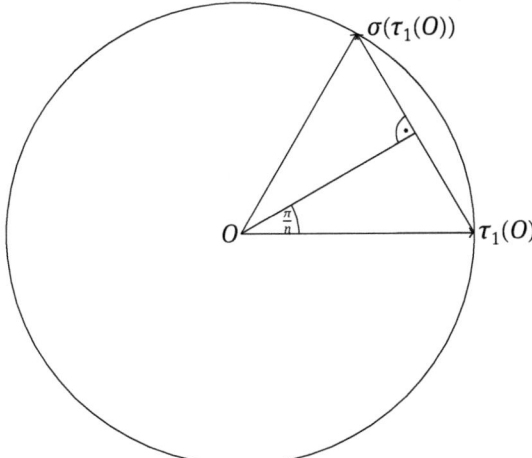

Figure 4.15: No rotation of order $n > 6$.

Let $\vec{v}_1$ be the vector from $\tau_1(O)$ to $\sigma(\tau_1(O))$, $\vec{v}_1$ is a period of T_G. But $\|\vec{v}_1\| = 2\|\vec{w}_1\| \sin(\frac{\pi}{n}) < \|\vec{w}_1\| = r$ because $\sin(\frac{\pi}{n}) < \frac{1}{2}$ for $n \geq 7$. This gives a contradiction to the minimality of $\|\vec{w}_1\|$. Hence, $n \leq 6$.

Now we assume that $n = 5$. Then we have the following situation given in Figure 4.16.

We have

$$\sigma^3 \circ \tau_1^{-1} \circ \sigma^{-3} \circ \tau_1^{-1}(\tau_1(O)) = \sigma^3 \circ \tau_1^{-1}(O),$$

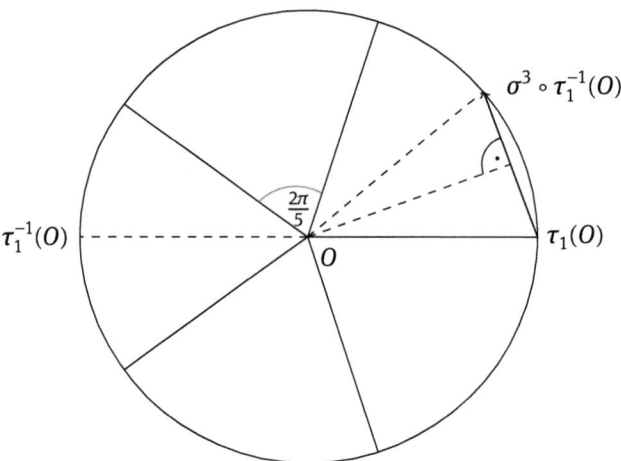

Figure 4.16: No rotation of order 5.

and $\sigma^3 \circ \tau_1^{-1} \circ \sigma^{-3} \circ \tau_1^{-1}$ is in T_G. Hence the vector $\vec{v}_1$ from $\tau_1(O)$ to $\sigma^3 \circ \tau_1^{-1}(O)$ is a period of T_G, and $\|\vec{v}_1\| = 2\|\vec{w}_1\| \sin(\frac{\pi}{10}) < \|\vec{w}_1\| = r$. This gives a contradiction to the minimality of $\|\vec{w}_1\|$. Hence, $n \neq 5$.

Therefore, σ can only have order 2, 3, 4 or 6 if $n \geq 2$, which means that all rotations in G^+ have order 1, 2, 3, 4 or 6.

If G^+ contains a rotation σ_1 with angle $\frac{2\pi}{2}$ and a rotation σ_2 with angle $\frac{2\pi}{3}$, then $\sigma_2^{-1} \circ \sigma_1$ is a rotation with angle $\frac{2\pi}{6}$. Thus if G^+ contains rotations of order 2 and 3, it contains rotations of order 6.

If G^+ contains a rotation σ_1 with angle $\frac{2\pi}{3}$ and a rotation σ_2 with angle $\frac{2\pi}{4}$, then $\sigma_2^{-1} \circ \sigma_1$ is a rotation with angle $\frac{2\pi}{12}$, and hence of order 12. But this gives a contradiction because the maximal order in G^+ is 6. Thus G^+ cannot contain rotations of order 3 and 4.

We conclude that if n is the greatest order of a rotation σ in G^+, then all other rotations σ_1 in G^+ have order n_1 dividing n. Let σ be a rotation with angle $\frac{2\pi}{n}$ and σ_1 be a rotation with angle $\frac{2\pi}{n_1}$ where $n_1 \cdot m = n$. Then σ^m has the angle $\frac{2\pi}{n_1}$, and hence $\sigma^{-m} \circ \sigma_1 = \tau$ is a translation. Therefore σ_1 is in the group generated by T_G together with σ. This shows that G^+ is generated by T_G together with σ.

We continue and let $n = 1, 2, 3, 4$ or 6 be the greatest order of a rotation in G^+, and write G_n for G^+ according to the case.

If $n = 1$ then $G^+ = G_1 = T_G$. If $n = 2$, then σ of order 2 has angle π, whence $\sigma \circ \tau \circ \sigma^{-1} = \tau^{-1}$ for all τ in T_G, and G_2 has the given presentation. The presence of σ of order 2 imposes no condition on the lattice L.

Let $n = 4$ and let σ be a rotation of order 4 and center O. Since σ has angle $\frac{\pi}{2}$, we get that $\sigma \circ \tau_1 \circ \sigma^{-1}$ is a translation in a direction orthogonal to that of τ_1. Since the period for $\sigma \circ \tau_1 \circ \sigma^{-1}$ has the same length $\|\vec{w}_1\|$, $\vec{w}_1$ being the period for τ_1, we may choose $\tau_2 = \sigma \circ \tau_1 \circ \sigma^{-1}$. Then $\sigma \circ \tau_2 \circ \sigma^{-1} = \tau_1^{-1}$ and G_4 has the given presentation.

We have the situation given in Figure 4.17 and the lattice is a square lattice.

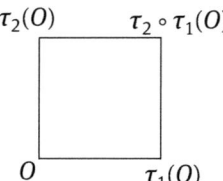

$\tau_2(O)$ $\tau_2 \circ \tau_1(O)$

O $\tau_1(O)$ **Figure 4.17:** Square lattice.

Suppose now that G^+ contains a rotation σ of maximal order 3 and center O. The angle between the vectors $\overrightarrow{O\tau_1(O)}$ and $\overrightarrow{O\sigma \circ \tau_1 \circ \sigma^{-1}(O)}$ is $\frac{2\pi}{3}$, and the length of the period of $\sigma \circ \tau_1 \circ \sigma^{-1}$ is the same as that of τ_1. Therefore the three points O, $\sigma \circ \tau_1 \circ \sigma^{-1}(O)$ and $\tau_1 \circ \sigma \circ \tau_1 \circ \sigma^{-1}(O)$ are the vertices of an equilateral triangle, see Figure 4.18.

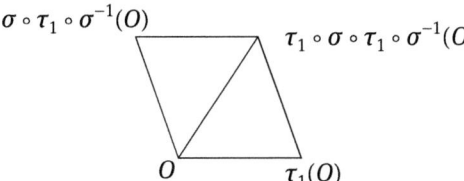

$\sigma \circ \tau_1 \circ \sigma^{-1}(O)$ $\tau_1 \circ \sigma \circ \tau_1 \circ \sigma^{-1}(O)$

O $\tau_1(O)$ **Figure 4.18:** Triangular lattice.

Thus the length of the period of $\tau_1 \circ \sigma \circ \tau_1 \circ \sigma^{-1}$ is the same as that of τ_1, and we may choose $\tau_2 = \tau_1 \circ \sigma \circ \tau_1 \circ \sigma^{-1}$, which gives $\sigma \circ \tau_1 \circ \sigma^{-1} = \tau_2 \circ \tau_1^{-1}$ and $\sigma \circ \tau_2 \circ \sigma^{-1} = \tau_1^{-1}$. In this case the lattice L is a triangular lattice.

If $n = 3$, then G_3 has the given presentation.

If $n = 6$, then σ^2 has order 3, whence the lattice is triangular. Now σ has angle $\frac{2\pi}{6}$, whence $\sigma \circ \tau_1 \circ \sigma^{-1} = \tau_2$ and $\sigma \circ \tau_2 \circ \sigma^{-1} = \tau_2 \circ \tau_1^{-1}$, which gives the presentation for G_6. □

From now on we suppose that $G \neq G^+$. Hence G is generated by G^+ together with any element ρ of G that is not in G^+. Now ρ is a reflection or a glide reflection, and in either case $\rho^2 \in T_G$.

Lemma 4.17. *If $\rho \in G$, $\rho \notin G^+$, then exactly one of the following conditions holds:*
(T1) T_G is generated by elements τ_1 and τ_2 such that $\rho \circ \tau_1 \circ \rho^{-1} = \tau_1$ and $\rho \circ \tau_2 \circ \rho^{-1} = \tau_2^{-1}$.
(T2) T_G is generated by elements τ_1 and τ_2 such that $\rho \circ \tau_1 \circ \rho^{-1} = \tau_2$ and $\rho \circ \tau_2 \circ \rho^{-1} = \tau_1$.

Proof. We first show that (T1) and (T2) are incompatible.

Suppose that (T2) holds and T_G is generated by τ_1 and τ_2 such that $\rho \circ \tau_1 \circ \rho^{-1} = \tau_2$ and $\rho \circ \tau_2 \circ \rho^{-1} = \tau_1$. Assume that α, β are elements of T_G such that $\rho \circ \alpha \circ \rho^{-1} = \alpha$ and $\rho \circ \beta \circ \rho^{-1} = \beta^{-1}$.

If $\alpha = \tau_1^h \circ \tau_2^k$ for some integers h and k, then $\rho \circ \alpha \circ \rho^{-1} = \tau_1^k \circ \tau_2^h$ and $\rho \circ \alpha \circ \rho^{-1} = \alpha$ implies $h = k$ and $\alpha = (\tau_1 \circ \tau_2)^h$.

If $\beta = \tau_1^m \circ \tau_2^n$ for some integers n and m, then $\rho \circ \beta \circ \rho^{-1} = \tau_1^n \circ \tau_2^m$, and $\rho \circ \beta \circ \rho^{-1} = \beta^{-1}$ implies $m = -n$ and $\beta = (\tau_1 \circ \tau_2^{-1})^m$.

Now α and β are both in the subgroup generated by $\tau_1 \circ \tau_2$ and $\tau_1 \circ \tau_2^{-1}$, which is a proper subgroup of T_G. Thus α and β do not generate T_G.

The other case can be handled analogously.

Now we want to show that (T1) or (T2) occurs.

Suppose $T_G = \langle \tau_1, \tau_2 \rangle$ with

$$\tau_1 : \mathbb{R}^2 \to \mathbb{R}^2, \quad \vec{v} \mapsto \vec{v} + \vec{w}_1$$

and

$$\tau_2 : \mathbb{R}^2 \to \mathbb{R}^2, \quad \vec{v} \mapsto \vec{v} + \vec{w}_2$$

as above. Again we may assume that $0 < \|\vec{w}_1\|$ is minimal among all $\|\vec{w}\|$ for nontrivial periods $\vec{w}$ of T_G, and $\|\vec{w}_2\|$ is minimal among all $\|\vec{w}\|$ for all nontrivial periods $\vec{w}$ which are not of the form $m\vec{w}_1$ for some $m \in \mathbb{Z}$.

We choose a base point $O \in \mathbb{R}^2$ for the lattice L for T_G. Without loss of generality, we may assume that $O = \vec{0}$, the zero vector. Then, if $y \in T_G$, we may consider both the line segment $\overline{Oy(O)}$ and $y(O) \in \mathbb{R}^2$ as a vector $\overrightarrow{y(O)} := \overrightarrow{Oy(O)}$. If $y \in T_G$, then we write $\overrightarrow{y(O)}^{\perp}$ and $\overrightarrow{y(O)}^{\parallel}$ for the components of $\overrightarrow{y(O)}$ orthogonal and parallel to $\overrightarrow{\tau_1(O)}$, respectively.

We now impose some supplementary conditions. After replacing τ_1 by τ_1^{-1}, if necessary, we may assume that the angle between the vectors $\overrightarrow{\tau_1(O)}$ and $\overrightarrow{\tau_2(O)}$ is some Θ with $\Theta \leq \frac{\pi}{2}$. The minimality of $\|\vec{w}_2\|$ requires that $\|\overrightarrow{\tau_2(O)}\| \leq \|\overrightarrow{\tau_2(O)} - \overrightarrow{\tau_1(O)}\| = \|\overrightarrow{\tau_1^{-1} \circ \tau_2(O)}\|$. Therefore $\|\overrightarrow{\tau_2(O)}^{\parallel}\| \leq \frac{1}{2}\|\overrightarrow{\tau_1(O)}\| = \frac{1}{2}\|\vec{w}_1\|$, that is, $\tau_2(O)$ is either on the perpendicular bisector of $\overline{O\tau_1(O)}$ or on the same side of it as O, see Figure 4.19.

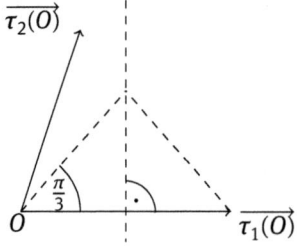

Figure 4.19: Location of $\tau_2(O)$.

This, together with the condition

$$\|\overrightarrow{\tau_1(O)}\| = \|\vec{w}_1\| \le \|\vec{w}_2\| = \|\overrightarrow{\tau_2(O)}\|,$$

implies that $\Theta \ge \frac{\pi}{3}$. Thus $\frac{\pi}{3} \le \Theta \le \frac{\pi}{2}$.

Suppose first that $\rho \circ \tau_1 \circ \rho^{-1} = \tau_1$, whence ρ has its axis parallel to $\overrightarrow{\tau_1(O)} = \vec{w}_1$.
If $\Theta = \frac{\pi}{2}$, then $\rho \circ \tau_2 \circ \rho^{-1} = \tau_2^{-1}$, and (T1) holds.
If $\Theta < \frac{\pi}{2}$, then

$$\overrightarrow{\rho \circ \tau_2 \circ \rho^{-1}(O)}^{\perp} = \overrightarrow{\tau_1 \circ \tau_2^{-1}(O)}^{\perp}$$

and

$$\overrightarrow{\rho \circ \tau_2 \circ \rho^{-1}(O)}^{\parallel} = \overrightarrow{\tau_2(O)}^{\parallel}.$$

Since $0 < \|\overrightarrow{\tau_2(O)}^{\parallel}\| \le \frac{1}{2}\|\overrightarrow{\tau_1(O)}\|$ implies that $\frac{1}{2}\|\overrightarrow{\tau_1(O)}\| \le \|\overrightarrow{\tau_1 \circ \tau_2^{-1}(O)}^{\parallel}\| < \|\overrightarrow{\tau_1(O)}\|$, we conclude that $\rho \circ \tau_2 \circ \rho^{-1}$ can only be

$$\rho \circ \tau_2 \circ \rho^{-1} = \tau_1 \circ \tau_2^{-1},$$

see Figure 4.20. Let $\alpha = \tau_2$ and $\beta = \tau_1 \circ \tau_2^{-1}$, then τ_1 and τ_2 generate T_G while ρ exchanges them, and (T2) holds.

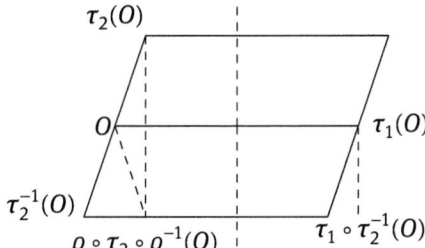

Figure 4.20: (T2) for $\rho \circ \tau_1 \circ \rho^{-1} = \tau_1$.

Suppose next that $\rho \circ \tau_1 \circ \rho^{-1} = \tau_1^{-1}$, whence ρ has its axis orthogonal to $\vec{w}_1 = \overrightarrow{\tau(O)}$.
If $\Theta = \frac{\pi}{2}$, then $\rho \circ \tau_2 \circ \rho^{-1} = \tau_2$, and (T1) holds with τ_1 and τ_2 exchanged.
If $\Theta < \frac{\pi}{2}$, then $\overrightarrow{\tau_2(O)} - \overrightarrow{\rho \circ \tau_2 \circ \rho^{-1}(O)}$ is parallel to $\overrightarrow{\tau_1(O)}$ and of length $2\|\overrightarrow{\tau_2(O)}^{\parallel}\| \le \|\tau_1(O)\| = \|\vec{w}_1\|$, whence by minimality of $\|\overrightarrow{\tau_1(O)}\|$, it must have length $\|\overrightarrow{\tau_1(O)}\|$, and $\rho \circ \tau_2 \circ \rho^{-1} = \tau_2 \circ \tau_1^{-1}$, see Figure 4.21. We have $T_G = \langle \tau_2, \tau_2 \circ \tau_1^{-1} \rangle$.
Now $\rho \circ \tau_2 \circ \rho^{-1} = \tau_2 \circ \tau_1^{-1}$ and $\rho \circ \tau_2 \circ \tau_1^{-1} \circ \rho^{-1} = \tau_2 \circ \tau_1^{-1} \circ \tau_1 = \tau_2$, and (T2) holds.
Suppose finally that $\rho \circ \tau_1 \circ \rho^{-1} \ne \tau_1, \tau_1^{-1}$, whence $\rho \circ \tau_1 \circ \rho^{-1}$ is not a power of τ_1, but $\|\overrightarrow{\rho \circ \tau_1 \circ \rho^{-1}(O)}\| = \|\overrightarrow{\tau_1(O)}\|$. We take $\alpha = \tau_1, \beta = \rho \circ \tau_1 \circ \rho^{-1}$. Even though the

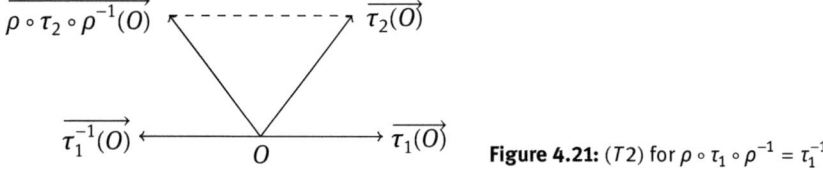

Figure 4.21: $(T2)$ for $\rho \circ \tau_1 \circ \rho^{-1} = \tau_1^{-1}$.

supplementary conditions from above need not be fulfilled, the minimality conditions are satisfied, whence α and β generate T_G and $(T2)$ holds. $\qquad\qquad\qquad\square$

Enumeration of the cases

We now enumerate the possible types of $G \neq G^+ = G_n$, according to $n = 1, 2, 3, 4, 6$. As for G_n we may take $O = \vec{0}$ as the center of the rotation σ (if $n \geq 2$).

Case $n = 1$

Here $G_1 = T_G$.

If ρ_1, ρ_2 are elements of G, not in G^+, then $\rho_2 = \tau \circ \rho_1$ for some $\tau \in T_G$, whence $\rho_1 \circ \tau \circ \rho_1^{-1} = \rho_2 \circ \tau \circ \rho_2^{-1}$ for all $\tau \in T_G$.

If G contains any reflection, we choose ρ to be a reflection, with $\rho^2 = 1$. It is clear that both cases $(T1)$ and $(T2)$ from Lemma 4.17 can be realized, for example (after stretching $\vec{w}_1$ and $\vec{w}_2$ if necessary), with a square lattice, see Figure 4.22.

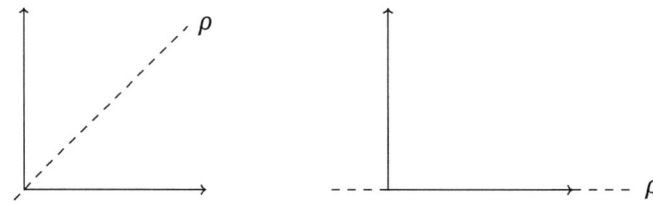

Figure 4.22: Realization of $(T1)$ and $(T2)$.

We obtain two groups G_1^1 and G_1^2 with presentation obtained from $G_1 = T_G$ by adjoining the new generator ρ, a relation $\rho^2 = 1$, and two relations giving the values of $\rho \circ \tau_1 \circ \rho^{-1}$ and $\rho \circ \tau_2 \circ \rho^{-1}$.

Suppose now that G contains no reflection, whence we must choose ρ as glide reflection with $\rho^2 = \tau \neq 1$ in T_G. If $(T1)$ holds, then $\rho \circ \tau \circ \rho^{-1} = \tau$ implies that $\tau = \tau_1^h$ for some integer h. If $\rho_1 = \tau_1^m \circ \rho$, then $\rho_1^2 = \tau_1^{h+2m}$, and replacing ρ by ρ_1 for suitable m we may suppose that $\rho^2 = 1$ or $\rho^2 = \tau_1$. Since $\rho^2 = 1$ implies that ρ is a reflection, we must have $\rho^2 = \tau_1$. Thus ρ is a reflection at an axis ℓ parallel to τ_1 followed by a translation along ℓ through a distance $\frac{1}{2}\|\overrightarrow{\tau_1(O)}\|$. We must verify that this group G_1^3 is new, that is, that it does not contain a reflection. Now every element of $G = G_1^3$, not in G^+, has the form $\rho_2 = \tau_1^m \circ \tau_2^n \circ \rho$ for some integers m and n, and we find that

$$\rho_2^2 = \tau_1^m \circ \tau_2^n \circ \rho \circ \tau_1^m \circ \tau_2^n \circ \rho$$
$$= \tau_1^m \circ \tau_2^n \circ \tau_1^m \circ \tau_2^{-n} \circ \rho^2$$
$$= \tau_1^{2m+1} \neq 1,$$

whence ρ_2 is not a reflection.

There remains the possibility that G contains no reflection and (T2) holds. Now $\rho^2 = \tau$ and $\rho \circ \tau_1 \circ \rho^{-1} = \tau_2$, $\rho \circ \tau_2 \circ \rho^{-1} = \tau_1$ implies that $\tau = (\tau_1 \circ \tau_2)^h$ for some integer h, and we can suppose that $\rho^2 = \tau_1 \circ \tau_2$. Now let $\rho_2 = \tau_1^{-1} \circ \rho$. Then

$$\rho_2^2 = \tau_1^{-1} \circ \rho \circ \tau_1^{-1} \circ \rho$$
$$= \tau_1^{-1} \circ \rho \circ \tau_1^{-1} \circ \rho^{-1} \circ \rho^2$$
$$= \tau_1^{-1} \circ \tau_2^{-1} \circ \tau_1 \circ \tau_2 = 1$$

and G contains a reflection, contrary to the hypothesis. We get the following theorem.

Theorem 4.18. *If $G \neq G^+ = G_1 = T_G$, then there are exactly three possible isomorphism types for G:*

$$G_1^1 = \langle \tau_1, \tau_2, \rho \mid \tau_1 \circ \tau_2 = \tau_2 \circ \tau_1, \rho^2 = 1, \rho \circ \tau_1 \circ \rho^{-1} = \tau_1, \rho \circ \tau_2 \circ \rho^{-1} = \tau_2^{-1} \rangle,$$
$$G_1^2 = \langle \tau_1, \tau_2, \rho \mid \tau_1 \circ \tau_2 = \tau_2 \circ \tau_1, \rho^2 = 1, \rho \circ \tau_1 \circ \rho^{-1} = \tau_2, \rho \circ \tau_2 \circ \rho^{-1} = \tau_1 \rangle,$$
$$G_1^3 = \langle \tau_1, \tau_2, \rho \mid \tau_1 \circ \tau_2 = \tau_2 \circ \tau_1, \rho^2 = \tau_1, \rho \circ \tau_1 \circ \rho^{-1} = \tau_1, \rho \circ \tau_2 \circ \rho^{-1} = \tau_2^{-1} \rangle.$$

Case $n = 2$

Here G_2 is generated by τ_1, τ_2, σ where $\sigma^2 = 1$ and $\sigma \circ \tau_1 \circ \sigma^{-1} = \tau_1^{-1}$, $\sigma \circ \tau_2 \circ \sigma^{-1} = \tau_2^{-1}$.

We first suppose that G contains a reflection ρ and that (T1) holds, with $\rho \circ \tau_1 \circ \rho^{-1} = \tau_1$, $\rho \circ \tau_2 \circ \rho^{-1} = \tau_2^{-1}$. This implies that $\overrightarrow{\tau_1(O)}$ and $\overrightarrow{\tau_2(O)}$ are orthogonal. Then ρ has its axis parallel to τ_1.

Let $\rho_1 = \rho \circ \sigma$; then $\rho_1 \circ \tau_2 \circ \rho_1^{-1} = \tau_2$, and ρ_1 has its axis parallel to $\overrightarrow{\tau_2(O)}$, and so $\rho_1^2 = \tau_2^k$ for some integer k. Replacing ρ_1 by $\tau_2^m \circ \rho_1$ for some integer m, we can suppose, according to the parity of h, that $\rho_1^2 = 1$ or $\rho_1^2 = \tau_2$. If ρ and ρ' are any two reflections with axis parallel to $\overrightarrow{\tau_1(O)}$, then $\rho \circ \rho' = \tau_2^m$ for some integer m and $\rho' = \rho \circ \tau_2^m$, whence different choices of ρ yield the same cases $\rho_1^2 = 1$ or $\rho_1^2 = \tau_2$ and the two cases are distinct.

The two cases G_2^1 and G_2^2 are easily realized with the square lattice L, see Figure 4.23. For the first we take ρ with axis ℓ, the line through O and $\tau_1(O)$. For the second we take the axis ℓ parallel to $\overrightarrow{\tau_1(O)}$, at a distance of $\frac{1}{4}\|\overrightarrow{\tau_2(O)}\|$ in the direction of $\overrightarrow{\tau_2(O)}$.

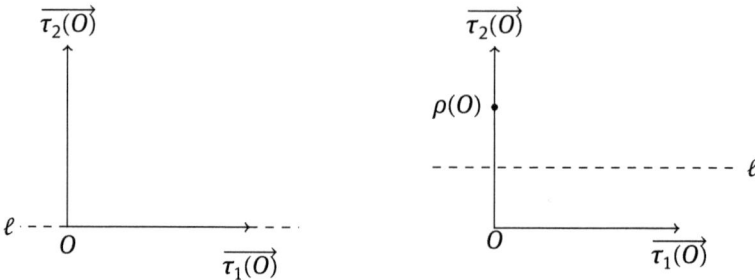

Figure 4.23: Types G_2^1 and G_2^2.

In the second $\rho_1(O) = \rho \circ \sigma(O) = \rho(O)$ is the midpoint of the line segment from O to $\overrightarrow{\tau_2(O)}$, and ρ_1, which has as axis the line through O and $\overrightarrow{\tau_2(O)}$, carries O to $\rho(O)$, whence $\rho_1^2 = \tau_2$.

We remark that G_2^1 is isomorphic to a semidirect product $T_G \rtimes D_2$.

Suppose now that G contains a reflection ρ and that $(T2)$ holds, with $\rho \circ \tau_1 \circ \rho^{-1} = \tau_2$, $\rho \circ \tau_2 \circ \rho^{-1} = \tau_1$, whence ρ has the axis parallel to $\overrightarrow{\tau_1 \circ \tau_2(O)}$.

If $\rho_1 = \rho \circ \sigma$, then $\rho_1 \circ \tau_1 \circ \rho_1^{-1} = \tau_2^{-1}$ and $\rho_1 \circ \tau_2 \circ \rho_1^{-1} = \tau_1^{-1}$ and $\rho_1 \circ \tau_2 \circ \tau_1^{-1} \circ \rho_1^{-1} = \tau_1^{-1} \circ \tau_2$.

Hence $\rho_1 = \rho \circ \sigma$ has the axis parallel to $\overrightarrow{\tau_1^{-1} \circ \tau_2(O)}$ and $\rho_1^2 = (\tau_1^{-1} \circ \tau_2)^h$ for some integer h. Replacing ρ_1 by $(\tau_1^{-1} \circ \tau_2)^m \circ \rho_1$ for some integer m, we may assume, according to the parity of h, that $\rho_1^2 = 1$ or $\rho_1^2 = \tau_1^{-1} \circ \tau_2$.

If $\rho_1^2 = \tau_1^{-1} \circ \tau_2$, then let $\rho_2 = \tau_1 \circ \rho_1$. Then $\rho_2^2 = \tau_1 \circ \rho_1 \circ \tau_1 \circ \rho_1 = \tau_1 \circ \tau_2^{-1} \circ \tau_1^{-1} \circ \tau_2 = 1$, and ρ_2 is a reflection.

In either case G contains two reflections with orthogonal axes parallel to $\overrightarrow{\tau_1 \circ \tau_2(O)}$ and $\overrightarrow{\tau_1^{-1} \circ \tau_2(O)}$ (the lines containing the diagonals of the rhombus in the figure) whose product is a rotation of order 2, see Figure 4.24.

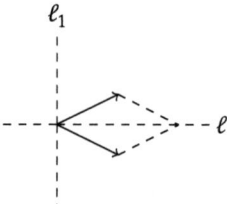

Figure 4.24: Type G_2^3.

We get the group G_2^3 which can be realized with a rhombus lattice.

In fact G_2^3 is a semi-direct product $T_G \rtimes G_0$ with $G_0 \cong D_2$, the Klein four group.

Finally, suppose that G contains no reflection. If $(T1)$ holds for τ_1, τ_2, we may suppose that $\rho^2 = \tau_1$. If $\rho_1 = \rho \circ \sigma$, then $\rho_1 \circ \tau_1 \circ \rho_1^{-1} = \tau_1^{-1}$ and $\rho_1 \circ \tau_2 \circ \rho_1^{-1} = \tau_2$, and we may likewise suppose that $\rho_1^2 = \tau_2$. To check that the group G_2^4, defined that way, does not contain reflections, we verify that

$$(\tau_1^m \circ \tau_2^n \circ \rho)^2 = \tau_1^{2m+1} \neq 1$$

and that

$$(\tau_1^m \circ \tau_2^n \circ \rho_1)^2 = \tau_2^{2n+1} \neq 1.$$

This group is easily realized in the square lattice L by taking ρ and ρ_1 with axes parallel to $\overrightarrow{\tau_1(O)}$ and $\overrightarrow{\tau_2(O)}$, respectively, at a distance $\frac{1}{2}\|\overrightarrow{\tau_1(O)}\| = \frac{1}{2}\|\overrightarrow{\tau_2(O)}\|$, and with translation displacement $\frac{1}{2}\|\overrightarrow{\tau_1(O)}\| = \frac{1}{2}\|\overrightarrow{\tau_2(O)}\|$, see Figure 4.25.

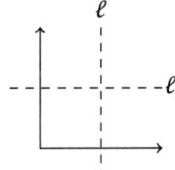

Figure 4.25: Type G_2^4.

If (T2) holds, with $\rho \circ \tau_1 \circ \rho^{-1} = \tau_2$ and $\rho \circ \tau_2 \circ \rho^{-1} = \tau_1$, we may suppose, as before, that $\rho^2 = \tau_1 \circ \tau_2$, and for $\rho_1 = \rho \circ \sigma$, that $\rho_1^2 = \tau_1^{-1} \circ \tau_2$.

As before, $\rho_2 = \tau_1 \circ \rho_1$ is then a reflection, contrary to the hypothesis. Altogether we get the following.

Theorem 4.19. *If $G \neq G^+ = G_2$, there are exactly four possible isomorphisms types for G:*

$$G_2^1 = \langle \tau_1, \tau_2, \rho, \sigma \mid \tau_1 \circ \tau_2 = \tau_2 \circ \tau_1, \sigma^2 = \rho^2 = (\rho \circ \sigma)^2 = 1, \sigma \circ \tau_1 \circ \sigma^{-1} = \tau_1^{-1},$$
$$\sigma \circ \tau_2 \circ \sigma^{-1} = \tau_2^{-1}, \rho \circ \tau_1 \circ \rho^{-1} = \tau_1, \rho \circ \tau_2 \circ \rho^{-1} = \tau_2^{-1}\rangle,$$

$$G_2^2 = \langle \tau_1, \tau_2, \rho, \sigma \mid \tau_1 \circ \tau_2 = \tau_2 \circ \tau_1, \sigma^2 = \rho^2 = 1, (\rho \circ \sigma)^2 = \tau_2, \sigma \circ \tau_1 \circ \sigma^{-1} = \tau_1^{-1},$$
$$\sigma \circ \tau_2 \circ \sigma^{-1} = \tau_2^{-1}, \rho \circ \tau_1 \circ \rho^{-1} = \tau_1, \rho \circ \tau_2 \circ \rho^{-1} = \tau_2^{-1}\rangle,$$

$$G_2^3 = \langle \tau_1, \tau_2, \rho, \sigma \mid \tau_1 \circ \tau_2 = \tau_2 \circ \tau_1, \sigma^2 = \rho^2 = (\rho \circ \sigma)^2 = 1, \sigma \circ \tau_1 \circ \sigma^{-1} = \tau_1^{-1},$$
$$\sigma \circ \tau_2 \circ \sigma^{-1} = \tau_2^{-1}, \rho \circ \tau_1 \circ \rho^{-1} = \tau_2, \rho \circ \tau_2 \circ \rho^{-1} = \tau_1\rangle,$$

$$G_2^4 = \langle \tau_1, \tau_2, \rho, \sigma \mid \tau_1 \circ \tau_2 = \tau_2 \circ \tau_1, \sigma^2 = 1, \rho^2 = \tau_1, (\rho \circ \sigma)^2 = \tau_2, \sigma \circ \tau_1 \circ \sigma^{-1} = \tau_1^{-1},$$
$$\sigma \circ \tau_2 \circ \sigma^{-1} = \tau_2^{-1}, \rho \circ \tau_1 \circ \rho^{-1} = \tau_1, \rho \circ \tau_2 \circ \rho^{-1} = \tau_2^{-1}\rangle.$$

Case $n = 4$
Here $G^+ = G_4$ is generated by τ_1, τ_2 and σ with $\sigma^4 = 1, \sigma \circ \tau_1 \circ \sigma^{-1} = \tau_2, \sigma \circ \tau_2 \circ \sigma^{-1} = \tau_1^{-1}$, and the lattice L is a square.

Let $\rho \in G$, $\rho \notin G_4$. If ρ satisfies (T1), that is, $\rho \circ \tau_1 \circ \rho^{-1} = \tau_1, \rho \circ \tau_2 \circ \rho^{-1} = \tau_2^{-1}$, and if $\rho_1 = \rho \circ \sigma$, then $\rho_1 \circ \tau_1 \circ \rho_1^{-1} = \tau_2^{-1}, \rho_1 \circ \tau_2 \circ \rho_1^{-1} = \tau_1^{-1}$, and whence ρ_1, which exchanges the two generators τ_1 and τ_2^{-1}, satisfies (T2). Thus the cases (T1) and (T2) coincide, and we may suppose that ρ satisfies (T1), where ρ is either a reflection or a glide reflection.

The set of four elements $\{\tau_1, \tau_1^{-1}, \tau_2, \tau_2^{-1}\}$ is uniquely determined by the facts that σ permutes them and that they generate T_G.

We show that if G contains a reflection with axes parallel to one of $\overrightarrow{\tau_1(O)}$, $\overrightarrow{\tau_2(O)}$, then it also contains a reflection with axes parallel to the other.

By symmetry, it suffices to consider the case that G contains a reflection ρ with axis parallel to $\overrightarrow{\tau_1(O)}$, hence satisfying (T1).

Since $\rho_1 = \rho \circ \sigma$ satisfies $\rho_1 \circ (\tau_1^{-1} \circ \tau_2) \circ \rho_1^{-1} = \tau_1^{-1} \circ \tau_2$, the axis of ρ_1 is parallel to $\overrightarrow{\tau_1^{-1} \circ \tau_2(O)}$, and $\rho_1^2 = (\tau_1^{-1} \circ \tau_2)^h$ for some integer h. Now, $\tau_2^{-h} \circ \rho_1$ has the axis parallel to that of ρ_1, hence to the axis of $\tau_1^{-1} \circ \tau_2$, and $(\tau_2^{-h} \circ \rho_1)^2 = \tau_1^h \circ \tau_2^{-h} \circ \rho_1^2 = (\tau_1^{-1} \circ \tau_2)^{h-h} = 1$. We have therefore two reflections with axes meeting at some point O_1 with an angle $\frac{2\pi}{8}$. Hence, their product σ_1 is a rotation about O_1 of order 4. The group G then is a semi-product $G_4^1 = T_G \rtimes G_{O_1}$ with $G_{O_1} \cong D_4$. This group is easily realized on the square lattice. In particular, it contains a reflection with an axis parallel to $\overrightarrow{\tau_2(O)}$.

We remark that G_4^1 is exactly the symmetry group of a regular tessellation of the plane by squares (see Section 4.2).

Suppose now that G contains a glide reflection ρ with axis parallel to $\overrightarrow{\tau_1(O)}$, but no reflection with axis parallel to $\overrightarrow{\tau_1(O)}$. Then $G = G_4^2$ is not isomorphic to G_4^1. To find a presentation for G_4^2, after replacing ρ by $\tau_1^m \circ \rho$ for some integer m, we can suppose, as before, that $\rho^2 = \tau_1$. If we now replace ρ by some $\tau_2^n \circ \rho$, the relation $\rho^2 = \tau_1$ remains valid because $(\tau_2^n \circ \rho)^2 = \rho^2 = \tau_1$, while, as before, for a proper choice of n we can make $\rho_1^2 = 1$, where again $\rho_1 = \rho \circ \sigma$. Thus we have the relation $(\rho \circ \sigma)^2 = 1$ or $\rho \circ \sigma \circ \rho^{-1} = \sigma^{-1}$.

This group can be realized on the square lattice as follows.

Let σ have center $O = \vec{0} = (0,0)$, and let $\overrightarrow{\tau_1(O)} = (1,0)$, $\overrightarrow{\tau_2(O)} = (0,1)$. If ρ is chosen with axis ℓ parallel to $\overrightarrow{\tau_1(O)}$ and through the point $(\frac{1}{4}, \frac{1}{4})$, then the axis ℓ_1 of $\rho_1 = \rho \circ \sigma$ will also pass through this point. In the figure, points marked $\cdot$ are centers of rotations of order 4, points marked $\times$ are centers of rotations of order 2, oblique solid lines $\backslash$; $/$ are axes of reflections, and horizontal or vertical broken lines $--$; $\vdots$ are axes of glide reflections, see Figure 4.26.

Theorem 4.20. *If $G \neq G^+ = G_4$, there are exactly two possible isomorphism types for G:*

$$G_4^1 = \langle \tau_1, \tau_2, \sigma, \rho \mid \tau_1 \circ \tau_2 = \tau_2 \circ \tau_1, \sigma^4 = \rho^2 = (\rho \circ \sigma)^2 = 1, \sigma \circ \tau_1 \circ \sigma^{-1} = \tau_2,$$

$$\sigma \circ \tau_2 \circ \sigma^{-1} = \tau_1^{-1}, \rho \circ \tau_1 \circ \rho^{-1} = \tau_1, \rho \circ \tau_2 \circ \rho^{-1} = \tau_2^{-1} \rangle,$$

$$G_4^2 = \langle \tau_1, \tau_2, \sigma, \rho \mid \tau_1 \circ \tau_2 = \tau_2 \circ \tau_1, \sigma^4 = 1, \rho^2 = \tau_1, (\rho \circ \sigma)^2 = 1, \sigma \circ \tau_1 \circ \sigma^{-1} = \tau_2,$$

$$\sigma \circ \tau_2 \circ \sigma^{-1} = \tau_1^{-1}, \rho \circ \tau_1 \circ \rho^{-1} = \tau_1, \rho \circ \tau_2 \circ \rho^{-1} = \tau_2^{-1} \rangle,$$

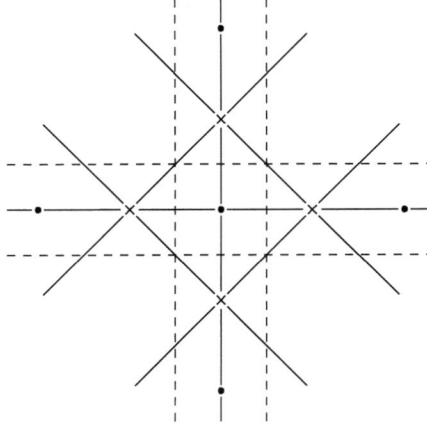

Figure 4.26: Type G_4^2.

Case $n = 3$

Here $G^+ = G_3$ is generated by τ_1, τ_2 and σ with $\sigma^3 = 1$, $\sigma \circ \tau_1 \circ \sigma^{-1} = \tau_1^{-1} \circ \tau_2$, $\sigma \circ \tau_2 \circ \sigma^{-1} = \tau_1^{-1}$, and the lattice L is triangular.

We first show that if G contains a glide reflection ρ with axis ℓ, then G also contains a reflection ρ_1 with an axis ℓ_1 parallel to ℓ. Let $\rho = \tau_0 \circ \rho_0$, where ρ_0 is a reflection at ℓ and τ_0 is a translation along ℓ; here τ_0 and ρ_0 are not necessarily in G.

Thus $\rho^2 = \tau_0^2 = \tau$ in T_G. Let ℓ_1 be a line parallel to ℓ and at a distance $\frac{\sqrt{3}}{2} \|\overrightarrow{\tau_0(O)}\|$ from ℓ, on the left side as one faces in the direction of $\overrightarrow{\tau(O)}$. Let P be a point on ℓ_1, considered also as a vector $\overrightarrow{OP} = \vec{P}$, analogously we regard $\rho_0(P)$ and $\rho(P)$ as vectors $\overrightarrow{\rho_0(P)}$ and $\overrightarrow{\rho(P)}$, respectively.

Then $\|\vec{P} - \overrightarrow{\rho_0(P)}\| = \sqrt{3}\|\overrightarrow{\tau_0(O)}\|$ and $\|\overrightarrow{\rho_0(P)} - \overrightarrow{\rho(P)}\| = \sqrt{3}\|\overrightarrow{\tau_0(O)}\|$. Hence, the points $P, \rho_0(P), \rho(P)$ form a right angle triangle with hypotenuse, the line segment from P to $\rho(P)$, of length $2\|\overrightarrow{\tau_0(O)}\| = \|\overrightarrow{\tau(O)}\|$ at an angle of $\frac{2\pi}{3}$ from the direction of $\overrightarrow{\tau(O)}$ along ℓ. Therefore $\sigma \circ \tau \circ \sigma^{-1}$ carries $\rho(P)$ to P, and $\sigma \circ \tau \circ \sigma^{-1} \circ \rho(P) = P$. We have shown that $\rho_1 = \sigma \circ \tau \circ \sigma^{-1} \circ \rho$ fixes every point P of ℓ_1, whence ρ_1 is a reflection with axis ℓ_1 parallel to ℓ, see Figure 4.27.

Let G be generated by $G^+ = G_3$ and some ρ that satisfies (T1), and let $\rho \circ \tau_1 \circ \rho^{-1} = \tau_1$. Then the axis of ρ is parallel to $\overrightarrow{\tau_1(O)}$ and the axis of $\rho \circ \sigma$ is parallel to $\overrightarrow{\tau_1 \circ \tau_2^{-1}(O)}$. We get $\rho \circ \tau_2 \circ \rho^{-1} = \tau_1 \circ \tau_2^{-1}$ because $\tau_2 = \sigma^{-1} \circ \tau_1^{-1} \circ \sigma$ and $(\rho \circ \sigma)^2 = 1$, and therefore $\rho \circ \tau_2 \circ \rho^{-1} = \sigma \circ \tau_1^{-1} \circ \sigma^{-1}$.

By the same argument above, G contains reflections ρ and ρ' with axes parallel to $\overrightarrow{\tau_1(O)}$ and $\overrightarrow{\tau_2(O)}$, respectively. The axes of ρ and ρ' meet at a point O_1, and $\sigma_1 = \rho' \circ \rho$ is a rotation about O_1 of order 3. Thus $G = T_G \rtimes G_{O_1}$, a semi-direct product, where $G_{O_1} \cong D_3$.

If ρ satisfies (T2), then it has an axis parallel to $\overrightarrow{\tau_1 \circ \tau_2(O)}$, and the same argument shows again $G = T_G \rtimes G_{O_1}$, a semi-direct product, but now with the three axes of reflections through O_1 in a different position relative to $\overrightarrow{\tau_1(O)}$ and $\overrightarrow{\tau_2(O)}$. We have thus two types of G_3^1 and G_3^2, see Figure 4.28.

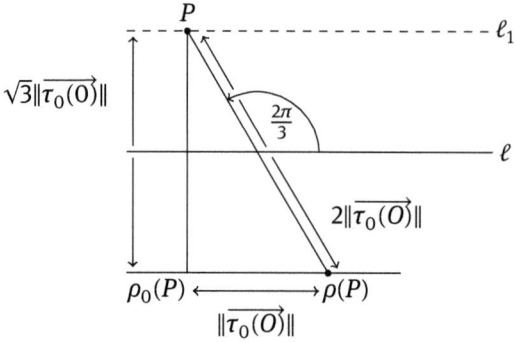

Figure 4.27: Existence of a reflection.

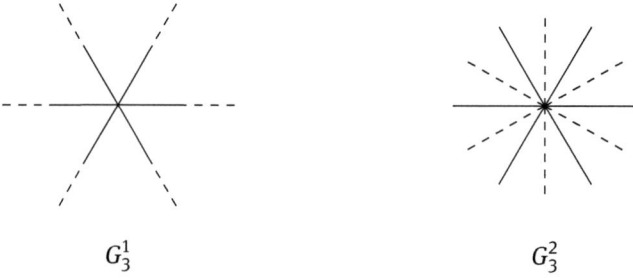

Figure 4.28: Two types of G_3^1 and G_3^2.

Theorem 4.21. *If $G \neq G^+ = G_3$, there are exactly two isomorphism types for G:*

$$G_3^1 = \langle \tau_1, \tau_2, \rho, \sigma \mid \tau_1 \circ \tau_2 = \tau_2 \circ \tau_1, \sigma^3 = \rho^2 = (\rho \circ \sigma)^2 = 1, \sigma \circ \tau_1 \circ \sigma^{-1} = \tau_1^{-1} \circ \tau_2,$$
$$\sigma \circ \tau_2 \circ \sigma^{-1} = \tau_1^{-1}, \rho \circ \tau_1 \circ \rho^{-1} = \tau_1, \rho \circ \tau_2 \circ \rho^{-1} = \tau_1 \circ \tau_2^{-1} \rangle,$$
$$G_3^2 = \langle \tau_1, \tau_2, \rho, \sigma \mid \tau_1 \circ \tau_2 = \tau_2 \circ \tau_1, \sigma^3 = \rho^2 = (\rho \circ \sigma)^2 = 1, \sigma \circ \tau_1 \circ \sigma^{-1} = \tau_1^{-1} \circ \tau_2,$$
$$\sigma \circ \tau_2 \circ \sigma^{-1} = \tau_1^{-1}, \rho \circ \tau_1 \circ \rho^{-1} = \tau_2, \rho \circ \tau_2 \circ \rho^{-1} = \tau_1 \rangle.$$

Case $n = 6$

If ρ is a reflection or glide reflection with axis ℓ, then $\rho_1 = \rho \circ \sigma$ has an axis ℓ_1 making an angle of $\frac{2\pi}{12}$ with ℓ.

Since G contains the rotation σ^2 of order 3, from the discussion of the case $n = 3$, we get that G contains reflections ρ and ρ' with axes making an angle of $\frac{2\pi}{12}$ whence $\sigma_1 = \rho' \circ \rho$ is a rotation of order 6 about their point O_1 of intersection. Thus $G_6^1 = T_G \rtimes G_{O_1}$, a semi-direct product, where $G_{O_1} \cong D_6$. Among the six reflections with axes passing through O_1, there is one satisfying (T1) and another one satisfying (T2). Hence, G is always of the form G_6^1.

Theorem 4.22. *If $G \neq G^+ = G_6$, then there is exactly one isomorphism type for G, namely*

$$G_6^1 = \langle \tau_1, \tau_2, \rho, \sigma \mid \tau_1 \circ \tau_2 = \tau_2 \circ \tau_1, \sigma^6 = \rho^2 = (\rho \circ \sigma)^2 = 1, \sigma \circ \tau_1 \circ \sigma^{-1} = \tau_2,$$

$$\sigma \circ \tau_2 \circ \sigma^{-1} = \tau_1^{-1} \circ \tau_2, \rho \circ \tau_1 \circ \rho^{-1} = \tau_1, \rho \circ \tau_2 \circ \rho^{-1} = \tau_1 \circ \tau_2^{-1} \rangle.$$

We remark that G_6^1 is exactly the symmetry group of a regular tessellation of the plane by equilateral triangles, and hence also by regular hexagons.

Summary. We have seen that there are in all 17 isomorphism types of wallpaper groups:

$$G_1, G_1^1, G_1^2, G_1^3, G_2, G_2^1, G_2^2, G_2^3, G_2^4, G_4, G_4^1, G_4^2, G_3, G_3^1, G_3^2, G_6, G_6^1.$$

From the proofs we realize that these 17 groups are also pairwise geometrically-different types. The following figures (see Figure 4.29), which have to be extended periodically in an obvious way, are figures in the plane which have the respective groups as symmetry groups.

Final remark. Crystallographic groups of dimensions 2 and 3 arose from their connections with chemistry and physics.

They came into mathematical prominence in 1900 with Hilbert's famous list of important outstanding mathematical problems. We have seen that in dimension 2 there are 17 types of cryptographic group. We remark that in dimension 3 there are 219 types of crystallographic group (see, for instance, [25]).

4.7 A Non-Periodic Tessellation of the Plane $\mathbb{R}^2$

So far we considered the periodic tessellation of the plane. As is usually done, we now use the word *tiling* for tessellation. Here, we describe one beautiful non-periodic tiling by Sir Roger Penrose (1931, Nobel Prize for Physics in 2000) which is his second tiling and uses only two tiling types called the *kite* and *dart* which may be combined to make a rhombus. However, the matching rules for periodic tilings (tessellations) prohibit such a combination.

The kite is a quadrilateral whose four interior angles are 72°, 72°, 72° and 144°. The kite may be bisected along its axis of symmetry to form a pair of triangles (with angles of 36°, 72° and 72°), see Figure 4.30.

The dart is a non-convex quadrilateral whose four interior angles are 36°, 72°, 72° and 144°. The dart may be bisected along its axis of symmetry to form a pair of obtuse triangles (with angles of 36°, 36° and 108°), which are smaller than the acute triangles, see Figure 4.31.

A *matching rule* is, for instance, to color the vertices with two colors (e.g., blue and red) and to require that adjacent tiles have matching vertices, see Figure 4.32.

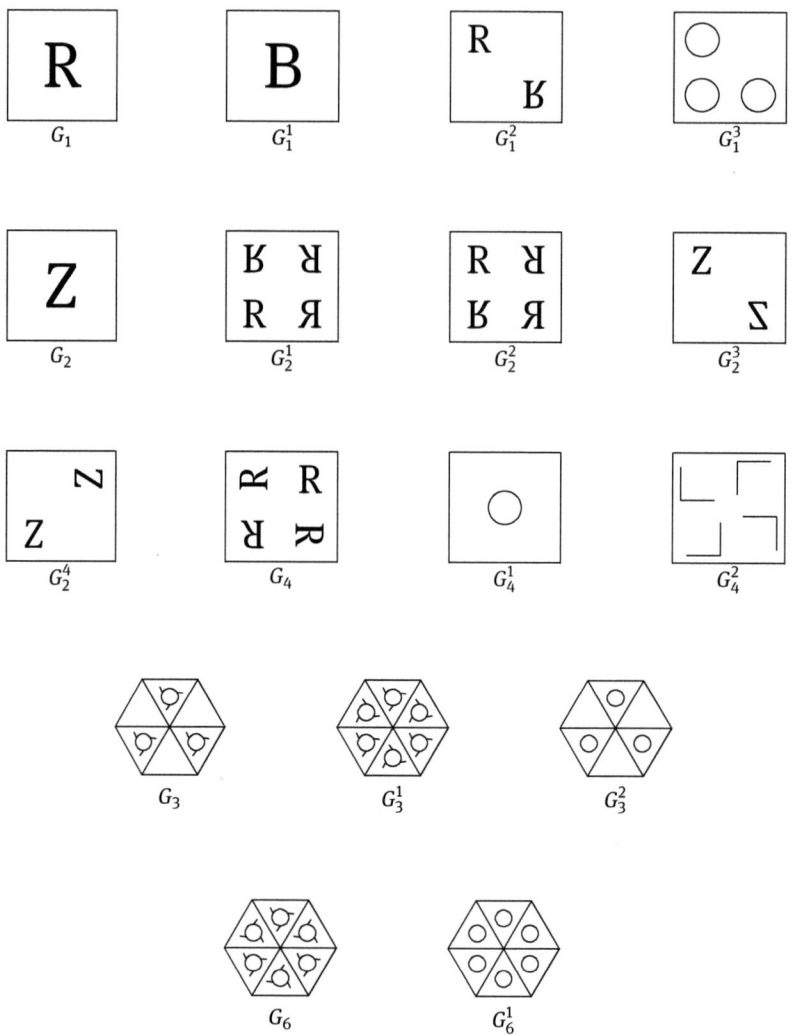

Figure 4.29: 17 isomorphism types of wallpaper groups.

Theorem 4.23 (Penrose).
1. *Taking into account the matching rules it is possible to tile the plane with kites and darts completely.*
2. *It is not possible to find a periodic tiling with kites and darts.*

Proof. We prove the theorem in four steps.
(a) If we pay attention to which angle of a kite and which angle of a dart together can result in 360°, we see that there are exactly seven possibilities to combine kites and darts to a complete vertex v, that is, the angle sum of the angles with the neighbors is 360°, see Figure 4.33.

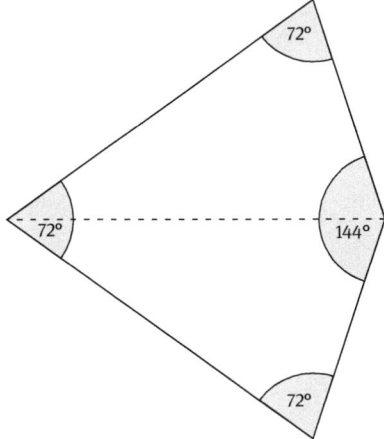

Figure 4.30: The kite.

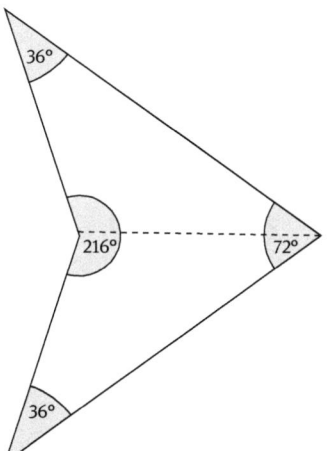

Figure 4.31: The dart.

(b) If we have a given tiling then we may produce from this a coarser tiling and a finer tiling which consist of kites and darts which are larger and smaller, respectively, by the factor $\phi = \frac{1}{2}(\sqrt{5} + 1)$ than in the original tiling. In coarsening, a bigger kite arises from two smaller kites and two half smaller darts, see Figure 4.34(a), and a bigger dart arises from a smaller kite and two half smaller darts, see Figure 4.34(b). The situation is opposite in the refinement. From the seven possibilities in step (a) we see that the smaller and bigger figures, respectively, come together to a new tiling by the defined matching rules. This already shows that we may tile the plane with kites and darts. If we start with the suitable starting configuration and then

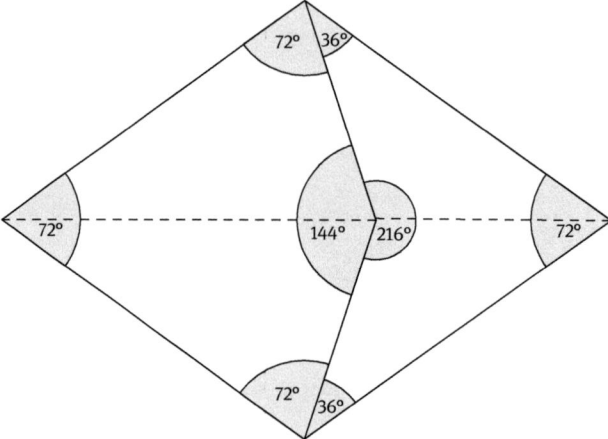

Figure 4.32: Matching rule.

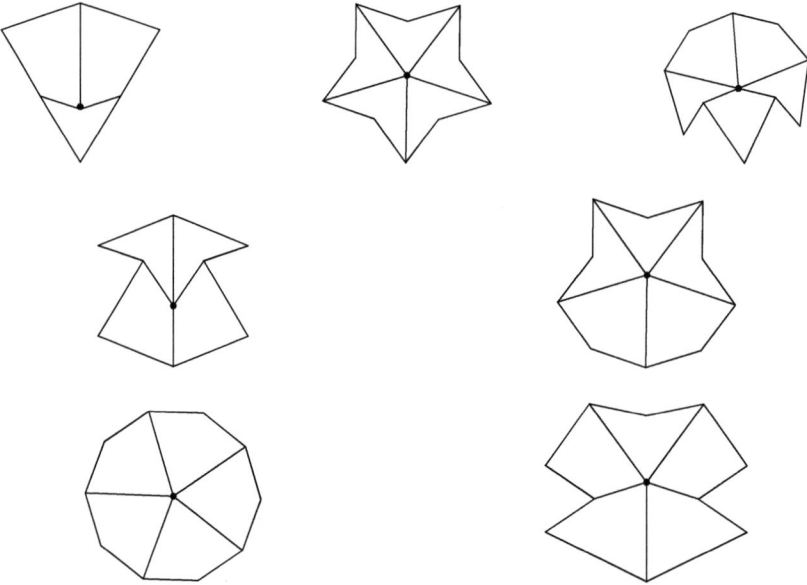

Figure 4.33: Possible vertices.

first refine and then enlarge by the stretch factor ϕ, and if we continue the process again and again, then we cover each point of the plane.

(c) If we consider an area with x kites and y darts and refine as in step (b) we get from one kite two new kites and $\frac{2}{2}$ new darts, altogether $2x$ new kites and x new darts. From one dart we get one new kite and $\frac{2}{2}$ new darts, altogether y new kites and y new darts.

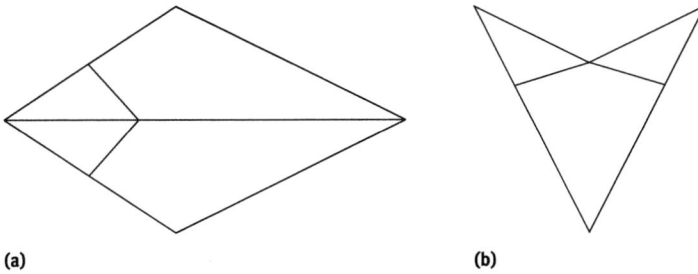

(a)   (b)

Figure 4.34: Coarsening of a tiling.

Altogether, we get $X = 2x + y$ new kites and $Y = x + y$ new darts. If we repeat this refinement process n times and with x_n kites and y_n darts, then we get the recursion formulas

$$X_{i+1} = 2x_i + y_i \quad \text{and} \quad Y_{i+1} = x_i + y_i$$

for $i = 0, 1, 2, \ldots$. Then

$$\lim_{i \to \infty} \frac{x_i}{y_i} = \phi = \frac{1}{2}(\sqrt{5} + 1) \approx 1.6180.$$

This fact is independent of the starting values x_i and y_i, if both are not zero.
Let $q_i = \phi + f_i$ with a rounding error for $i \geq 0$. Then, using the recursion formulas, for $i \geq 0$, we get

$$
\begin{aligned}
f_{i+1} &= -\phi + q_{i+1} \\
&= -\phi + \frac{x_{i+1}}{y_{i+1}} \\
&= -\phi + \frac{2x_i + y_i}{x_i + y_i} \\
&= 1 - \phi + \frac{x_i}{x_i + y_i} \\
&= 1 - \phi + \frac{1}{1 + \frac{y_i}{x_i}} \\
&= 1 - \phi + \frac{1}{1 + \frac{1}{q_i}} \\
&= 1 - \phi + \frac{1}{1 + \frac{1}{\phi + f_i}} \\
&= 1 - \phi + \frac{\phi + f_i}{1 + \phi + f_i} \\
&= \frac{1 + \phi - \phi^2 + (2 - \phi)f_i}{1 + \phi + f_i}
\end{aligned}
$$

$$= \frac{2 - \phi}{1 + \phi + f_i} f_i$$

because $1 + \phi - \phi^2 = 0$ by the golden equation (see [10]). Hence, we have

$$0 < \frac{2 - \phi}{1 + \phi + f_i} \leq \frac{2 - \phi}{1 + \phi - \phi + 1} < \frac{2 - \phi}{2} < 0.5$$

where we recall that $1 - \phi \approx -0.4$. Hence,

$$\lim_{i \to \infty} f_i = 0$$

which gives

$$\lim_{i \to \infty} \frac{x_i}{y_i} = 0.$$

(d) We have finally to show that there is no periodic tiling with kites and darts. We first assume that such a tiling exists. Then the assumed periodicity is based on some basic tile G with u kites and v darts, which is mapped over the entire plane by the two generating tessellations. Each figure, which will be enlarged by a centric stretching with stretch factor $d \to \infty$ contains a square-increasing number of complete basic tiles whereby the number of the basic tiles out from the edge of the figure increases only linearly. For the quotient of the numbers $x(d)$ of kites and $y(d)$ of darts, which are inside the figure, we get

$$\frac{x(d)}{y(d)} \to \frac{u}{v} \quad \text{for } d \to \infty.$$

The limit is a rational number. Now, in step (c) we have shown that the quotient of these numbers converges to the golden number ϕ which is irrational. Hence, a periodic tiling with kites and darts does not exist. □

Remark 4.24.
1. There is a fascinating ongoing work done in non-periodic tilings, and there exists several examples. Some of them can be considered as realistic models for the growth of quadrilaterals.
2. The described Penrose tiling is realized at the TU Dortmund where the foyer of the Auditorium Maximum is designed with a non-periodic Penrose tiling, see Figure 4.35.

Figure 4.35: Penrose tiling at TU Dortmund.

Exercises

1. Let G be a group and $H < G$ be a subgroup of G of index 2. Show that H is a normal subgroup of G.

2. Let G be a finite group and $H < G$ be a subgroup of index $[G : H]$. Show the Theorem of Lagrange:

$$|G| = [G : H] \cdot |H|.$$

3. Let $G = \langle g \mid g^n = 1 \rangle$, $n \geq 2$, be a cyclic group of order n. Show that:
   (a) For each divisor m of n, $1 \leq m \leq n$, there exists exactly one subgroup U of G with order m, and U is cyclic.
   (b) An element $g^k \neq 1$ generates G if and only if $\gcd(k, n) = 1$.
   (c) For each divisor m of n, $1 \leq m \leq n$, there exists exactly one factor group of G of order m.
   (d) Let $f : \mathbb{R}^2 \to \mathbb{R}^2$ be a rotation with center $\vec{0} = (0,0)$ and rotation angle $\alpha = \frac{2\pi}{n}$. Show that G is isomorphic to $\langle f \rangle$.

4. Let $D_n = \langle a, b \mid a^n = b^2 = (ab)^2 = 1 \rangle$, $n \geq 3$, be the dihedral group of order $2n$.
   (a) Let $U < D_n$ be a subgroup of D_n. Show that U is either cyclic, a dihedral group D_m with $m \mid n$, $m \geq 3$, or the Klein four group if n is even.
   (b) Show that D_3 is isomorphic to the symmetric group S_3 on $\{1, 2, 3\}$.
   (c) Show that D_n, $n \geq 4$, is isomorphic to a proper subgroup of the symmetric group S_n on $\{1, 2, \ldots, n\}$.
   (*Hint*: Consider in S_n the two permutations (written as cycles)

$$g = (1, 2, \ldots, n) \quad \text{and} \quad h = (2, n)(3, (n-1)) \cdots (i, (n+2-i)) \cdots.$$

   For the required details on symmetric groups, see, for instance, [12].)

5. Show that the set of all symmetries of a figure F forms a group.

6. Let G be a subgroup of the group E of the planar isometries. We call G discrete if there is no sequence $(\varphi_n)_{n \in \mathbb{N}}$ of pairwise distinct $\varphi_n \in G$ such that the sequence

$(\|\varphi_n(P)\|)_{n\in\mathbb{N}}$ converges to P for each $P \in \mathbb{R}^2$ (considered as local vectors $\overrightarrow{OP}$). Show that a discontinuous subgroup of E is discrete.

7. Given the two planar isometries

$$\tau : \mathbb{R}^2 \to \mathbb{R}^2, \quad \begin{pmatrix} x \\ y \end{pmatrix} \mapsto \begin{pmatrix} x+y \\ y \end{pmatrix}$$

and

$$\rho : \mathbb{R}^2 \to \mathbb{R}^2, \quad \begin{pmatrix} x \\ y \end{pmatrix} \mapsto \begin{pmatrix} x+y \\ -y \end{pmatrix}.$$

Let $G_1 = \langle \tau \rangle$ and $G_2 = \langle \rho \rangle$. Show that G_1 and G_2 are algebraically isomorphic but geometrically distinct. Find in $\mathbb{R}^2$ figures F_1 and F_2 such that $G_1 = \mathrm{Sym}(F_1)$ and $G_2 = \mathrm{Sym}(F_2)$.

8. (a) In the plane $\mathbb{R}^2$ consider the rotation

$$\sigma : \begin{pmatrix} x \\ y \end{pmatrix} \mapsto \begin{pmatrix} -1 & 0 \\ 0 & -1 \end{pmatrix} \begin{pmatrix} x \\ y \end{pmatrix}$$

and the translation

$$\tau : \begin{pmatrix} x \\ y \end{pmatrix} \mapsto \begin{pmatrix} x \\ y \end{pmatrix} + \begin{pmatrix} 1 \\ 0 \end{pmatrix}.$$

Let $G = \langle \sigma, \tau \rangle$. Describe G by generators and relations.
Find a figure F in $\mathbb{R}^2$ with $G = \mathrm{Sym}(F)$. Mark the rotation centers in F.

(b) In the plane $\mathbb{R}^2$ consider the reflection

$$\rho : \begin{pmatrix} x \\ y \end{pmatrix} \mapsto \begin{pmatrix} -1 & 0 \\ 0 & 1 \end{pmatrix} \begin{pmatrix} x \\ y \end{pmatrix}$$

and the translation

$$\tau : \begin{pmatrix} x \\ y \end{pmatrix} \mapsto \begin{pmatrix} x \\ y \end{pmatrix} + \begin{pmatrix} 1 \\ 0 \end{pmatrix}.$$

Let $G = \langle \rho, \tau \rangle$. Describe G by generators and relations.
Find a figure F in $\mathbb{R}^2$ with $G = \mathrm{Sym}(F)$. Mark the reflection lines in F.

9. In the plane $\mathbb{R}^2$ consider the translations

$$\tau_1 : \begin{pmatrix} x \\ y \end{pmatrix} \mapsto \begin{pmatrix} x \\ y \end{pmatrix} + \begin{pmatrix} 1 \\ 0 \end{pmatrix}$$

and

$$\tau_2 : \begin{pmatrix} x \\ y \end{pmatrix} \mapsto \begin{pmatrix} x \\ y \end{pmatrix} + \begin{pmatrix} 0 \\ 1 \end{pmatrix}.$$

Let $G = \langle \tau_1, \tau_2 \rangle$. Describe G by generators and relations and find a figure F in $\mathbb{R}^2$ with $G = \text{Sym}(F)$.

10. (a) In the plane $\mathbb{R}^2$ consider the reflection

$$\rho : \begin{pmatrix} x \\ y \end{pmatrix} \mapsto \begin{pmatrix} 0 & 1 \\ 1 & 0 \end{pmatrix} \begin{pmatrix} x \\ y \end{pmatrix}$$

and the two translations

$$\tau_1 : \begin{pmatrix} x \\ y \end{pmatrix} \mapsto \begin{pmatrix} x \\ y \end{pmatrix} + \begin{pmatrix} 1 \\ 0 \end{pmatrix}$$

and

$$\tau_2 : \begin{pmatrix} x \\ y \end{pmatrix} \mapsto \begin{pmatrix} x \\ y \end{pmatrix} + \begin{pmatrix} 0 \\ 1 \end{pmatrix}.$$

Let $G = \langle \rho, \tau_1, \tau_2 \rangle$. Describe G by generators and relations and find a figure F in $\mathbb{R}^2$ with $G = \text{Sym}(F)$.

(b) In the plane $\mathbb{R}^2$ consider the rotation

$$\sigma : \begin{pmatrix} x \\ y \end{pmatrix} \mapsto \begin{pmatrix} -1 & 0 \\ 0 & -1 \end{pmatrix} \begin{pmatrix} x \\ y \end{pmatrix}$$

and the two translations

$$\tau_1 : \begin{pmatrix} x \\ y \end{pmatrix} \mapsto \begin{pmatrix} x \\ y \end{pmatrix} + \begin{pmatrix} 1 \\ 0 \end{pmatrix}$$

and

$$\tau_2 : \begin{pmatrix} x \\ y \end{pmatrix} \mapsto \begin{pmatrix} x \\ y \end{pmatrix} + \begin{pmatrix} 0 \\ 1 \end{pmatrix}.$$

Let $G = \langle \sigma, \tau_1, \tau_2 \rangle$. Describe G by generators and relations and find a figure F in $\mathbb{R}^2$ with $G = \text{Sym}(F)$.

5 Graph Theory and Graph Theoretical Problems

In this chapter we give an introduction to graph theory. Our aim is to present some basic material together with a variety of applications. We remark that graph theoretical results and techniques are often used in the forthcoming chapters. Graphs serve the description and illustration of relations between objects. The idea is to represent the objects by points, and if there is a connection between two objects, we draw a line between them. The objects are denoted as vertices and the connections between them as edges.

The sets of vertices and edges form a graph. Through abstraction we may represent many coherences using graphs. For instance, we may take all cities as vertices and the road network as edges, that is, we draw an edge between two cities if they are connected via a road. We will use this model when we discuss the Taveling Salesman Problem in the last section of this chapter. Another example appears with movies and actors as set of vertices. We draw an edge between an actor x and a movie y, if x played a part in y. We also may regard the game positions of a game as vertex set. We draw an edge from a game position x to a game position y if x gets converted to y by one move.

In such a manner, different facts can be modeled by graphs. In general, any relation $R \subset A \times B$ can be considered as a graph with a vertex set $A \cup B$ and an edge between $x, y \in A \cup B$ if $(x, y) \in R$. This type of presentation has several advantages. First, graphs can be illustrated figuratively. Second, we can use existing results and methods from graph theory for the solution of problems. Last but not least, graph theory allows uniform terminology.

5.1 Graph Theory

We begin with the definition of a graph. The graphs we define here are called *simple* and *undirected* in the literature.

Definition 5.1. A (simple) *graph* is a pair $G = (V, E)$, consisting of a nonempty set V of *vertices* and a set E of unordered pairs of distinct elements of V, which we call *edges* and denote by $k = \{x, y\}, x, y \in V, x \neq y$. Typically, a graph is depicted in diagrammatic form as a set of dots for the vertices, joined by lines or curves for the edges.

Remarks 5.2.
(1) In simple graphs, we do not allow loops and we do not allow more than one edge between two elements of V. In an undirected graph an edge has no direction.
(2) An edge $k = \{x, y\}$ is called a *connecting edge* from x to y, and x, y are named as the termini (or end points) of k. We say that k connects the vertices x and y.
(3) Examples for graphs are certainly the known geometric figures like triangles, rectangles, n-polygons, tetrahedrons, cubes and so on, but also the *Hasse diagrams*

https://doi.org/10.1515/9783110740783-005

of finite partial orders over a set M. Recall that a partial order over M is a binary relation $\leq$ over M which is reflexive, antisymmetric and transitive, that is, which satisfies for all $a, b, c \in M$:

(a) $a \leq a$,

(b) $a \leq b$ and $b \leq a$, then $a = b$,

(c) $a \leq b$ and $b \leq c$, then $a \leq c$.

The *Hasse diagram*, named after H. Hasse (1898–1979), is a type of mathematical diagram to represent a finite partially ordered set. For a finite partially ordered set $(M, \leq)$, one represents each element of M as a vertex in the plane $\mathbb{R}^2$ and draws a line segment upwards from a to b whenever $a < b$ and there is no c such that $a < c < b$. These line segments may cross each other but must not touch any vertices other than their endpoints. Such a diagram, with labeled vertices, uniquely determines its partial order. We consider examples of Hasse diagrams:

(1) Let $M = \{1, 2, 3\}$. Then we have the following possible Hasse diagrams, see Figure 5.1.

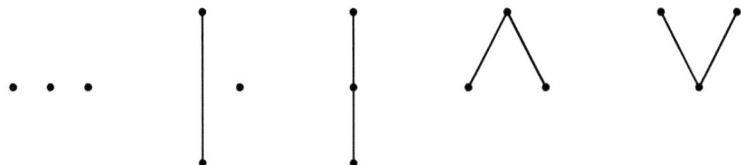

Figure 5.1: Hasse diagrams.

(2) Let $M = \{1, 2, \ldots, 10\}$ be equipped with the divisor relation as the partial order, see Figure 5.2.

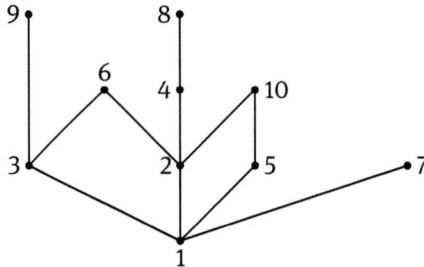

Figure 5.2: Diagram with divisor relation as partial order.

(3) Let B^n, $n \geq 1$, be the n-fold Cartesian product of $B = \{0, 1\}$, that is, the set of the n-digit 0–1-sequences $(a_1, a_2, \ldots, a_n)$ with $a_i = 0$ or 1 for $i = 1, 2, \ldots, n$.

B^n can be partially ordered by

$$(a_1, a_2, \ldots, a_n) \leq (b_1, b_2, \ldots, b_n) \quad \Leftrightarrow \quad a_i \leq b_i \quad \text{for } i = 1, 2, \ldots, n$$

and

$$(a_1, a_2, \ldots, a_n) < (b_1, b_2, \ldots, b_n) \quad \Leftrightarrow \quad a_i \leq b_i \quad \text{for } i = 1, 2, \ldots, n \quad \text{and}$$
$$a_i < b_i \quad \text{for at least one } i.$$

The Hasse diagrams for B^1, B^2 and B^3 are shown in Figure 5.3.

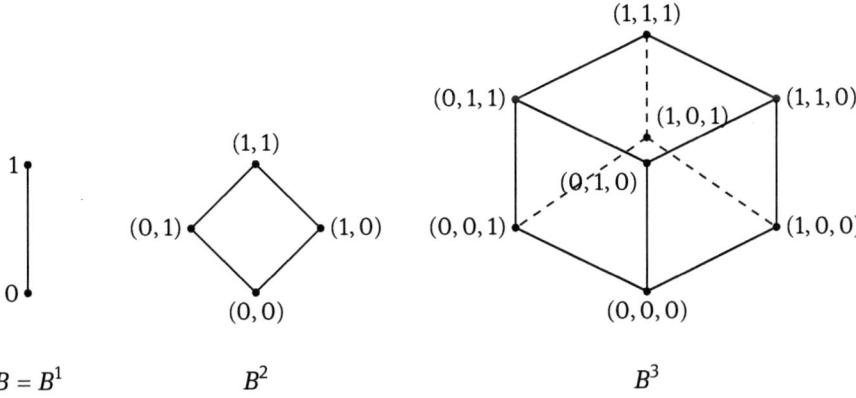

Figure 5.3: Hasse diagrams for B^1, B^2 and B^3.

These examples already show that it is custom to realize graphs in the plane $\mathbb{R}^2$ or the space $\mathbb{R}^3$.

Definition 5.3.
(1) Two edges k, k' are called *adjacent* (neighboring) in G, if they have a common terminus (end point), see Figure 5.4.
(2) $x \in V$ is called *isolated* if it is not incident to an edge.
(3) A graph $G = (V, E)$ is called *finite* if both V and E are finite.

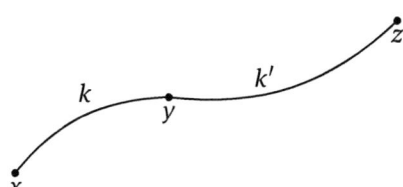

Figure 5.4: Adjacent edges.

Remark 5.4. Since we only consider simple graphs, then certainly $|E| < \infty$ if $|V| < \infty$.

Agreement. From now on all graphs are finite.

Definition 5.5. Let $G = (V, E)$ be a graph and $x \in V$. Then

$$d(x) = \left| \{ u \in E \mid x \in u, \text{ that is, } u = \{x, y\} \text{ for some } y \in V \} \right|$$

is called the *degree of x*.

Remark 5.6. If $x \in V$ is isolated then $d(x) = 0$, if $d(x) = 1$ then we call x a *final vertex*. Certainly we have

$$|E| = \frac{1}{2} \sum_{x \in V} d(x).$$

Definition 5.7. Let $G = (V, E)$ be a graph. An *edge sequence* with start vertex x_0 and end vertex x_n is a sequence of edges of the form $u_1 = \{x_0, x_1\}, u_2 = \{x_1, x_2\}, \ldots, u_n = \{x_{n-1}, x_n\}$.

An *edge line* is an edge sequence where all edges are distinct.

An *edge path* is an edge line $u_1 = \{x_0, x_1\}, u_2 = \{x_1, x_2\}, \ldots, u_n = \{x_{n-1}, x_n\}$ with $x_i \neq x_j$ for $i \neq j$, $0 \leq i, j \leq n$ with the possible exception $x_0 = x_n$. If $x_0 = x_n$ then the edge sequence (line or path, respectively) is called *closed* or an *edge circle*.

Remarks 5.8.
(1) From an edge sequence from x_0 to x_n we get by reduction an edge line, and finally an edge path from x_0 to x_n.
(2) By the definition

$$x_0 \sim_P x_n \quad \Leftrightarrow \quad \text{there exists an edge path from } x_0 \text{ to } x_n$$

we get an equivalence relation $\sim_P$ on the vertex set V of $G = (V, E)$.

Definition 5.9. Let $G = (V, E)$ be a graph.
(1) A graph (V', E') is called a *subgraph* of G if $V' \subset V$ and $E' \subset E$.
    A subgraph (V', E') of G is called a *section graph* of G if E' is composed exactly of those edges of G for which both termini are in V'.
    The section graph is uniquely determined by V' and is called the *spanning section graph of V'*.
(2) A *connected component* of G is the spanning section graph of an equivalence class of V with respect to the equivalence relation $\sim_P$ on V.

Example 5.10. Figure 5.5 shows a graph with three connected components.

Definition 5.11. G is called *connected* if G has exactly one connected component.

Figure 5.5: Graph with three connected components.

Theorem 5.12. *If $G = (V, E)$ is connected, then $|E| \geq |V| - 1$.*

Proof. We proof the theorem by induction on $n = |V|$.
  If $n = 1$, then $V = \{x\}$ and hence $|E| = 0 \geq |V| - 1 = 0$.
  If $n \geq 2$ and $d(x) \geq 2$ for all $x \in V$, then $|E| \geq n$ because

$$|E| = \frac{1}{2} \sum_{x \in V} d(x).$$

Now let $n \geq 2$ and x_0 be a vertex with $d(x_0) = 1$ (the case $d(x_0) = 0$ cannot hold, because G is connected). Let u be an edge with x_0 as a terminus vertex.
  The subgraph $(V \setminus \{x_0\}, E \setminus \{u\})$ is connected. By the induction hypothesis, we have

$$|E \setminus \{u\}| \geq |V \setminus \{x_0\}| - 1,$$

that is,

$$|E| - 1 \geq |V| - 2, \quad \text{and hence} \quad |E| \geq |V| - 1. \qquad \square$$

Definition 5.13. Let $G = (V, E)$ be a graph. G is called a *tree* if G is connected and has no edge circle. A vertex of a tree is called a *leaf* if it has at most degree 1. Vertices which are not leaves are called the *inner vertices*.

In trees with more than one vertex the leaves are exactly those vertices of degree 1.

Example 5.14.
(1) All trees with $|V| = 5$, see Figure 5.6.

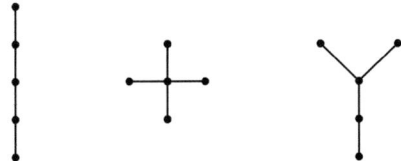

Figure 5.6: All trees with $|V| = 5$.

(2) Binary trees (V, E).
  These are trees with the following properties:

(a) There exists a distinguished vertex x_0 with $d(x_0) \leq 2$; x_0 is called a root of the tree, and

(b) $d(x) \leq 3$ for all $x \in V$. These may occur in several situations:

(i) Decision tree; see Figure 5.7.

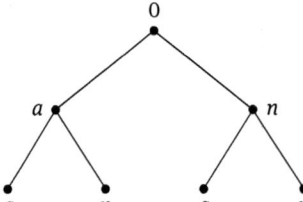

Figure 5.7: Decision tree.

(ii) Probability tree; see Figure 5.8.

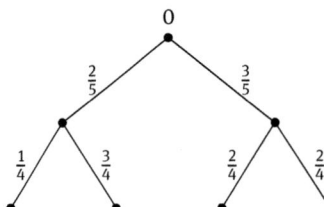

Figure 5.8: Probability tree.

(iii) Binary search tree of an ordered set $M = \{x_1, x_2, \ldots, x_n\}$. For $M = \{21, 17, 6, 3, 7, 25, 12, 11, 19, 30, 29, 27, 14, 35\}$ we see an example (with the given ordering) in Figure 5.9:

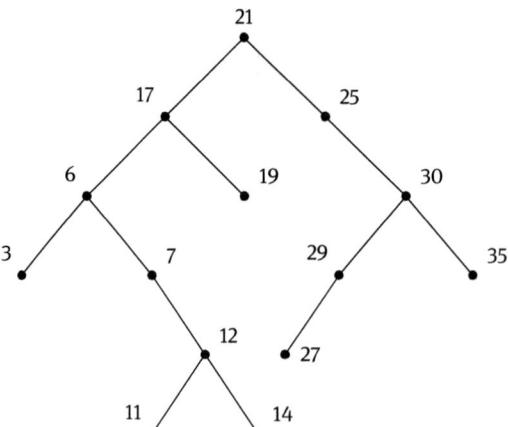

Figure 5.9: Example of a binary search tree.

(iv) Brackets for multiple products, see Figures 5.10 and 5.11.

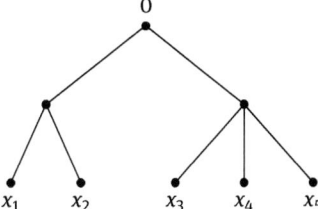

Figure 5.10: Brackets for multiple products $(x_1 x_2)(x_3 x_4 x_5)$.

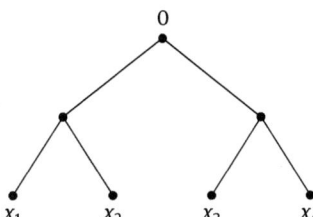

Figure 5.11: Brackets for multiple products $(x_1 x_2)(x_3 x_4)$.

Theorem 5.15. *Let $G = (V, E)$ be a graph. Then the following are equivalent:*
(1) *G is a tree.*
(2) *G is connected and $|E| = |V| - 1$.*
(3) *Any two distinct vertices in G are connected by exactly one path.*

Proof. (1) $\Rightarrow$ (2) A tree is connected by definition. Let n, m be the number of vertices and edges of G, respectively. We show $m = n - 1$ by induction. If $n = 1$ then $m = 0$, hence, $m = n - 1$.

Now let $n \geq 2$ and $u = \{a, b\}$ be an edge. Let $G' = (V, E \setminus \{u\})$. Then G' is not connected, and G' comprises two connected components, $G'_1 = (V_1, E_1)$ and $G'_2 = (V_2, E_2)$, see Figure 5.12.

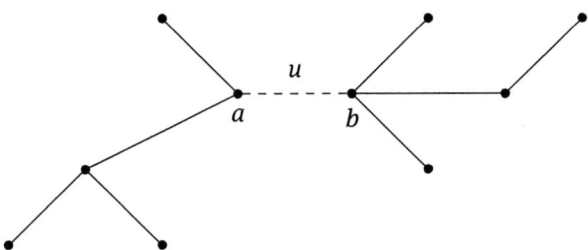

Figure 5.12: Components G'_1 and G'_2.

G_i' is a tree with, say $a \in G_1'$, $b \in G_2'$. By induction hypothesis, $|E_i| = |V_i| - 1$, hence

$$|E| = m = |E_1| + |E_2| + 1 = |V_1| + |V_2| - 2 + 1,$$

that is, $m = n - 1$.

(2) $\Rightarrow$ (3) We have

$$2|E| = \sum_{x \in V} d(x) = 2|V| - 2.$$

If $|V| = 1$, there is nothing to prove. Now let $|V| \geq 2$. Then there exists a final vertex $x \in V$ with $d(x) = 1$. Let $u = \{x, y\}$ be the edge to which x is incident.

Then $G' = (V \setminus \{x\}, E \setminus \{u\})$ also is connected and satisfies the condition in (2).

Let $a, b \in V \setminus \{x\}$. A path from a to b in G cannot contain u because $d(x) = 1$. Then, inductively, there exists exactly one path in G' which connects a and b. A path from x to a must start with u, which gives the uniqueness of the path from x to a in G.

(3) $\Rightarrow$ (1) From the existence of a closed path, which contains $a, b \in V$, we get the existence of two distinct paths between a and b. $\qquad\square$

Remark 5.16. From Theorem 5.15 we automatically get the following statements.
(1) Let $G = (V, E)$ be a graph with $|V| \geq 2$. Then the following are equivalent:
   (a) G is a tree.
   (b) G contains no closed path, and for any $x, y \in V$, $x \neq y$, with $\{x, y\} \notin E$, the graph $(V, E \cup \{x, y\})$ contains a closed path.
(2) Each tree has a leaf. Especially, each tree with at least two vertices has at least two leaves.

Definition 5.17. Let $G = (V, E)$ be a graph. A subgraph of G which is a tree and which contains all vertices of G is called a *spanning tree* of G.

Theorem 5.18. *A graph $G = (V, E)$ has a spanning tree if and only if G is connected.*

Proof. "$\Rightarrow$" This is clear, because a spanning tree is connected.

"$\Leftarrow$" Let G be connected. Starting with $G_0 = (V, \emptyset)$, we construct a chain

$$G_0 \subset G_1 \subset \cdots$$

of subgraphs, where $G_k = (V, E_k)$ for $k \geq 1$ is formed from G_{k-1} by adding an edge u in such a manner that G_k does not contain an edge circle. After $|V|$ steps, we get a spanning tree. $\qquad\square$

Definition 5.19. The graph $K_n = (V, E)$ with $|V| = n \geq 2$ and $|E| = \binom{n}{2}$ edges, that is, every pair of distinct vertices is connected by a unique edge, is called the *complete graph on n vertices*.

The Cayley formula (after A. Cayley, 1821–1895) determines the number of spanning trees in K_n, $n \geq 2$.

Theorem 5.20 (Cayley formula). *Let $n \geq 2$. The number of spanning trees in K_n is n^{n-2}.*

Proof. Let $K_n = (V, E)$ be the complete graph on n vertices. We assume that the vertices are arranged linearly. Let (V, T) with $T \subset E$ be a spanning tree of K_n. We encode T by a sequence in V^{n-2} using a method of E. P. H. Prüfer (1896–1934). If $n = 2$, then this is the empty sequence, and this matches the only spanning tree with $T = E$.

Now, let $b_1 \in V$ be the smallest leaf of (V, T). The essential trick is to note down the neighbor p_1 of b_1 and not the leaf itself. Hence, $\{b_1, p_1\} \in T$. Let $V' = V \setminus \{b_1\}$ and $T' = T \setminus \{\{b_1, p_1\}\}$. Then (V', T') is a spanning tree of a complete graph with $n - 1$ vertices. By induction there exists a sequence $(p_2, \ldots, p_{n-2})$ which encodes (V', T'). We define the encoding of (V, T) by the sequence $(p_1, p_2, \ldots, p_{n-2})$ and call this sequence the Prüfer code of (V, T). We see by induction that $\{p_1, \ldots, p_{n-2}\}$ is exactly the set of the inner vertices of T. Especially, some of the p_i's may be equal. We now can discern the leaf b_1 from the sequence $(p_1, \ldots, p_{n-2})$, it is the smallest element in $V \setminus \{p_1, \ldots, p_{n-2}\}$. We know that $\{b_1, p_1\} \in T$. Inductively, we may reconstruct from $\{p_2, \ldots, p_{n-2}\}$ the spanning tree T' of $V' = V \setminus \{b_1\}$. We get $T = T' \cup \{\{b_1, p_1\}\}$.

The assignment which assigns each T the Prüfer code is therefore an injective map from the set of all spanning trees of (V, E) in the set V^{n-2}.

It remains to show that each sequence $(p_1, \ldots, p_{n-2})$ is a Prüfer code of some spanning tree. Let b_1 be the smallest element of $V \setminus \{p_1, \ldots, p_{n-2}\}$. By induction, the residual sequence $(p_2, \ldots, p_{n-2})$ is the Prüfer code of a spanning tree T' of $V' \setminus \{b_1\}$. Therefore, (V', T') is connected and T' has $n - 2$ edges. We set $T = T' \cup \{\{b_1, p_1\}\}$, then (V, T) is connected and T has $n - 1$ vertices. Hence, (V, T) is a spanning tree with Prüfer code $(p_1, \ldots, p_{n-2})$. □

Examples 5.21.
(1) We consider the tree with Prüfer code (2, 7, 7, 1, 7, 1), see Figure 5.13.

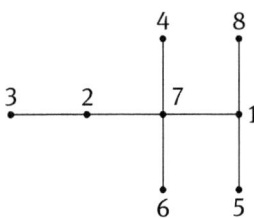

Figure 5.13: Tree with Prüfer code (2, 7, 7, 1, 7, 1).

(2) All spanning trees with edge set $\{1, 2, 3\}$, see Figure 5.14.

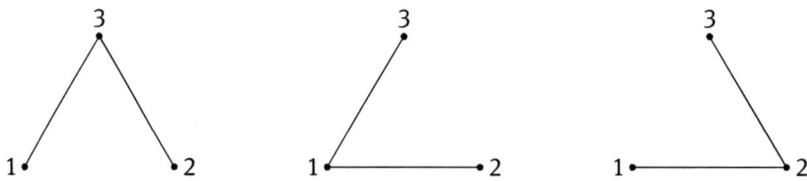

Figure 5.14: Spanning trees with edge set $\{1, 2, 3\}$.

Remark 5.22. If a graph $G = (V, E)$ is realized in the plane $\mathbb{R}^2$, then in general intersections of edges occur at points which are not vertices of G.

Definition 5.23.
(1) A graph $G = (V, E)$ in the plane $\mathbb{R}^2$ is called *crossing-free* if intersections of edges are only at their endpoints. In other words, no edges cross each other.
(2) Two graphs $G = (V, E)$ and $G' = (V', E')$ are called *isomorphic* if their exists a bijection $f : V \to V'$ such that

$$\{x, y\} \in E \quad \Leftrightarrow \quad \{f(x), f(y)\} \in E',$$

f then is called a graph isomorphism.
(3) A graph $G = (V, E)$ is called *planar* if it is isomorphic to a crossing-free graph in the plane $\mathbb{R}^2$.

Example 5.24. Let $K_n = (V, E)$ be the complete graph with $|V| = n$ and $|E| = \binom{n}{2}$. K_2 and K_3 are certainly planar, see Figure 5.15.

Figure 5.15: K_2 and K_3.

K_4 is also planar, see Figure 5.16.

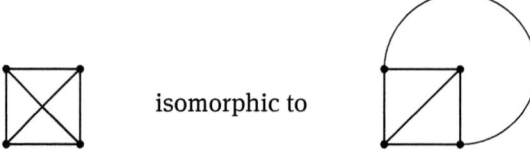

isomorphic to

Figure 5.16: K_4.

We will show that K_5 is not planar. Then in general K_n, $n \geq 5$, is not planar.

To prove this, we need Euler's formula for planar, connected graphs.

Agreement. If in what follows we call a graph $G = (V, E)$ planar then we mean that G already is realized in $\mathbb{R}^2$ as crossing-free.

Definition 5.25. Let $G = (V, E)$ be a planar graph in the plane $\mathbb{R}^2$.
(1) $x \in \mathbb{R}^2$ is called *distinct from G* if x is neither a vertex of G nor an element of an edge of G. Diagrammatically, x is not a vertex and not on a line or curve joining two vertices.
(2) Let $x \in \mathbb{R}^2$ be different from G. A face $F(x)$ containing x is the set of all points from $\mathbb{R}^2$ which can be reached from x by a finite polygonal chain, whose points are all distinct from G.

Remark 5.26. This defines an equivalence relation:
$x \sim y \Leftrightarrow x$ and y are distinct from G and can be connected by a finite polygonal chain, whose points are all distinct from G.
The equivalence classes are called the *faces of G*. There is always the external or unbounded face.

Example 5.27. We give an example in Figure 5.17.

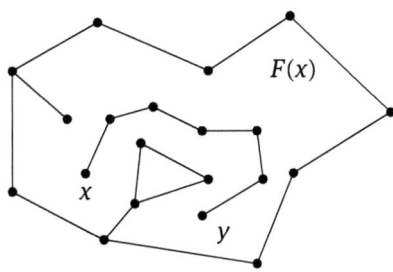

Figure 5.17: Face $F(x)$.

We are now prepared to prove Euler's polyhedra formula (L. Euler, 1707–1783).

Theorem 5.28 (Euler's formula). *Let G be a connected, planar graph in the plane $\mathbb{R}^2$. Let n, m and f be the number of vertices, edges and faces of G, respectively.*
Then

$$n - m + f = 2.$$

Proof. We give the proof by induction on m. If $m = 0$, then $n = 1$ because G is connected, and $f = 1$ (there is only the unbounded face). Therefore

$$n - m + f = 1 - 0 + 1 = 2$$

for $m = 0$.

Now let $m \geq 1$ and assume that Euler's formula holds for all respective graphs with $m - 1$ edges.

First, if G is a tree, then $f = 1$ because we only have the unbounded face, and $m = n - 1$ by Theorem 5.15, which gives $n - m + f = n - (n - 1) + 1 = 2$. Therefore Theorem 5.28 holds if G is a tree.

Now assume G is not a tree. Let e be an edge of an edge circle of G, see Figure 5.18.

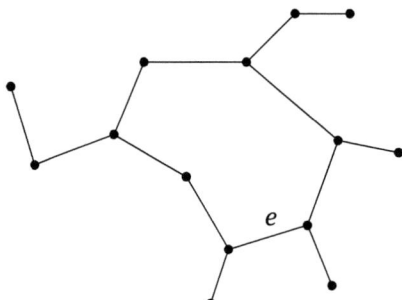

Figure 5.18: An edge circle and an edge e.

Then $G \setminus \{e\} := (V, E \setminus \{e\})$ is a connected planar graph with n vertices, $m - 1$ edges and $f - 1$ faces, and hence,

$$n - (m - 1) + (f - 1) = n - m + f = 2,$$

by the induction hypothesis.

Altogether $n - m + f = 2$. □

Corollary 5.29. *Let G be a connected, planar graph with n, n ≥ 3, vertices and m edges. Then m ≤ 3n − 6.*

Proof. Since each face is bounded by at least three edges, and each edge bounds at most two faces, we get $3f \leq 2m$. This gives

$$6 = 3n - 3m + 3f \leq 3n - m,$$

and hence $m \leq 3n - 6$. □

Corollary 5.30. K_5 *is not planar.*

Proof. Assume that K_5 is planar. Then, by Corollary 5.29, we get

$$10 = m \leq 3n - 6 = 15 - 6 = 9$$

which gives a contradiction. Hence, K_5 is not planar. □

Corollary 5.31. *Each planar graph has a vertex x with d(x) ≤ 5.*

Proof. Without loss of generality, let the graph be connected and have at least 3 vertices.

If $d(x) \geq 6$ for all vertices x, then $6n \leq 2m$ with n and m the number of vertices and edges, respectively. By Corollary 5.29, we have then $3n \leq 3n - 6$ which gives a contradiction. □

5.2 Coloring of Planar Graphs

Definition 5.32. A *C-coloring* of a graph $G = (V, E)$ is a map $c : V \to C$ with $c(x) \neq c(y)$ for all $\{x, y\} \in E$. Here C is the set of colors. We say that G is *k-colorable* if there exists a C-coloring with $|C| = k$.

Remark 5.33. We give the following application for graph coloring in the context of school timetables. We consider the graph $G = (V, E)$ where the vertices present the lessons and two lessons are connected by an edge if they cannot take place at the same time. There are many practical reasons for this: A teacher cannot teach two lessons at the same time and a class cannot attend two lessons simultaneously, and many more. If the graph admits a k-coloring, then there exists a time table for all courses with k time slots.

The famous Four Color Theorem by K. Appel (1932–2013) and W. Haken (born 1928) says that each planar graph is 4-colorable, see [2]. In the context of geographic maps this means that four colors are enough to color the regions of a map so that adjacent regions have different colors. Here we consider a planar graph $G = (V, E)$ where the vertices present the regions and two regions are connected by an edge if they have a common border.

We only show here that each planar graph is 5-colorable. First, we check that each planar graph is 6-colorable. Let $v \in V$ be a vertex of G with $d(v) \leq 5$. Such a $v \in V$ exists by Corollary 5.31. The graph without the vertex v (and the edges $\{v, x\} \in E$) is assumed to be 6-colorable by induction. Since v has 5 neighbors x with $\{v, x\} \in E$, we need for these neighbors x at most five colors. If we add again v, then one of the six colors is left for v. We use this idea for the proof of the next theorem.

Theorem 5.34. *Each planar graph is 5-colorable.*

Proof. We make induction on the number of vertices and show that each planar graph is colorable using colors C with $|C| = 5$.

Let $G = (V, E)$ be a planar graph. By Corollary 5.31, there exists a vertex $v \in V$ with degree $d(v) \leq 5$. If $d(v) \leq 4$ then we consider the spanning section graph of $V \setminus \{v\}$ of G. This has by induction a C-coloring with $|C| = 5$. We add again v to get G and may take for v one of the five colors which was not used. The colors of the remaining vertices are not changed. This gives a 5-coloring for G.

Now, let $d(v) = 5$, and let a, b, c, d, e be the neighbors of v ordered clockwise, see Figure 5.19. It is impossible for both edges $\{a, c\}$ and $\{b, d\}$ to exist because otherwise they would intersect. Without loss of generality, let $\{a, c\}$ not be in E, that is, $\{a, c\}$ does not exist.

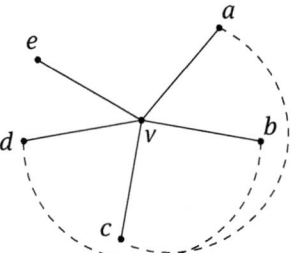

Figure 5.19: Five neighbors of v.

We remove from G the vertex v and all the edges $\{v, a\}$, $\{v, b\}$, $\{v, c\}$, $\{v, d\}$ and $\{v, e\}$.

Now we shift the vertices a and c together and join them up to a single vertex $z_{ac} \notin V$.

This gives a planar graph without two vertices. By induction, there exists a 5-coloring c' for the resulting graph. We now construct a coloring $c : V \rightarrow C$ of G in such a way that the vertices a and c both get the color $c'(z_{ac})$. The remaining vertices of G keep their color. We only have to color the vertex v. Since the five neighbors of v need at most four colors, there is one color left, which we may use for v. □

5.3 The Marriage Theorem

Let A and B be non-empty, disjoint sets and let $G = (A \cup B, E)$ be a *bipartite graph*, that is, each edge in E connects a vertex in A with a vertex in B. A subset $M \subset E$ is a *matching* if no two edges in M have a common vertex. M is called a *perfect matching for A* if each vertex from A is on an edge from M.

In Figure 5.20 we present an example of a matching and a perfect matching for a bipartite graph.

We remark that in the exercises, we will also study complete bipartite graphs. A graph $G = (V, E)$ is called *complete bipartite* if $V = A \cup B$ with $A \cap B = \emptyset$ and $E = \{\{a, b\} \mid a \in A, b \in B\}$. If $|A| = m$ and $|B| = n$, then we write $K_{m,n}$ for G.

The marriage theorem specifies a condition when exactly a perfect matching exists for A. Let $N_G(a) = \{b \in B \mid \{a, b\} \in E\}$ be the set of neighbors of $a \in A$ in G. This notation can be extended by $N_G(X) = \bigcup_{a \in X} N_G(a)$ to subsets $X \subset A$. The *marriage condition* holds if $|N_G(X)| \geq |X|$ for all subsets $X \subset A$.

If $|B| < |A|$ then there cannot exist a perfect matching. Also the marriage condition must hold. The marriage theorem says that the reverse is true: If the marriage

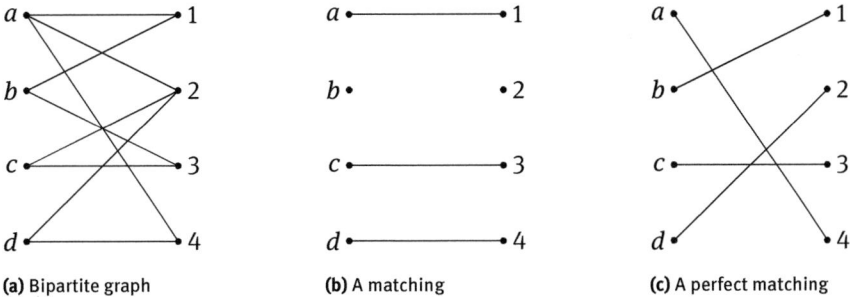

(a) Bipartite graph **(b)** A matching **(c)** A perfect matching

Figure 5.20: A matching and a perfect matching for a bipartite graph.

condition holds, then there exists a perfect matching for A. In this form, the marriage theorem was proved in 1935 by P. Hall (1904–1982). In a slightly different form it was proved in 1931 independently by D. König (1884–1944) and J. Egerváry (1891–1958). Nevertheless, it is now customary to attribute the marriage theorem to Hall.

The name giver for the marriage theorem is the following situation: We may imagine the set A as a set of women and B as a set of men. An edge between a woman and a man exists if a marriage is possible. A perfect matching then means that it is possible for all women in A to get married under the condition that no man gets married with more than one women.

Theorem 5.35 (Marriage theorem). *Let $G = (A \cup B, E)$ be a bipartite graph. Then there exists a perfect matching for A if and only if $|N_G(X)| \geq |X|$ for all $X \subset A$.*

Proof. If M is a perfect matching, then the graph $(A \cup B, M)$ satisfies the marriage condition, and, hence, also G. Now, let G satisfy the marriage condition. If for each non-empty proper subset X of A the inequality $|N_G(X)| > |X|$ holds, then we may remove any arbitrary edge e from G. The remaining graph still satisfies the marriage condition and has, by induction on the number of edges, a perfect matching. This is also a perfect matching of G. If the above case does not hold then there exists a non-empty proper subset X of A with $|N_G(X)| = |X|$.

Let G_1 be the subgraph of G induced by $X \cup N_G(X)$, and let G_2 be the subgraph induced by the remaining vertices (outside of $X \cup N_G(X)$). The graph G_1 satisfies the marriage condition because $N_{G_1}(X') = N_G(X')$ for all $X' \subset X$. We have to show that G_2 satisfies the marriage condition. Then by induction, both G_1 and G_2 have perfect matchings, and so their union is a perfect matching for A. Let $G_2 = (A' \cup B', E')$ with $A' \subset A$ and $B' \subset B$. For $X' \subset A'$ we have

$$\left|N_{G_2}(X')\right| + \left|N_G(X)\right| \geq \left|N_G(X' \cup X)\right| \geq |X' \cup X| = |X'| + |X|,$$

and hence, $|N_{G_2}(X')| \geq |X'|$. Therefore G_2 satisfies the marriage condition. □

A frequent application of the marriage theorem is the case when $|X| = |B|$. Then each perfect matching for A in the bipartite graph $G = (A \cup B, E)$ is also a perfect matching for B. Compare the example for a perfect matching above.

5.4 Stable Marriage Problem

Let two equally sized sets A and B be given, with an ordering of preference for each element. A matching between A and B is *stable* when there does not exist any match (a, b), $a \in A$, $b \in B$, for which both a and b prefer each other to their current partners under the matching. A *stable marriage problem* is the problem of finding a stable matching between A and B.

Originally the stable marriage problem is as follows: Given n men and n women, where each person has ranked all members of the opposite sex in order of preference. When there are no pairs of people that prefer each other to their current partners, the set of marriages is considered stable.

Algorithms for finding solutions to the stable marriage problem have many applications, for instance, in the assignment of graduating medical students to their first hospital appointments. In 2012, the Nobel Prize in Economics was given to L. S. Shapley (1923–2016) and A. E. Roth (born 1951) for the theory of stable allocations and the design of certain markets. In this section we want to describe the underlying algorithm of D. Gale (1921–2008) and Shapley which is central for Roth's empiric work and applications in real world situations. Let A and B be sets of n persons each. Without loss of generality let A be the women and let B be the men. For each $a \in A$ we define a linear relation P_a on the set B by $b_i > b_j$ if and only if a prefers b_i over b_j. In this case we also write $P_a(b_i) > P_a(b_j)$. Analogously, each man $b \in B$ has a list P_b of preferences. A marriage (or the matching $M \subset A \times B$) is stable if all women are married and there are no divorces. If there is one divorce then there exist two couples $(a, b'), (a', b)$ with $P_a(b) > P_a(b')$ and $P_b(a) > P_b(a')$. Then a and b are getting divorced from their partners and form a new married couple. If afterwards a' and b' get together then again all are married. Then the satisfaction of a and b is increased while those of a' and b' may be decreased.

The situation of two couples is easy to analyze. There are two women a, a' and two men b, b'. Let us first consider the case in which there exists a couple which gives each other the highest preference. Without loss of generality, let (a, b) be this couple. Hence $P_a(b) > P_a(b')$ and $P_b(a) > P_b(a')$. Then $(a, b), (a', b')$ is stable, and this is the only stable marriage. If there is no such couple, then the preferences cross each other. Say, a has favorite b, but b favors a' which now prefers b', for which finally the woman a is of the highest preference. We get a circular arrangement of highest preferences, see Figure 5.21.

In this case both possible pairings are stable; the difference between them is that either both men or both women marry their favorites. Anyway, also in more compli-

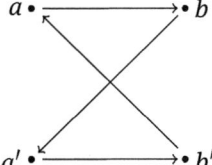

Figure 5.21: Circular arrangement of highest preferences.

cated situations there is a stable marriage possible, however only under the preferences of one party. In general, a stable marriage does not appear if partners meet each other in random order and, with appropriate preferences, split up from their current partner and remarry. In the previous situation first (a, b) could be married, then (a', b), then (a', b'), then (a, b') and finally (a, b) again. Here, the remaining two persons are unmarried at time. In our consideration of a stable marriage, there are no unmarried persons, so we add an additional woman a_0 and an additional man b_0 which always have the lowest preference among the earlier men and women, respectively. Unmarried persons in the above example get married now with a_0 or b_0. This provides an infinite sequence of unstable marriages in each of which all persons are married. Figure 5.22 illustrates a run; the dashed edge always contradicts the stability.

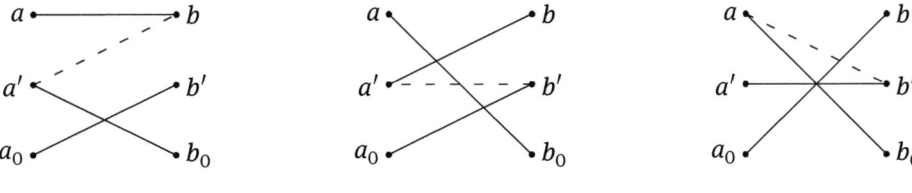

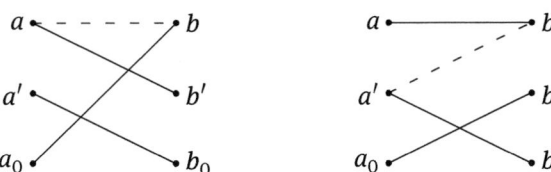

Figure 5.22: Illustration of a run.

When calculating a stable marriage it is customary to differentiate getting engaged and getting married from each other, so as not to have divorced couples. Engaging means provisional selection of a partner while getting married is final.

The Gale–Shapley algorithm calculates a marriage as follows:

(1) At the beginning nobody is engaged or married.

(2) As long as there is an unengaged man $b \in B$, he makes a proposal to that woman $a \in A$ he had not yet proposed to and that is for him of the highest preference among these women. The woman a accepts the proposal and gets engaged to b if she does not have a partner or if she prefers b to her current fiancé. If necessary, an engagement is broken off in order to enter into another.

(3) If all men are engaged, then everyone marries their fiancée.

In what follows we denote each pass through of step (2) as a circuit.

Theorem 5.36. *The Gale–Shapley algorithm calculates a stable marriage in at most n^2 circuits.*

Proof. Engaged women remain engaged in each circuit. A woman only gets engaged again if she improves. Especially each woman gets engaged maximally n times. At the latest, each woman gets a proposal after n^2 circuits, and all women (and therefore also all men) are engaged. Assume that the marriage, calculated by the Gale–Shapley algorithm, is unstable, that is, there are two married couples (a, b') and (a', b) with $P_a(b) > P_a(b')$ and $P_b(a) > P_b(a')$. But then a had, before the engagement of b with a', either rejected an application from b or had left him. The reason for this was an engagement with a man b'' with $P_a(b'') > P_a(b)$. But since women only improved as the process progresses, we have $P_a(b') > P_a(b'')$. This contradicts $P_a(b) > P_a(b')$. □

Remark 5.37. Especially, there exists always a stable marriage. The marriage calculated by the Gale–Shapley algorithm can be described a little more precisely. The Gale–Shapley algorithm is optimal for the men. They each receive the one partner who has the highest preference among all women with which a stable pairing is even possible. For this, it is enough to show that a couple $(a, b') \in A \times B$ is in no marriage realizable, if the woman a has either an application from b' rejected or left him in the Gale–Shapley procedure. To get a contradiction, we consider the first moment t at which either a woman a or a man is rejected or left, although there exists a stable marriage M with $(a, b') \in M$. In both cases the reason is a man b with $P_a(b) > P_a(b')$. We have $(a', b) \in M$ for a woman $a' \neq a$. If we would have $P_b(a) < P_b(a')$, then b had received a negative reply from a' or would be left by a' already before the moment t. But this is not possible by the choice of (a, b'). Hence, $P_b(a) > P_b(a')$. Thus, if (a, b') and (a', b) meet each other, then a and b leave their partners and form a new couple (a, b). As a result M is not stable, which gives a contradiction.

5.5 Euler Line

With Euler's formula we are now essentially prepared for the classification of the Platonic solids. But before we do this in the next chapter, we want, for historical reasons, to talk about a problem which was solved by Euler. His solution is the reason why Euler often is considered as the founder of graph theory.

Definition 5.38. A (closed) edge line C in the graph $G = (V, E)$ is called an *Euler line* (*Euler cycle, respectively*) if C contains each edge of G exactly one time.

Remarks 5.39.
(1) This is certainly only meaningful in connected graphs.
(2) The naming is after Euler who in 1736 solved the famous problem of the Seven Bridges of Königsberg. Its negative solution by Euler laid the foundation of graph theory.

  The city of Königsberg was set on both sides of the Pregel River, and included two large islands which where connected to each other and to the two mainland portions of the city by seven bridges, see Figure 5.23.

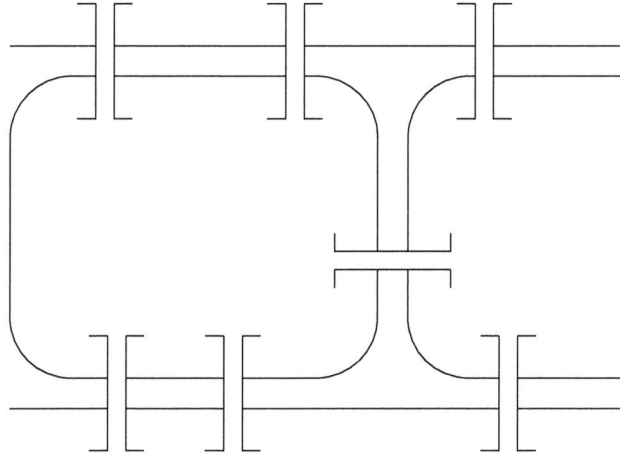

Figure 5.23: Seven bridges of Königsberg.

  The problem was to devise a walk through the city that would cross each of those bridges once and only once.

  By way of specifying the logical task unambiguously, solutions involving either
  – reaching an island or mainland bank other than via one of the bridges, or
  – accessing any bridge without crossing to its other end
  are unacceptable.

  Euler pointed out that the choice of route inside each land mass is irrelevant. The only important feature of a route is the sequence of bridges crossed.

  In modern terms, or in graph-theoretical interpretations, one replaces each land mass with an abstract vertex (or node), and each bridge with an abstract edge, which only serves to record which pair of vertices (land masses) are connected by that bridge. The resulting structure is shown in Figure 5.24.

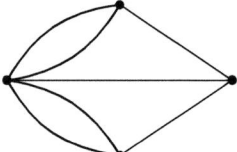

Figure 5.24: Resulting graph with multiple edges.

This structure has multiple edges between two vertices. To bring this in line with our definition of a graph, we may introduce additional vertices and edges so that the problem now is equivalent to the original one, see Figure 5.25.

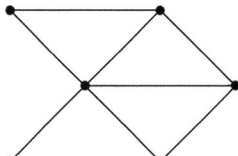

Figure 5.25: Resulting simple graph.

The task is now to find an Euler cycle in this extended graph. The following Theorem 5.40 tells us that this is not possible.

(3) For an example of an Euler line, see Figure 5.26.

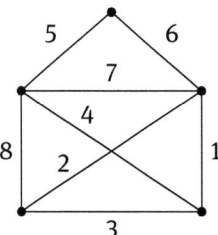

Figure 5.26: House of Nicolaus.

This is the house of Nicolaus (one edge for each syllable).

Theorem 5.40. *A graph $G = (V, E)$ contains an Euler line (Euler cycle) if and only if G is connected and the number of vertices with an odd degree is less than or equal to 2 (= 0, respectively).*

Before we prove Theorem 5.40 we describe a necessary lemma.

Lemma 5.41. *Let $G = (V, E)$ be a connected graph with $d(x)$ even for all $x \in V$. If $a \in V$ and $u \in E$ with a as one terminus, then $G' = (V, E \setminus \{u\})$ is connected.*

Proof. Let b be the second terminus of u. Assume that G' is not connected. Then G' comprises two connected components, one of which contains a and the other contains b.

Let $G_1 = (V_1, E_1)$ be that component which contains a, and let d_1 be the degree function of G_1.

We have that $d_1(a)$ is odd, and $d_1(x)$ is even for all $x \in V_1$, $x \neq a$, from the condition on d. This contradicts

$$|E_1| = \frac{1}{2} \sum_{x \in V_1} d_1(x).$$

Hence, G' is connected. □

We now give the proof of Theorem 5.40.

Proof. "$\Rightarrow$" Let G contain an Euler line (Euler cycle). Each passage through one vertex x gives a contribution of 2 to the value $d(x)$, with a possible exception of the beginning vertex and the ending vertex of the line, these two vertices are equal in a cycle.

"$\Leftarrow$" Let G be connected, and let the number of vertices with an odd degree be ≤ 2 or 0. Then the number of vertices with an odd degree is 0 or 2 because $|E| = \frac{1}{2} \sum_{x \in V} d(x)$. We may reduce the case with two vertices with an odd degree to the case where all vertices have even degree. This we can see as follows. Let a and b be two vertices with an odd degree. We extend G by adding an additional vertex c and two edges $\{a, c\}$ and $\{b, c\}$ to a graph $G' = (V \cup \{c\}, E \cup \{a, c\} \cup \{b, c\})$; and each Euler cycle in G' corresponds to an Euler line in G from a to b. Now let the number of vertices with an odd degree be 0.

The following algorithm provides an Euler cycle:
(1) Choose any arbitrary vertex a as the start vertex.
(2) Choose any arbitrary edge u_1 from the set of edges with a as an end vertex. By Lemma 5.41, the graph $(V, E \setminus \{u_1\})$ is connected.
(3) If $u_1, u_2, \ldots, u_n$ are iteratively already chosen with final vertex x_n, then choose from the set of unused edges with x_n as terminus an edge u_{n+1} in such a way that the remaining graph G_{n+1}, formed from a and those edges which are distinct from $u_1, u_2, \ldots, u_{n+1}$ and their vertices, is connected.

An example for (3) is given in Figure 5.27.

In Figure 5.27, $u_1, u_2, \ldots, u_6$ are chosen. The remaining graph G_6 after the choice of $u_1, u_2, \ldots, u_6$ is given in Figure 5.28.

Edge u' is not allowed to be chosen as u_7, because then G_7 is not connected; u'' can be chosen. □

This procedure gives an Euler cycle.

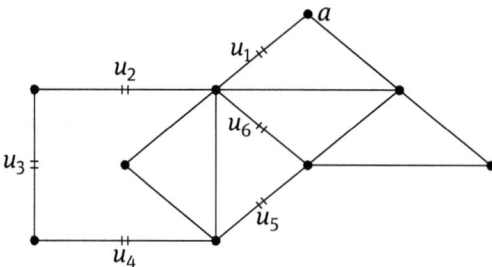

Figure 5.27: Example for (3).

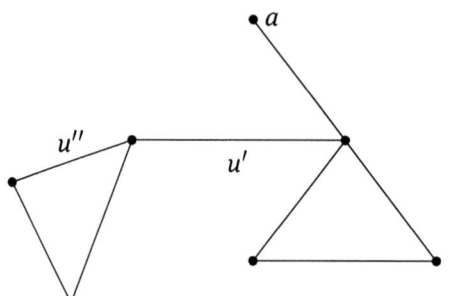

Figure 5.28: Remaining graph.

Proof. *Case 1.* $x_n = a$

If all edges are used then we have an Euler cycle. If G_n still contains edges, then among them there is one (for degree reasons, even two) with a as terminus, since G_n is connected, and among those we may choose any as u_{n+1} by Lemma 5.41.

Case 2. $x_n \neq a$

Then $d_n(x_n)$ and $d_n(a)$ are odd (here d_n is the degree function for G_n), and in any case G_n still contains edges.

If $d_n(x_n) = 1$, then this one edge is acceptable as u_{n+1}. Now let $d_n(x_n) \geq 3$ and let $u = \{x_n, y\}$ be an edge which is not acceptable as u_{n+1}.

By removing u, we get that G_n comprises two connected components, one of which contains x_n and the other contains y. For degree reasons, a and y are in the same component, and the component which contains x_n has only vertices of even degree. Hence, each edge with terminus x_n and different from u is acceptable as u_{n+1}. □

5.6 Hamiltonian Line

In an analogous manner to the Euler line we may introduce Hamiltonian lines (or Hamiltonian paths) and Hamiltonian cycles. They are named after W. R. Hamilton (1805–1865).

Definition 5.42. A *Hamiltonian line* in a graph is an edge line that visits each vertex exactly once. A *Hamiltonian cycle* is a Hamiltonian line that is a cycle.

Determining whether such lines or cycles exist in graphs is the Hamiltonian line problem, which is considered to be a hard problem that can be used in cryptology.

Examples 5.43.
1. The complete graph $K_n, n \geq 3$, always has a Hamiltonian cycle. Certainly, if $n = 2$ then we have a Hamiltonian line.
2. The edge graph of a dodecahedron (see the next chapter on Platonic solids) has a Hamiltonian cycle. It is indicated in Figure 5.29 by thick edges.

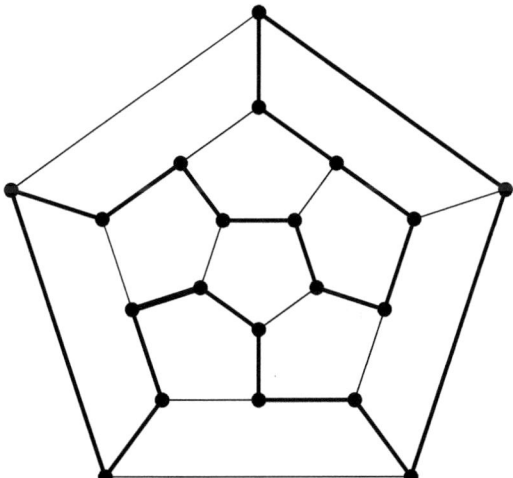

Figure 5.29: The edge graph of a dodecahedron.

Theorem 5.44 (Theorem of Ore). *If $d(x) + d(y) \geq n = |V|$ for all pairs of non-adjacent vertices x, y of the connected graph $G = (V, E)$, then G is Hamiltonian, that is, G has a Hamiltonian cycle.*

Proof. We consider cyclic arrangements of the vertices

$$x_1, x_2, \ldots, x_{n-1}, x_n \quad (x_{n+1} = x_1) \tag{Z}$$

and call a consecutive pair x_k, x_{k+1} a gap, if x_k and x_{k+1} are not adjacent in the graph. Assume that G does not have a Hamiltonian cycle. Then each cycle arrangement has at least one gap. Now, let (Z) be a cyclic arrangement with a minimal number of gaps, and let x_k, x_{k+1} be herein a gap. We claim that then $d(x_k) + d(x_{k+1}) \leq n - 1$. If $d(x_k) = 0$, then this is correct because the degree of each vertex is at most $n-1$. Now, let $d(x_k) > 0$ and x_j be a neighbor of x_k. Then x_{k+1} and x_{j+1} cannot be adjacent because otherwise the cyclic arrangement

$$x_1, \ldots, x_k, x_j, x_{j-1}, \ldots, x_{k+1}, x_{j+1}, \ldots, x_n \tag{Z'}$$

if $j > k$, respectively

$$x_1 \ldots, x_j, x_k, x_{k-1}, \ldots, x_{j+1}, x_{k+1}, \ldots, x_n \tag{Z"}$$

if $j < k$ has at least one gap less. Hence, if x_k has m neighbors ($m \leq n - 1$), then it has m forbidden adjacencies as a consequence. From this we get $d(x_{k+1}) \leq n - 1 - m$ and therefore the statement. □

Palmer [24] describes an algorithm for constructing a Hamiltonian cycle which reflects the arguments in the proof of Ore's theorem:
1. Take a cycle arrangement of the vertices, ignoring adjacencies in the graph.
2. If the cycle arrangement contains a gap x_k, x_{k+1}, perform the following two steps:
   (a) Search for an index i such that the four vertices x_k, x_{k+1}, x_i and x_{i+1} are all distinct and such that the graph contains edges from x_k to x_i and from x_{i+1} to x_{k+1}.
   (b) Reverse the part of the cycle between x_{k+1} and x_{j+1} (inclusive).
   Each step increases the number of consecutive pairs in the cyclic arrangement that are adjacent, by one or two pairs (depending on whether x_i and x_{i+1} are already adjacent). The desired index i must exist, or else the non-adjacent vertices x_i and x_{i+1} would have a too small total degree.

This algorithm certainly terminates. We close this section with the sufficient condition by G. A. Dirac (1925–1984).

Theorem 5.45. *Let $G = (V, E)$ be a graph with $|V| \geq 3$ and let $\delta = \delta(G)$ be the minimum degree. If $\delta \geq \frac{|V|}{2}$, then G has a Hamiltonian cycle.*

Proof. Suppose that the statement is false, and let G be a maximal graph with $|V| \geq 3$ and $\delta \geq \frac{|V|}{2}$ which has no Hamiltonian cycle. Since $|V| \geq 3$, G cannot be complete. Let u and v be non-adjacent vertices in G. By the choice of G then $G' = (V, E \cup \{u, v\})$ has a Hamiltonian cycle. Moreover each Hamiltonian cycle of G' must contain the edge $\{u, v\}$. Thus there is a Hamiltonian line $\{v_1, v_2\}, \ldots, \{v_{k-1}, v_k\}$ in G from $u = v_1$ to $v = v_k$.

Let $S = \{v_i \mid \{u, v_{i+1}\} \in E\}$ and $T = \{v_i \mid \{v_i, v\} \in E\}$. Since $v_k \notin S \cup T$, we have $|S \cup T| < |V|$. Furthermore, $|S \cap T| = 0$, since if $S \cap T$ contained some vertex v_i, the G would have a Hamiltonian cycle $\{v_1, v_2\}, \{v_2, v_3\}, \ldots, \{v_i, v_k\}, \{v_k, v_{k-1}\}, \ldots, \{v_{i+1}, v_1\}$ contrary to the assumption, see Figure 5.30.

Hence, we obtain

$$d(u) + d(v) = |S| + |T| = |S \cup T| < |V|.$$

But this contradicts the hypothesis that $\delta \geq \frac{|V|}{2}$. □

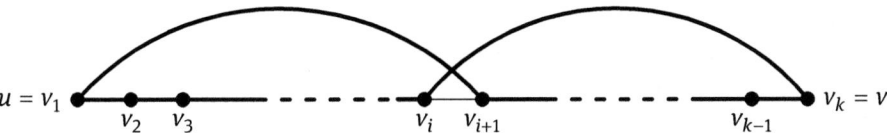

Figure 5.30: Hamiltonian cycle for Dirac's theorem.

5.7 The Traveling Salesman Problem

In this section we discuss the traveling salesman problem, its connection to Hamiltonian cycles, and have a glance at algorithms and applications. One considers the following situation: A salesman needs to visit a list of cities connected by routes of certain lengths and asks for the shortest route that connects these cities. As we have pointed out in the beginning of this chapter, it is customary to model this geographic data as a graph G with the cities as vertices and edges connecting two cities if there is a road between these cities. In order to model the lengths of the roads we introduce weighted graphs.

Definition 5.46. A *weighted graph* $G = (V, E, w)$ is a graph (V, E) together with a function $w \colon E \to \mathbb{R}_+$.

We remark that we interpret the weight of an edge as the length of a road. Of course, depending on the application, this could also be regarded as the travel time or costs between the respective cities. Let $p = (e_1, \ldots, e_n)$ be a sequence in a weighted graph $G = (V, E, w)$. Then we understand that $w(p) = \sum_{i=1}^{n} w(e_i)$ is the weight of p.

With these notions at hand, we can formulate the traveling salesman problem in a graph-theoretic manner. We just take the list of cities as the vertex set V; if there is a road between cities A and B we add $\{A, B\}$ to our edge set E and encode the lengths of the road as a weight function $w \colon E \to \mathbb{R}_+$. Hence, a *traveling salesman problem* (*TSP*) is completey determined by a weighted graph $G = (V, E, w)$.

Definition 5.47. Let $G = (V, E, w)$ be a TSP. A *solution* of the TSP is a sequence

$$p = (\{v_1, v_2\}, \{v_2, v_3\}, \ldots, \{v_{n-1}, v_n\})$$

in G with $v_1 = v_n$ that contains all vertices and that is of minimal length among all those sequences.

We restrict ourselves to the following kind of TSPs.

Remark 5.48. We remark that what we consider in the following is called a *metric, symmetric TSP* in the literature. That is, we assume that the distance from city A to B is always the same as from B to A (this yields an undirected graph) and that it is always longer to take a detour through a city C than to use the direct connection between A and B (in fact, we assume the Euclidean metric, that is the Euclidean distance).

For theoretical considerations and for practical computations it is customary to give an alternative description of the traveling salesman problem in terms of complete graphs and Hamiltonian cycles. However, we have not found a rigorous treatment of this equivalence in the literature, so we discuss it in the following.

We first consider a new graph associated to a TSP.

Definition 5.49. Let $G = (V, E, w)$ be a TSP. We then consider the graph

$$G' = (V, E \cup E', w')$$

where E' is the set of edges that completes the graph G and w' is the weight function with $w'(e) = w(e)$ for $e \in E$ and for an edge $e = \{u, v\} \in E'$ we set $w'(e) = w(p)$ where p is any sequence of minimal weight in G that connects u and v.

That is, in G' we add the necessary edges such that G becomes a complete graph. Note that these edges are 'artifical' in the respect that they do not represent real roads. We then extent the weight function in such a way that a new edge e gets assigned the minimal length of a sequence of old edges (the actual roads) that connects the start and end point of e.

We now discuss how to translate between solutions p of a TSP and Hamiltonian cycles of least weight in the graph G'.

Definition 5.50. Let $G = (V, E, w)$ be a TSP.
1. For a solution p of the TSP we define a Hamiltonian cycle $\varphi(p)$ in G' as follows: Read p from left to right and replace any subsequence

$$(\{v, v_1\}, \{v_1, v_2\}, \dots, \{v_{n-1}, v_n\}, \{v_n, v'\}),$$

where $v_1, \dots, v_n$ have already been part of the sequence but v, v' have not or v has not and v' is the end vertex of p, by the edge $\{v, v'\}$ in E'.
2. For a Hamiltonian cycle h in G' we define the sequence $\psi(h)$ in G as follows: Replace any edge $\{u, u'\} \in E'$ by a sequence $p' = (\{u, u_1\}, \{u_1, u_2\}, \dots, \{u_n, u'\})$ of least weight.

We remark that in Definition 5.50.1 the edge $\{v, v'\}$ is indeed in E' because if the edge already existed in E, replacing the subsequence $(\{v, v_1\}, \{v_1, v_2\}, \dots, \{v_{n-1}, v_n\}, \{v_n, v'\})$ by the edge $\{v, v'\}$ would yield a path that is of less weight (metric TSP) but that still contains all cities because $v_1, \dots, v_n$ have already been visited before. We also remark that the choice of p' in the definition of ψ is not unique and that φ is not injective.

We observe the following:

Lemma 5.51. *Let $G = (V, E, w)$ be a TSP. For a solution p of the TSP and a Hamiltonian cycle h of least weight in G' we have*
1. *$w(p) = w'(\varphi(p))$ and*
2. *$w'(h) = w(\psi(h))$.*

Proof. For the first equation we have already seen that the subsequences

$$(\{v, v_1\}, \{v_1, v_2\}, \dots, \{v_{n-1}, v_n\}, \{v_n, v'\})$$

we replace are paths of minimal weight from v to v'. Hence, by definition of $w'(\{v, v'\})$, the Hamiltonian cycle $\varphi(p)$ is of the same weight as p. For the second equation observe that the artificial edges have by definition the weight of any sequence of least weight that connects their vertices. Hence replacing them by such a sequence does not change the weight. $\square$

We now give our main theorem that shows the equivalence of the approaches.

Theorem 5.52. *Let $G = (V, E, w)$ be a TSP.*
1. *If p is a solution of the TSP, then $\varphi(p)$ is a Hamiltonian cycle of least weight in the associated graph G'.*
2. *If h is a Hamiltonian cycle in G' of least weight, then $\psi(h)$ is a solution of the TSP.*

Proof.
1. If p is a solution of the TSP, then $\varphi(p)$ is certainly a Hamiltonian cycle in G'. Now assume that there is a Hamiltonian cycle h of less weight than $\varphi(p)$, that is,

$$w'(h) < w'(\varphi(p)).$$

Then by Lemma 5.51 we have

$$w(\psi(h)) = w(h) < w'(\varphi(p)) = w(p)$$

contradicting the minimality of p.
2. On the other hand, let h be a Hamiltonian cycle of least weight, then $\psi(h)$ certainly visits all cities and has identical start and end point. Now assume that there is such a sequence p of less weight, that is, $w(p) < w(\psi(h))$. Then again by Lemma 5.51 we have

$$w'(\varphi(p)) = w(p) < w(\psi(h)) = w'(h)$$

contradicting the minimality of h. $\square$

We illustrate these correspondences in Figure 5.31 that could model the situation of two islands connected by a bridge (we remark that the arrows do not describe edges, they just mean possible flows). Note that in the visualization of the graph G' we will omit the artificial edges that are not part of $\varphi(p)$.

An optimal solution of the traveling salesman problem is often out of sight as the number of possible routes one has to consider depends overexponentially on the number of cities. The characterization with respect to Hamiltonian cycles shows that in a symmetric TSP in $n > 2$ cities there are $(n-1)!/2$ cycles to consider. Even for eleven cities there are 1814400 possible cycles. If we now consider the 275 cities in America

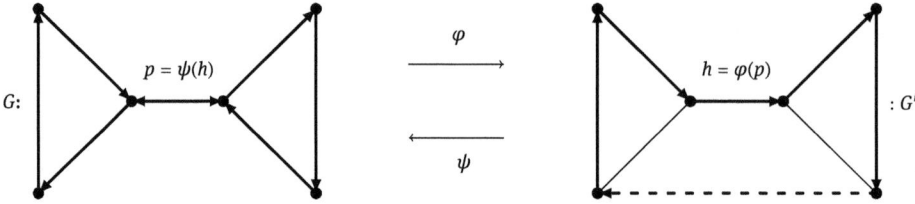

Figure 5.31: Example of the graphs G, G' and the correspondences φ, ψ.

with a population of more than 100000, we already get approx. $1.8 \cdot 10^{550}$ possible routes. For the reader who knows about complexity classes we mention that the traveling salesman problem is NP-hard. Under the assumption NP≠P this means that there is no algorithm that computes every instance of a TSP in polynomial time.

This is why in applications one aims for approximation algorithms with reasonable running time. Here, one distinguishes between heuristic opening or construction algorithms and post-optimization algorithms. In the following we will present an example for each one of them and give a sample calculation.

Example 5.53 (Nearest neighbor algorithm). As an opening method we present the *nearest neighbor algorithm*. Here, one starts with an arbitrary city and then continues with the nearest unvisited city and so forth. This algorithm may result in arbitrarily bad solutions. As a sample calculation we consider the cities Chicago (C), Houston (H), Los Angeles (LA), New York City (NY) and Philadelphia (P), and ask for a short traffic route connecting all of them.

As a starting point we choose New York City and proceed according to the nearest neighbour algorithm. We finally find a route of 6262 miles that goes through NY–P–C–H–LA–NY (see Figure 5.32, distances in miles).

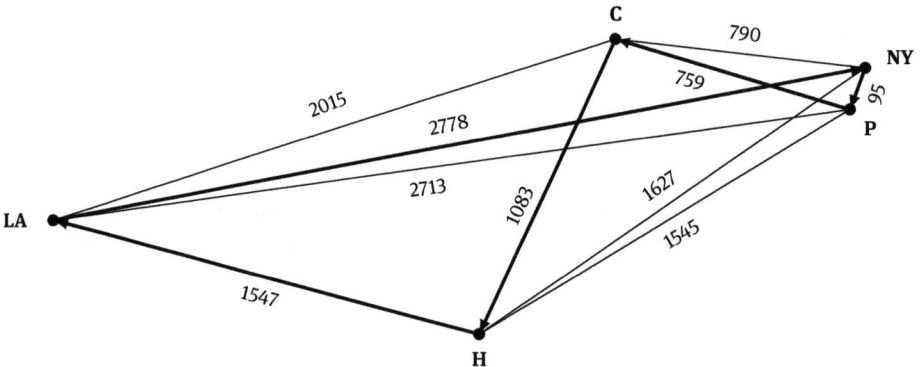

Figure 5.32: Sample calculation for the nearest neighbour algorithm.

We now try to post-optimize this route with the so-called *2-opt algorithm*.

Example 5.54 (2-opt algorithm). As a post-optimization method we present the 2-opt algorithm. Here, one has to realize the graph G as a Euclidean graph, that is, one takes the vertices as points with suitable coordinates in the real plane and draw the edges according to their weight with respect to the Euclidean metric. Now we step-by-step delete two edges $\{A, C\}$, $\{B, D\}$ that cross each other and replace them by edges $\{A, B\}$, $\{C, D\}$ that do not intersect.

We apply this algorithm to our running example. We first delete the crossing P–C, LA–NY (see Figure 5.32), and replace them with edges P–LA and C–NY (see Figure 5.33).

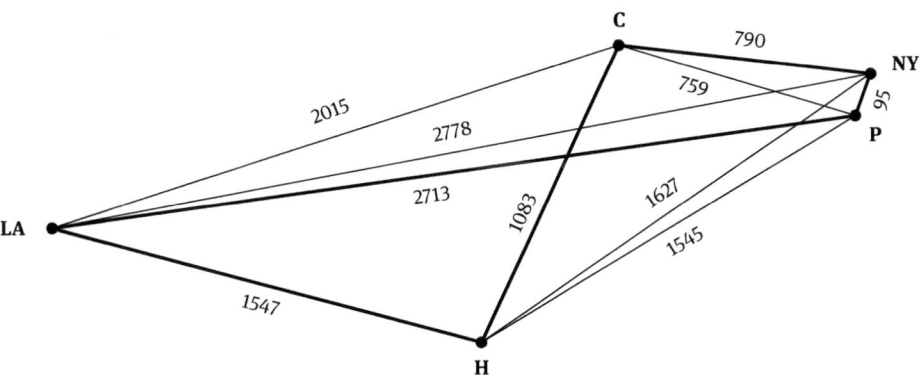

Figure 5.33: Sample calculation for the 2-opt algorithm (step 1).

Then we replace the new crossing P–LA, H–C by the edges P–H and LA–C. Our algorithm terminates at step 2 because there are no further crossings. We obtain the route NY–P–H–LA–C–NY with the length of 5992 miles, see Figure 5.34.

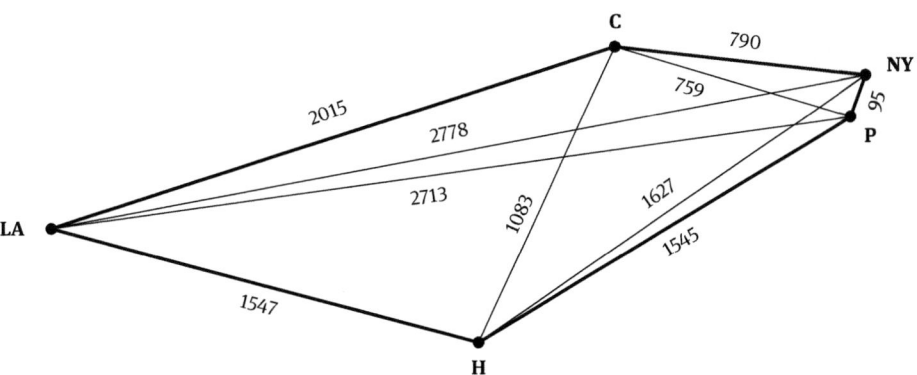

Figure 5.34: Sample calculation for the 2-opt algorithm (step 2).

In this small example one can in fact apply a brute-force method and give the optimal solution which indeed is the one we have obtained in Example 5.54 after two steps of the 2-opt method. In general, the solution found by an iterative application of 2-opt moves for n cities could be up to $(4\sqrt{n})$-times longer than the optimal solution. If 2-opt moves do not improve the route, there might be more general k-opt moves that do so. For completeness we finally mention that the worst route in our running example has the length of 8960 miles and is realized by the route NY–LA–P–C–H–NY.

Apart from the applications of finding shortest routes between cities that we discussed above, there are many useful real world applications for a TSP, for instance, in DNA sequencing, the design of microchips and shortest routes between stars in the context of astronomy. For general reading, more applications and algorithms we refer the reader to [9].

We once again mention that we restricted ourselves to metric, symmetric traveling salesman problems. Depending on the application context there are several variants of the TSP, for instance, the asymmetric TSP where one assumes that the route from A to B takes longer than from B to A (this yields a directed graph), or the multiple TSP with more than one salesman. Another variant is to consider transport capacities or time windows. Both of these are relevant, e. g., for breakdown and parcel services and remain an important and lively field of scientific research.

Exercises

1. Show that there are (up to isomorphisms) exactly eleven graphs with four vertices.
2. Draw all sixteen spanning trees of K_4.
3. Let $\delta = \delta(G)$ be the minimum degree of a graph G. Show that G contains a cycle if $\delta \geq 2$.
4. How many graphs with n vertices do exist?
5. Let $G = (V, E)$ be a connected graph. We call an edge $e \in E$ a bridge if $G' = (V, E \setminus \{e\})$ is non-connected.
   Show that:
   (a) If e is a bridge then $G' = (V, E \setminus \{e\})$ has two connected components.
   (b) A graph $G = (V, E)$ with $d(x)$ even for all $x \in V$ does not have a bridge.
6. (a) Let $G = (V, E)$ be a graph with $d(x) \geq 4$ for each $x \in V$. Show that $|E| \geq 2|V|$.
   (b) Does each graph with $n \geq 2$ vertices have at least two vertices with the same degree?
7. Let V_n be the set of all subsets of $\{1, 2, \ldots, n\}$, $n \geq 1$. Let G_n be the graph with V_n as the set of vertices and let vertices $A, B \in V_n$ be connected by an edge if and only if $A \cap B = \emptyset$. Draw the graph G_3.
   Is G_3 connected or planar?
8. How many edges does the complete bipartite graph $K_{m,n}$ have? Draw the graphs $K_{1,n}$, $K_{2,n}$ and $K_{3,3}$.

9. Let G be a connected planar graph with n, $n \geq 3$, vertices and m edges.
    Suppose that G does not contain a triangle, that is, no edge circle of the form

    $$u_1 = \{x_0, x_1\}, u_2 = \{x_1, x_2\}, u_3 = \{x_2, x_0\}$$

    with $x_0 \neq x_1 \neq x_2 \neq x_0$. Show that $m \leq 2n - 4$.
    Use this to show that the complete bipartite graph $K_{m,n}$, $3 \leq m, n$, is not planar.
10. Prove Remark 5.16 in detail.
11. Determine the number of perfect matchings with k edges in the complete graph
    K_{2n} and the complete bipartite graph $K_{n,n}$.
12. Show that:
    (a) The complete graph K_n, $n \geq 3$, is a Hamiltonian graph.
    (b) The complete bipartite graph $K_{m,n}$, $m, n \geq 2$, has a Hamiltonian cycle if and
        only if $m = n$.

6 Spherical Geometry and Platonic Solids

In three-dimensional space, a Platonic solid is a regular, convex polyhedron. It is constructed by congruent (identical in shape and size) regular (all angles equal and all sides equal) polygonal faces with the same number of faces meeting at each vertex. There are only five solids that meet these criteria: the *tetrahedron* which has four faces, the *cube* which has six faces, the *octahedron* which has eight faces, the *dodecahedron* which has 12 faces and the *icosahedron* which has 20 faces.

The ancient Greek geometers extensively studied the Platonic solids. They are named after Plato (428–348 BC) who wrote about them in the dialogue Timaeus around 360 BC. In this dialogue Plato associated each of the four classical elements (earth, air, water and fire) with one of the Platonic solids.

Euclid (ca. 300 BC) completely described the Platonic solids in the Elements. Book XIII is devoted to their properties. Much of the information by Euclid is probably derived from the work of Theaitetus (415–369 BC) who first described all five Platonic solids and may have proved that they are the only regular solids.

When Kepler (1571–1630) in the seventeenth century began to study the Solar System, he attempted to relate the five extraterrestrial planets known at that time (Mercury, Venus, Mars, Saturn and Jupiter) to the five Platonic solids. In Mysterium Cosmographicum, published in 1596, Kepler proposed a model of the Solar System, in which the five solids were set inside one another and separated by a series of inscribed and circumscribed spheres.

The purpose of this chapter is to describe and to classify the Platonic solids, that is, to show that the five we have mentioned are the only ones. For the proof of this classification, we use Euler's formula for planar, connected graphs via the stereographic projection from the sphere S^2 to the extended complex plane. Hence, we take this as an opportunity to give some principles of spherical geometry of the sphere S^2 and show some beautiful results in this area.

6.1 Stereographic Projection

Before we start with the classification of the Platonic solids, we need as a technical tool Riemann's number sphere and the stereographic projection.

B. Riemann (1826–1866) introduced a model of the complex numbers (see [12, Chapter 10]), which allows a visualization of the point ∞. The complex numbers $\mathbb{C}$ here will be identified with the points of the equator plane $x_3 = 0$ of the space $\mathbb{R}^3$. The complex numbers $z = x + iy \in \mathbb{C}$ with $x, y \in \mathbb{R}$ will be mapped onto the unit sphere

$$S^2 = \left\{ (x_1, x_2, x_3) \in \mathbb{R}^3 \mid \sum_{i=1}^{3} x_i^2 = 1 \right\} \subset \mathbb{R}^3$$

https://doi.org/10.1515/9783110740783-006

by doing the intersection Z of S^2 with the line through the north pole

$$N = \begin{pmatrix} 0 \\ 0 \\ 1 \end{pmatrix} \quad \text{and} \quad P = \begin{pmatrix} x \\ y \\ 0 \end{pmatrix},$$

see Figure 6.1.

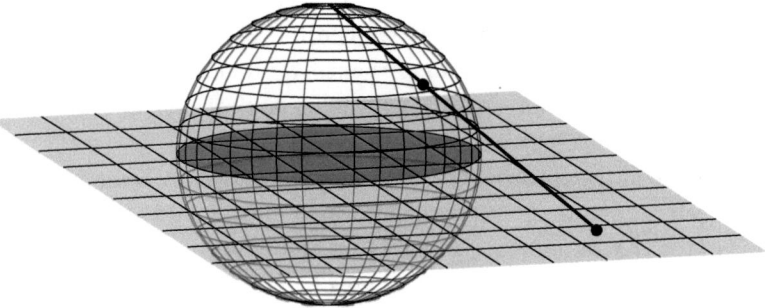

Figure 6.1: Stereographic projection.

Applying

$$Z = \begin{pmatrix} x_1 \\ x_2 \\ x_3 \end{pmatrix} = \lambda \begin{pmatrix} x \\ y \\ 0 \end{pmatrix} + (1 - \lambda) \begin{pmatrix} 0 \\ 0 \\ 1 \end{pmatrix}, \quad \lambda \in \mathbb{R}, \, \lambda > 0,$$

to the equation $x_1^2 + x_2^2 + x_3^2 = 1$ of the unit sphere gives $\lambda = \frac{2}{x^2 + y^2 + 1}$. This leads to

$$x_1 = \frac{2x}{x^2 + y^2 + 1} = \frac{z + \bar{z}}{z\bar{z} + 1},$$

$$x_2 = \frac{2y}{x^2 + y^2 + 1} = \frac{i(\bar{z} - z)}{z\bar{z} + 1},$$

$$x_3 = \frac{x^2 + y^2 - 1}{x^2 + y^2 + 1} = \frac{z\bar{z} - 1}{z\bar{z} + 1},$$

and we get the stereographic projection τ from $\mathbb{C}$ onto $S^2 \setminus \{N\}$. This is a bijection. The inverse map is given by

$$x = \frac{x_1}{1 - x_3}, \quad y = \frac{x_2}{1 - x_3},$$

that is,

$$z = \frac{x_1 + ix_2}{1 - x_3}.$$

By construction, the stereographic projection maps

$$E = \{z \in \mathbb{C} \mid |z| < 1\} \subset \mathbb{C}$$

onto the southern hemisphere and

$$F = \mathbb{C} \setminus \{z \in \mathbb{C} \mid |z| \le 1\} \subset \mathbb{C}$$

onto the northern hemisphere.

The points from $K = \{z \in \mathbb{C} \mid |z| = 1\}$ are mapped onto itself, that is, they are fixed points of τ. Further, as $|z| \to \infty$, the images $Z = \tau(z)$ tend to the north pole N. If we extend $\mathbb{C}$ by ∞ to be the extended, closed complex plane $\hat{\mathbb{C}} := \mathbb{C} \cup \{\infty\}$ (one-point compactification), then we may extend the stereographic projection τ via $\infty \mapsto N$ to a bijective map from $\hat{\mathbb{C}}$ to S^2, which we also name by τ. We mention the usual rules

- $z + \infty = \infty, z \in \mathbb{C}$;
- $z \cdot \infty = \infty, z \in \mathbb{C} \setminus \{0\}$;
- $\frac{z}{\infty} = 0, z \in \mathbb{C}$;
- $\frac{z}{0} = \infty, z \in \mathbb{C} \setminus \{0\}$;
- $\infty + \infty = \infty$ in $\hat{\mathbb{C}}$.

6.2 Platonic Solids

Recall that in the three-dimensional space $\mathbb{R}^3$, a Platonic solid is a regular, convex polyhedron. It is constructed by congruent regular polygonal faces with the same number of faces meeting at each vertex, and none of its faces intersect except at their edges. There are five Platonic solids which meet these criteria: tetrahedron (four faces), cube (six faces), octahedron (eight faces), dodecahedron (12 faces) and icosahedron (20 faces). The most important aim in this section is to show that these five Platonic solids are all one can get. The Platonic solids have been known since antiquity.

The attraction and fascination of the Platonic solids can be seen not only in ancient treatises or in more modern publications of, for instance, Pythagoras (570–510 BC), Plato (428–348 BC), Archimedes (287–212 BC), Euclid (ca. 300 BC), Leonardo da Vinci (1452–1519), Kepler (1571–1630) and Euler (1707–1783), but also in the frequent treatment in classes at schools and universities.

The ancient Greeks studied the Platonic solids extensively. Some sources credit Pythagoras (580–500 BC) to be familiar with the tetrahedron, cube and dodecahedron. The discovery of the octahedron and icosahedron is due to Theaitetos (415–369 BC). He gave a mathematical description of all five and may have been responsible for the first known proof that no other convex regular polyhedra exist.

The Platonic solids are prominent in the philosophy of Plato, their namesake. He wrote about them in his dialogue Timaios (360 BC), in which he associated each of the five Platonic solids to the five elementary elements of the world (earth, air, water, fire and heaven). The dodecahedron, in fact, was in a sense obscurely related by Plato to the heaven. A formal connection between the dodecahedron and the heaven was made by Aristotle (384–322 BC).

Euclid completely mathematically described the Platonic solids in the *Elements*, the last Book XIII, which is devoted to their properties. Euclid also argues that there are no further convex regular polyhedra. Much of the information in Book XIII is probably derived from the work of Theaitetos. Kepler (1571–1630) attempted to relate the five extraterrestrial planets known at that time to the five Platonic solids. He postulated a model of the Solar System in which the five Platonic solids were set inside one another and separated by a series of inscribed and circumscribed spheres, see Figure 6.2.[1] Kepler proposed that the distance relationship between the six planets known at that time could be understood in terms of the five Platonic solids enclosed within a sphere that represented the orbit of Saturn. The six spheres correspond to each of the planets: Mercury, Venus, Earth, Mars, Jupiter and Saturn.

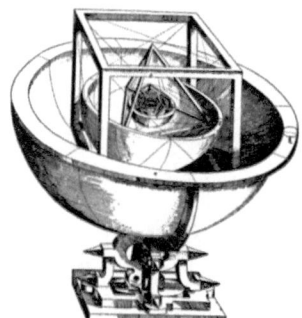

Figure 6.2: Kepler's Platonic solid model of the Solar System, from Mysterium Cosmographicum (1596).

The solids were ordered with the innermost being the octahedron, followed by the icosahedron, dodecahedron, tetrahedron and cube, thereby dictating the structure of the solar system and the distance relationship between the planets by the Platonic solids. In the end, after some discrepancies found by Brahe (1546–1601) between the reality and the model, Kepler's original idea had to be abandoned.

We now describe the five known Platonic solids, which are the cube, tetrahedron, octahedron, icosahedron and dodecahedron, before we show that there are exactly five Platonic solids.

Definition 6.1. As in the plane $\mathbb{R}^2$, a *figure* F in the space $\mathbb{R}^3$ is just a subset of $\mathbb{R}^3$. A symmetry of F is an isometry α of $\mathbb{R}^3$ with the property that $\alpha(F) = F$. The set of all symmetries of F forms a group which we again denote by Sym(F).

---

1 The figure is from Wikipedia, see https://en.wikipedia.org/wiki/Johannes_Kepler

Recall that $\alpha = \tau \circ f$ with τ a translation and f a linear isometry; α is called *oriented* if $\det(A) = 1$ and non-oriented if $\det(A) = -1$, where $A \in \mathrm{GL}(3, \mathbb{R})$ is the matrix which corresponds to f (with respect to an orthonormal basis of $\mathbb{R}^3$).

6.2.1 Cube (C)

The cube (or hexahedron) is a three-dimensional regular solid object bounded by six congruent square faces, with three meeting at each vertex. It has 6 faces, 12 edges and 8 vertices, see Figure 6.3.

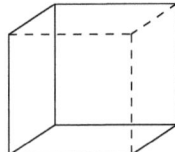

Figure 6.3: Cube, six faces.

We now consider the symmetries of the cube. First, there is the identity. Then, the cube contains three different types of symmetry axes for rotations:
-   three 4-fold axes, each of which passes through the centers of opposite faces,
-   four 3-fold axes, each of which passes through two opposite vertices,
-   six 2-fold axes, each of which passes through the midpoint of two opposite edges.

Hence, altogether, we have 24 oriented symmetries. Certainly, we have a reflection, for instance, at a plane through four vertices, given by the opposite diagonals of two opposite faces. Hence, altogether $|\mathrm{Sym}(C)| = 48$.

6.2.2 Tetrahedron (T)

A tetrahedron is a three-dimensional regular solid object bounded by four regular triangular faces with three meeting at each vertex. It has 4 faces, 6 edges and 4 vertices, see Figure 6.4.

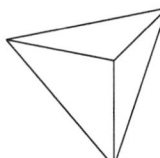

Figure 6.4: Tetrahedron, four faces.

We now consider the symmetries of the tetrahedron. First, there is the identity. The tetrahedron has only two different types of symmetry axes for rotations:
- four 3-fold axes, each of which passes through one vertex and the center of the opposite face,
- three 2-fold axes, each of which passes through midpoints of two edges.
  Hence, altogether, we have 12 oriented symmetries.

The latter three 2-fold axes give rise to three rotation-reflection planes. Hence, $|\mathrm{Sym}(T)| = 24$.

6.2.3 Octahedron (*O*)

An octahedron is a three-dimensional regular solid object bounded by eight regular triangular faces with four meeting at each vertex. It has 8 faces, 12 edges and 6 vertices, see Figure 6.5.

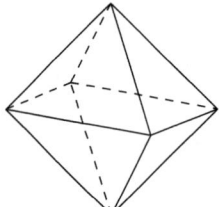

Figure 6.5: Octahedron, eight faces.

We now consider the symmetries of the octahedron. First, there is the identity. The octahedron has three different types of axes for rotations:
- three 4-fold axes, each of which passes through two opposite vertices,
- four 3-fold axes, each of which passes through the centers of two opposite faces, and
- six 2-fold axes, each of which passes through the midpoints of two opposite edges.

Hence altogether we have 24 oriented symmetries.

Certainly, we have a reflection, for instance, at a plane through any four vertices.

Hence, we get $|\mathrm{Sym}(O)| = 48$. The octahedron is dual to the cube in the following sense.

Starting with any regular polyhedron, its dual can be constructed in the following manner:
(1) Place a point in the center of each face of the original polyhedron.
(2) Connect each new point with the new points of its neighboring faces.
(3) Erase the original polyhedron, see Figures 6.6 and 6.7.

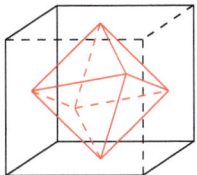

Figure 6.6: Octahedron in a cube.

This gives in particular $\mathrm{Sym}(C) \cong \mathrm{Sym}(O)$. We remark that the tetrahedron is dual to itself.

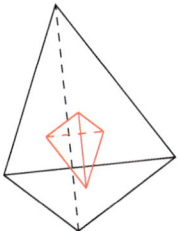

Figure 6.7: Tetrahedron in tetrahedron.

6.2.4 Icosahedron (*I*)

An icosahedron is a three-dimensional regular solid object bounded by 20 regular triangular faces with five meeting at each vertex. It has 20 faces, 30 edges and 12 vertices, see Figure 6.8.

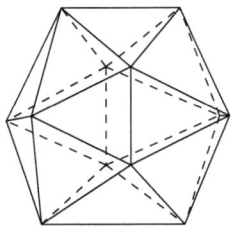

Figure 6.8: Icosahedron, 20 faces.

We now consider the symmetries of the icosahedron.

First, there is the identity. The icosahedron has three different types of axes for rotations:
- ten 3-fold axes, each of which passes through the centers of two opposite faces,
- six 5-fold axes, each of which passes through two opposite vertices, and
- fifteen 2-fold axes, each of which passes through the midpoints of two opposite edges.

Hence altogether we have 60 oriented symmetries.

The 2-fold axes give rise to rotation-reflection planes. Hence, $|\mathrm{Sym}(I)| = 120$.

6.2.5 Dodecahedron (*D*)

The dodecahedron is a three-dimensional regular solid object bounded by 12 regular pentagon faces with three meeting at each vertex. It has 12 faces, 30 edges and 20 vertices, see Figure 6.9.

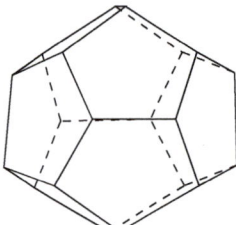

Figure 6.9: Dodecahedron, 12 faces.

We now consider the symmetries of the dodecahedron.

First, there is the identity. The dodecahedron has three types of axes for rotations:
- ten 3-fold axes, each of which passes through two opposite vertices,
- six 5-fold axes, each of which passes through the center of two opposite faces, and
- fifteen 2-fold axes, each of which passes through the midpoints of two opposite edges.

Hence altogether we have 60 oriented symmetries.

Again, the 2-fold axes give rise to rotation-reflection planes. Hence, |Sym(*D*)| = 120. The dodecahedron is dual to the icosahedron, see Figure 6.10.

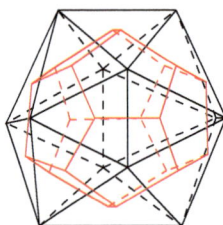

Figure 6.10: Dodecahedron in icosahedron.

This gives in particular Sym(*D*) ≅ Sym(*I*).

In Section 6.4 we will classify the Platonic solids and show that we have exactly those five solids just described.

Also we will give presentations of the symmetry groups by generators and relations.

For both proposes we need some geometry on the sphere S^2, the spherical geometry on S^2.

6.3 The Spherical Geometry of the Sphere S^2

In this section, spherical geometry is the geometry of the two-dimensional surface

$$S^2 = \{\vec{v} \in \mathbb{R}^3 \mid \|\vec{v}\| = 1\} \subset \mathbb{R}^3,$$

where $\vec{v} = (v_1, v_2, v_3)$ and $\|\vec{v}\| = \sqrt{v_1^2 + v_2^2 + v_3^2}$ as usual.

It is an example of a non-Euclidean geometry, in fact, a non-neutral geometry. In planar geometry, the basic concepts are points and lines. On the sphere S^2 points are defined in the usual sense, and the lines are the great circles.

Definition 6.2.
(1) A *point* is a vertex $\vec{v} \in \mathbb{R}^3$ with $\|\vec{v}\| = 1$, that is, the intersection of S^2 with a 1-dimensional half-subspace of $\mathbb{R}^3$, that is, the intersection of S^2 with a half-line starting at the center $O = \vec{0}$.
(2) A *line* is the intersection of S^2 with a 2-dimensional subspace of $\mathbb{R}^3$.

In this sense, we consider points P on S^2 as vectors $\vec{v} = \overrightarrow{OP} \in \mathbb{R}^3$ with $O = \vec{0}$ the center of the ball $B = \{\vec{x} \mid \|x\| \leq 1\}$; and a great circle on S^2 determines a plane through O in $\mathbb{R}^3$.

Remarks 6.3.
(1) All lines have finite length, and the lengths are all equal. Since the circumference of the unit circle is 2π, we define the length of a line as 2π.
(2) Other geometric concepts are defined as in planar geometry but with straight lines replaced by great circles. The spherical geometry has all the important axiomatic properties of the planar geometry in $\mathbb{R}^2$ except for the parallel axiom which holds in $\mathbb{R}^2$: If g is a line in $\mathbb{R}^2$ and $P \in \mathbb{R}^2$ a point with $P \notin g$, then there exists exactly one line h in $\mathbb{R}^2$ with $g \cap h = \emptyset$, and the incidence axiom 1: For any two distinct points P and Q in $\mathbb{R}^2$ there exists a unique line through P and Q.
 In spherical geometry any two lines (great circles) have a nonempty intersection, and there are infinitely many lines through two distinct points which are antipodal to each other. Recall that the antipodal point of a point on S^2 is the point which is diametrically opposite to it.
(3) A line segment between two non-antipodal points on S^2 is a segment between these points on the great circle through these points.
 Each spherical figure can be decomposed into spherical triangles. Hence, it is often enough just to consider spherical triangles.

Definition 6.4.
(1) Let $P, Q \in S^2$. The *(spherical) distance* $d(P, Q)$ is given by the angle between the vectors $\overrightarrow{OP}$ and $\overrightarrow{OQ}$, given in radian measure, see Figure 6.11.

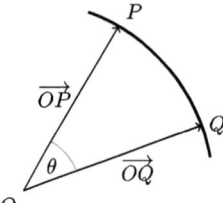

Figure 6.11: Spherical distance.

(2) The *(spherical) angle* between two great circles (lines) on S^2 is the dihedral angle between the planes in $\mathbb{R}^3$ determined by the great circles.
This is the Euclidean angle formed by the two tangents on the great circles in a cusp.

Remark 6.5. This defines also the angle between two line segments of S^2.

Theorem 6.6. *Let $P, Q \in S^2$. Then*

$$\cos(d(P, Q)) = \langle \overrightarrow{OP}, \overrightarrow{OQ} \rangle$$

where $\langle \overrightarrow{OP}, \overrightarrow{OQ} \rangle$ is the canonical scalar product of $\overrightarrow{OP}$ and $\overrightarrow{OQ}$.

Proof. Let k be the great circle in S^2 through P and Q, see Figure 6.12.

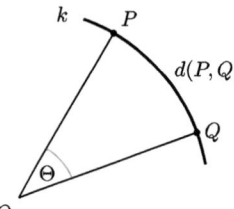

Figure 6.12: Great circle through P and Q.

By the Cauchy–Schwarz inequality (see [12, Chapter 14]), we get

$$\cos(\Theta) = \cos(d(P, Q)) = \langle \overrightarrow{OP}, \overrightarrow{OQ} \rangle$$

because $\|\overrightarrow{OP}\| = \|\overrightarrow{OQ}\| = 1$. □

Definition 6.7. A *spherical move* is the restriction $L : S^2 \to S^2$ of a linear isometry $f : \mathbb{R}^3 \to \mathbb{R}^3$, that is, $L = f \mid_{S^2}$.

We considered linear isometries $f : \mathbb{R}^3 \to \mathbb{R}^3$ in Chapter 2 in detail and classified them. From the definition of the distance and the angle in S^2, we get that spherical moves are length- and angle-preserving. From this definition it is also clear what we mean by a reflection at a spherical line (great circle) in S^2 and a rotation around a point of S^2. Moreover, we may distinguish between oriented and non-oriented moves.

Theorem 6.8. *The area F of a spherical triangle with interior angles α, β, γ is given as $F = \alpha + \beta + \gamma - \pi$.*

Proof. The area of the unit sphere S^2 is $4\pi = 2 \cdot 2\pi$. Also the area of a spherical bigon with interior angle α is 2α. To each angle of the spherical triangle there is a pair of spherical bigons because two great circles cut themselves in two points, that is, they form two bigons, one ahead and one behind.

The great circles, which define the triangle PQR, induce a triangulation of S^2 with 8 triangles, four in each hemisphere, see Figure 6.13.

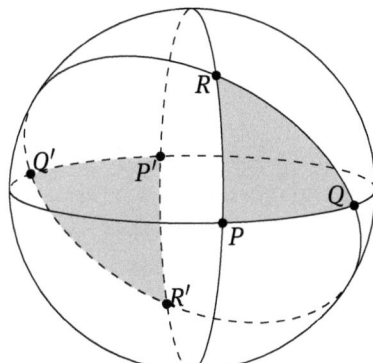

Figure 6.13: Induced triangulation.

The triangles PQR and $P'Q'R'$ have the same area.

We have pairs of spherical bigons with angular apertures α, β, γ, hence, altogether 6 bigons. These overlap the triangle PQR three times. They also overlap the opposite triangle $P'Q'R'$ three times. Otherwise the bigons overlap the whole sphere.

The area of the sphere is therefore the sum of the areas of all bigons reduced by the four extra triangles that form the overlappings. It follows that

$$4\pi = 2 \cdot 2\alpha + 2 \cdot 2\beta + 2 \cdot 2\gamma - 4F,$$

where F is the area of the spherical triangle.

Altogether $F = \alpha + \beta + \gamma - \pi$. □

Corollary 6.9. *The sum of the interior angles of a spherical triangle is bigger than π.*

Proof. This follows from Theorem 6.8 because $F > 0$. □

Corollary 6.10. *Spherical moves are area-preserving.*

Proof. If $P, Q \in S^2$, then we write $\overline{PQ}$ for the spherical line segment from P to Q. Let $L : S^2 \to S^2$ be a spherical move. Let $\Delta = ABC$ be a spherical triangle and $\Delta' = A'B'C'$ with $A' = L(A)$, $B' = L(B)$, $C' = L(C)$.

Then $\Delta' = L(\Delta)$ and

$$\sphericalangle(\overline{AB}, \overline{AC}) = \sphericalangle(\overline{A'B'}, \overline{A'C'}), \quad \sphericalangle(\overline{BA}, \overline{BC}) = \sphericalangle(\overline{B'A'}, \overline{B'C'}),$$
$$\sphericalangle(\overline{CA}, \overline{CB}) = \sphericalangle(\overline{C'A'}, \overline{C'B'}),$$

because L is length- and angle-preserving.

By Theorem 6.8, we get that Δ and Δ' have the same area.

Since all polygons may be decomposed into triangles, we get the statement. □

6.4 Classification of the Platonic Solids

The key here is the translation of Euler's formula for planar, connected graphs to an Euler formula for Platonic solids. For this we give two proofs, one which uses the spherical geometry for S^2 and one more general proof for convex polyhedra in $\mathbb{R}^3$.

Definition 6.11. A *polyhedron* P in $\mathbb{R}^3$ is a figure in the Euclidean space $\mathbb{R}^3$ which is bounded by a finite set of faces, and these faces are bounded by line segments as edges such that none of its faces intersect except at their edges. The intersections of the edges are the vertices of P.

A polyhedron P is *convex* if for any two points within P there is a line segment within P between these two points.

Theorem 6.12. *Let P be a convex polyhedron with n vertices, m edges and f faces. Then*

$$n - m + f = 2.$$

Proof. We describe the convex polyhedron as a planar, connected graph G in $\mathbb{R}^2$ with n vertices, m edges and $f - 1$ bounded faces.

This we do in the following manner. We remove one face from the polyhedron, and then we pull it straight so, that the edges do not overlap. The vertices of the polyhedron correspond to the vertices of the graph, and the edges of the polyhedron correspond to the edge of the graph. The face, removed from P, corresponds to the unbounded (external) face, and the other faces of the polyhedron correspond to the bounded faces of the graph.

By Theorem 5.28 (Euler's formula), we get

$$n - m + f = 2$$

as stated. □

We now give a different proof for Platonic solids which also gives some insight into their symmetry groups.

We now give again a definition for Platonic solids (in a more concrete manner).

Definition 6.13. A *Platonic solid* is a convex polyhedron whose faces are congruent regular polygons with the same number of faces meeting at each vertex.

Hence, the combinatorial properties are described as follows.

A convex polyhedron is a Platonic solid if and only if

(1) all its faces are congruent regular polygons and

(2) the same number of faces meet at each of its vertices.

The Platonic solids all posses three concentric spheres:

- the circumscribed sphere that passes through all the vertices,
- the midsphere that is tangent to each edge at the midpoint of the edge, and
- the inscribed sphere that is tangent to each face at the center of the face.

Important for us is the circumscribed sphere, its radius is called the circumradius. We now give a direct proof of Theorem 6.12 for Platonic solids.

Theorem 6.14. *Let P be a Platonic solid with n vertices, m edges and f faces. Then*

$$n - m + f = 2.$$

Proof. We project P onto its circumscribed sphere, starting from its center. We may assume, without loss of generality, that its circumradius is 1. The image of P on S^2 (this is now the circumscribed sphere) is a tessellation of S^2 by spherical faces bounded by spherical line segments. We choose the midpoint of one of these faces as the north pole of S^2. Then we apply the stereographic projection to this image of P and project it into the Euclidean plane. The image now is a planar, connected graph whose vertices, edges and faces correspond exactly to the vertices, edges and faces, respectively, of the tessellation. The unbounded face of the graph corresponds here to the spherical face whose midpoint is the north pole of S^2. Now we may apply Euler's formula as given in Theorem 5.28. □

Theorem 6.15. *There are exactly five Platonic solids: cube, tetrahedron, octahedron, icosahedron and dodecahedron.*

Proof. Certainly, the mentioned solids are Platonic solids in the sense of the definition.

Now let P be a Platonic solid. Let p be the number of vertices (which is also the number of edges) of one face, and let q be the number of edges at one vertex.

Recall that n, m, f denote the number of vertices, edges, faces, respectively, of the Platonic solid.

If we sum up over all faces, then we get $p \cdot f$ edges. Here each edge is counted twice. Hence, $p \cdot f = 2m$.

If we sum up over all vertices then we get $q \cdot n$ edges. Here each vertex is counted twice. Hence, $qn = 2m$.

From Theorem 6.14 we get therefore

$$\frac{1}{p} + \frac{1}{q} = \frac{1}{2} + \frac{1}{m}.$$

If $p, q \geq 3$, then this equation has exactly five solutions (p, q) which describe Platonic solids. These are

$$(q, p) = (3, 3), (3, 4), (3, 5), (4, 3), (5, 3).$$

If $(q, p) = (3, 3)$, we have the tetrahedron with $(n, m, f) = (4, 6, 4)$. If $(q, p) = (3, 4)$, we have the cube with $(n, m, f) = (8, 12, 6)$. If $(q, p) = (4, 3)$, we have the octahedron with $(n, m, f) = (6, 12, 8)$. If $(q, p) = (3, 5)$, we have the icosahedron with $(n, m, f) = (12, 30, 20)$. Finally, if $(q, p) = (5, 3)$, we have the dodecahedron with $(n, m, f) = (20, 30, 12)$. □

We now give a group-theoretical description of the symmetry groups of the five platonic solids.

We remark that dual Platonic solids have isomorphic symmetry groups. Hence we get three non-isomorphic group presentations. The faces of the Platonic solids are regular polygons in the plane $\mathbb{R}^2$. We described the symmetry groups of regular polygons in Chapter 4 and showed that these are dihedral groups.

With respect to the action of the dihedral groups on a planar regular polygon, a regular k-gon gets decomposed into $2k$ triangles. We now come back to the Platonic solids.

Let P be a Platonic solid. As in the proof of Theorem 6.14, we project P, starting from its center, onto its circumscribed sphere. Without loss of generality, we may assume that this is the sphere S^2.

The image of P gives a regular spherical tessellation of S^2 by regular spherical polygons which are the image of the faces of P. We remark that a regular spherical tessellation S^2 is analogously defined as in the plane $\mathbb{R}^2$. A regular spherical tessellation of S^2 is a division of S^2 into non-overlapping congruent regular spherical polygons.

We fix one of these regular spherical polygons, and in this one we label one of the spherical triangles (these triangles are images of the Euclidean triangles of the face of P as described above).

Let Δ be the labeled spherical triangle. Then the symmetry group of the regular spherical tessellation is generated by the reflections at the edges of Δ, and it is isomorphic to the symmetry groups Sym(P) of the Platonic solid P. Hence, we get for Sym(P) a presentation of the form

$$\text{Sym}(P) = \langle \alpha, \beta, \gamma \mid \alpha^2 = \beta^2 = \gamma^2 = (\beta \circ \gamma)^p = (\gamma \circ \alpha)^q = (\alpha \circ \beta)^r = 1 \rangle$$

with $2 \le p,q,r$, $\frac{1}{p} + \frac{1}{q} + \frac{1}{r} > 1$ and at most one of p, q, r equal to 2. The proof and the arguments are exactly as in the case of a regular tessellation of the Euclidean plane $\mathbb{R}^2$ using the Poincaré method. Also, as in the case of Euclidean plane $\mathbb{R}^2$, the subgroup $\mathrm{Sym}^+(P)$ of the oriented symmetries then is the triangle group

$$\mathrm{Sym}^+(P) = \langle x, y \mid x^p = y^q = (x \circ y)^r = 1 \rangle$$

with $x = \beta \circ y$ and $y = y \circ \alpha$.

We know that x and y are rotations. When we described the single Platonic solids, we always determined the orders of the rotations and the orders of the symmetry groups.

Hence, we get the following.

Theorem 6.16.

(1) *Let P be a tetrahedron. Then*

$$\mathrm{Sym}(P) = \langle \alpha, \beta, y \mid \alpha^2 = \beta^2 = y^2 = (\beta \circ y)^2 = (y \circ \alpha)^3 = (\alpha \circ \beta)^3 = 1 \rangle$$

and

$$\mathrm{Sym}^+(P) = \langle x, y \mid x^2 = y^3 = (x \circ y)^3 = 1 \rangle.$$

Further, $|\mathrm{Sym}(P)| = 24$ *and* $|\mathrm{Sym}^+(P)| = 12.$

(2) *Let P be a cube or an octahedron. Then*

$$\mathrm{Sym}(P) = \langle \alpha, \beta, y \mid \alpha^2 = \beta^2 = y^2 = (\beta \circ y)^2 = (y \circ \alpha)^3 = (\alpha \circ \beta)^4 = 1 \rangle$$

and

$$\mathrm{Sym}^+(P) = \langle x, y \mid x^2 = y^3 = (x \circ y)^4 = 1 \rangle.$$

Further, $|\mathrm{Sym}(P)| = 48$ *and* $|\mathrm{Sym}^+(P)| = 24.$

(3) *Let P be an icosahedron or dodecahedron. Then*

$$\mathrm{Sym}(P) = \langle \alpha, \beta, y \mid \alpha^2 = \beta^2 = y^2 = (\beta \circ y)^2 = (y \circ \alpha)^3 = (\alpha \circ \beta)^5 = 1 \rangle$$

and

$$\mathrm{Sym}^+(P) = \langle x, y \mid x^2 = y^3 = (x \circ y)^5 = 1 \rangle.$$

Further, $|\mathrm{Sym}(P)| = 120$ *and* $|\mathrm{Sym}^+(P)| = 60.$

Remarks 6.17.

(1) With the help of a computer algebra system like GAP it is easy to write down in each case a group table (or to do this by hand). Comparing these tables with those of the alternating groups A_4 and A_5, and the symmetric group S_4 (see [12, Chapter 8 and the respective exercise]), one gets:

Theorem 6.18.
(a) $\langle x, y \mid x^2 = y^3 = (x \circ y)^3 = 1 \rangle \cong A_4$,
(b) $\langle x, y \mid x^2 = y^3 = (x \circ y)^4 = 1 \rangle \cong S_4$,
(c) $\langle x, y \mid x^2 = y^3 = (x \circ y)^5 = 1 \rangle \cong A_5$.

(2) From the definition of the spherical moves, we also get that the groups A_4, S_4 and A_5 may be realized as subgroups of the special orthogonal group $SO(3) = O^+(3)$. We also may realize the dihedral group $D_n = \langle x, y \mid x^2 = y^2 = (x \circ y)^n = 1 \rangle$, $n \geq 2$, as a subgroup of $SO(3)$.

We take a spherical triangle Δ on S^2 with spherical sides a, b, c meeting with angles $\frac{\pi}{2}, \frac{\pi}{2}, \frac{\pi}{n}$. We get such a triangle if we choose one side on the equator and the other two sides orthogonal to the first one and meeting at the north pole with angle $\frac{\pi}{n}$. Let G be the group generated by the reflections at the sides a, b, c. It is clear that then $g(\Delta)$, $g \in G$, provides a tessellation of S^2. Analogously as in the Euclidean case, we get a presentation

$$G = \langle \alpha, \beta, \gamma \mid \alpha^2 = \beta^2 = \gamma^2 = (\beta \circ \gamma)^2 = (\gamma \circ \alpha)^2 = (\alpha \circ \beta)^n = 1 \rangle,$$

and $G^+ = G \cap SO(3) \cong D_n$, $n \geq 2$.
Certainly, all finite cyclic groups can be realized as subgroups of $SO(3)$.

Remark 6.19. In Chapter 8 we show the following classification.

Theorem 6.20. *Let G be a finite subgroup of $SO(3)$. Then G is isomorphic to a finite cyclic group, a dihedral group D_n, $n \geq 2$, the alternating group A_4, the symmetric group S_4 or the alternating group A_5.*

Exercises

1. Show that Platonic solids, considered as graphs, are Hamiltonian. Are they also Eulerian?
2. Given the ellipse

$$CS = \left\{ (x, y) \in \mathbb{R}^2 \mid \frac{x^2}{16} + \frac{y^2}{4} = 1 \right\}.$$

Describe the image of CS in S^2 under the stereographic projection.
(*Hint:* Write CS parametrically as $(x, y) = (4\cos(\Theta), 2\sin(\Theta))$.)

3. Given a spherical triangle as in Figure 6.14.

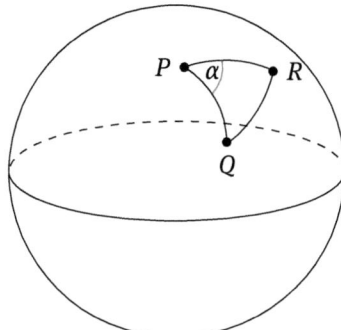

Figure 6.14: Spherical triangle.

Show the *spherical law of cosines*, namely that

$$\cos(d(Q,R)) = \cos(d(P,Q)) \cdot \cos(d(P,R)) + \sin(d(P,Q)) \cdot \sin(d(P,R)) \cdot \cos(\alpha).$$

Conclude from this the *spherical theorem of Pythagoras*, namely that

$$\cos(d(Q,R)) = \cos(d(P,Q)) \cdot \cos(d(P,R)) \quad \text{for } \alpha = \frac{\pi}{2}.$$

(*Hint*: If $\alpha \equiv 0 \mod 2\pi$ or $\alpha \equiv \pi \mod 2\pi$, then the spherical law of cosines follows from the trigonometric addition formula for cosine. If $\alpha \not\equiv 0 \mod 2\pi$ or $\alpha \not\equiv \pi \mod 2\pi$, then the vectors $\overrightarrow{QR}$, $\overrightarrow{PQ}$ and $\overrightarrow{PR}$ are linearly independent, and we may argue as for the cosine rule in $\mathbb{R}^3$ as given in [12].)

4. Deduce a spherical law of sines from the spherical law of cosines.
5. Show that the angle sum of a spherical triangle is greater than π and less than 3π.
6. Show that there is an upper bound for the area of a spherical triangle.
7. Prove Theorem 6.18 in detail. Realize the groups $\langle x,y \mid x^2 = y^3 = (x \circ y)^3 = 1 \rangle$, $\langle x,y \mid x^2 = y^3 = (x \circ y)^4 = 1 \rangle$ and $\langle x,y \mid x^2 = y^3 = (x \circ y)^5 = 1 \rangle$ as groups of permutations.
8. Show that:
   (a) S_4 contains the Klein four group as a normal subgroup.
   (b) If $N \triangleleft A_5$, then $N = \{1\}$ or $N = A_5$.
9. Determine (eventually with help of a computer algebra system) the subgroups of A_4, S_4 and A_5.
10. If we just cancel the condition that at each vertex the same number of faces meet in Definition 6.13, then there are five additional solids which are bounded by congruent equilateral triangles, see Figure 6.15.
    Find these solids by a suitable modification of Theorem 6.15.

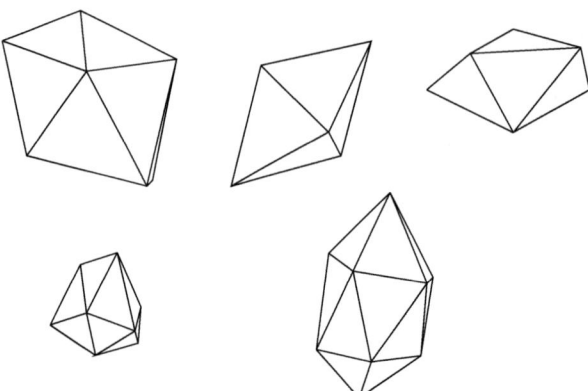

Figure 6.15: Five additional solids.

7 Linear Fractional Transformation and Planar Hyperbolic Geometry

As we saw in Chapter 1, given a neutral geometry, there are only two possibilities for the parallel postulate. These are the *Euclidean parallel postulate,* or EPP, which states that given a line ℓ and a point $P \notin \ell$ there exists a unique parallel to ℓ through point P and the *hyperbolic parallel postulate,* or HPP, which states that given a line ℓ and a point $P \notin \ell$ there exist more than one parallel to ℓ through point P. In the latter case we saw that then there are infinitely many parallels through P. Neutral geometry with EPP is standard Euclidean geometry while neutral geometry with the HPP is *hyperbolic geometry*. Hyperbolic geometry was introduced in the early part of the nineteenth century by C. F. Gauss (1777–1855), N. I. Lobachevsky (1792–1856) and J. Bolyai (1802–1860) all working independently. The discovery of hyperbolic geometry was a major step in the development of mathematics since so much work has gone into trying to prove the uniqueness of Euclidean geometry, that is, proving that the EPP follows from the other Euclidean axioms.

The main reason for this chapter is to introduce a model for a planar hyperbolic geometry.

As introduced in Chapter 5, let $\hat{\mathbb{C}} = \mathbb{C} \cup \{\infty\}$ be the one-point compactification of the complex numbers $\mathbb{C}$, realized in the Gaussian plane. We repeat the rules for calculations with ∞; these are

- $z + \infty = \infty, z \in \mathbb{C}$;
- $\infty + \infty = \infty$;
- $z \cdot \infty = \infty, z \in \mathbb{C}^* = \mathbb{C} \setminus \{0\}$;
- $\frac{z}{\infty} = 0, z \in \mathbb{C}$; and
- $\frac{z}{0} = \infty, z \in \mathbb{C}^*$.

Terms as $0 \cdot \infty, \frac{0}{0}, \infty - \infty$ and $\frac{\infty}{\infty}$ are not defined.

7.1 Linear Fractional Transformations

Definition 7.1. A *linear fractional transformation (LFT)* is a map $f : \hat{\mathbb{C}} \to \hat{\mathbb{C}}, z \mapsto \frac{az+b}{cz+d}$ with $a, b, c, d \in \mathbb{C}, ad - bc \neq 0$.

Proposition 7.2. *If $ad - bc = 0$, then f is a constant map.*

Proof. Let $ad - bc = 0$. Since f is not defined for $a = b = c = d = 0$, at least one of a, b, c, and d is not zero.

If $c \neq 0$, then we may rewrite f as $f(z) = \frac{bc-ad}{c^2}(z + \frac{d}{c})^{-1} + \frac{a}{c}$, that is, $f(z) = \frac{a}{c}$. Now let $c = 0$. Then $a = 0$ or $d = 0$.

If $a = d = 0$, then $b \neq 0$ and $f(z) = \infty$.

https://doi.org/10.1515/9783110740783-007

If $a = 0$ and $d \neq 0$, then $f(z) = \frac{b}{d}$.

If $d = 0$ and $a \neq 0$, then $f(z) = \infty$. □

The LFTs form a group $\text{Aut}(\hat{\mathbb{C}})$ of bijective maps from $\hat{\mathbb{C}}$ to $\hat{\mathbb{C}}$ with the composition as operation.

Remark 7.3. We remark that in general an automorphism of $\hat{\mathbb{C}}$ is defined to be a meromorphic bijection $f : \hat{\mathbb{C}} \to \hat{\mathbb{C}}$. In fact, an automorphism in this general sense is indeed an LFT $f : \hat{\mathbb{C}} \to \hat{\mathbb{C}}$ (see [1] or [16] for this and also for the discussion on meromorphic functions).

If $f(z) = \frac{az+b}{cz+d}$ and $g(z) = \frac{\alpha z+\beta}{\gamma z+\delta}$ are LFTs, then $f^{-1}(z) = \frac{dz-b}{-cz+a}$ and

$$g \circ f(z) = \frac{(\alpha a + \beta c)z + (\alpha b + \beta d)}{(\gamma a + \delta c)z + (\gamma b + \delta d)}. \tag{7.1}$$

The coefficients of $g \circ f$ arise from (7.1) by the multiplications of the matrices $\left(\begin{smallmatrix} \alpha & \beta \\ \gamma & \delta \end{smallmatrix}\right)$ and $\left(\begin{smallmatrix} a & b \\ c & d \end{smallmatrix}\right)$ corresponding to g and f, respectively.

Each matrix $\left(\begin{smallmatrix} a & b \\ c & d \end{smallmatrix}\right) \in \text{GL}(2,\mathbb{C})$ is assigned to the LFT $f(z) = \frac{az+b}{cz+d}$, and we have a homomorphism $\varphi : \text{GL}(2,\mathbb{C}) \to \text{Aut}(\hat{\mathbb{C}})$, that is, a map $\varphi : \text{GL}(2,\mathbb{C}) \to \text{Aut}(\hat{\mathbb{C}})$ with $\varphi(AB) = \varphi(A) \circ \varphi(B)$ where $A, B \in \text{GL}(2,\mathbb{C})$.

We define

$$\ker(\varphi) = \left\{ \begin{pmatrix} a & b \\ c & d \end{pmatrix} \in \text{GL}(2,\mathbb{C}) \,\middle|\, \varphi\left(\begin{pmatrix} a & b \\ c & d \end{pmatrix}\right) = \text{id}_{\hat{\mathbb{C}}} \right\}.$$

Let $\frac{az+b}{cz+d} = z$ for all $z \in \hat{\mathbb{C}}$. Then $cz^2 + (d-a)z = b$ for all $z \in \hat{\mathbb{C}}$. Thus, necessarily, $b = c = 0$ and $a = d$, that is,

$$\ker(\varphi) = \left\{ \begin{pmatrix} a & 0 \\ 0 & a \end{pmatrix} \,\middle|\, a \in \mathbb{C} \setminus \{0\} \right\}.$$

We consider some special LFTs. Let $z \in \hat{\mathbb{C}}$. The map $f(z)$ is a *homothety* if $f(z) = az$, $a \in \mathbb{R} \setminus \{0\}$; a *rotation* if $f(z) = cz$, $c \in \mathbb{C}$, $|c| = 1$; a *translation* if $f(z) = z + b$, $b \in \mathbb{C}$, the *inversion* if $f(z) = \frac{1}{z}$; and a *spiral similarity*, if $f(z) = az$, $a \in \mathbb{C} \setminus \mathbb{R}$. We call these elementary LFTs.

Theorem 7.4. *Every LFT is a composition of elementary ones.*

We note that this theorem is the analog of the classification of planar Euclidean isometries and the fact that any planar Euclidean isometry is the product of three or fewer reflections (see Chapters 3 and 4).

Proof. This is clear for $c = 0$. Now, let $c \neq 0$. Then we consider the chain

$$z \mapsto z + \frac{d}{c} \mapsto \left(z + \frac{d}{c}\right)^{-1} \mapsto \frac{bc - ad}{c^2}\left(z + \frac{d}{c}\right)^{-1}$$

$$\mapsto \frac{bc - ad}{c^2}\left(z + \frac{d}{c}\right)^{-1} + \frac{a}{c} = \frac{az + b}{cz + d} = f(z). \qquad \square$$

Corollary 7.5. *An LFT is angle-preserving and oriented, that is, preserves oriented angles between curves.*

Recall that the angle between two curves is the angle subtended by tangent lines where the curves intersect. Here, as usual, a curve is defined by a continuous function $y : I \to \mathbb{C}$ from an interval $I \subset \mathbb{R}$ (see [12, Chapter 12] for this type of functions), and we call $y(I)$ a curve.

Proof. We need to prove this only for the elementary LFTs. But for the elementary LFTs the statement is clear. $\qquad \square$

Theorem 7.6. *An LFT takes lines or circles in $\mathbb{C}$ onto lines or circles.*

Here we agree that each (Euclidean) line always contains ∞.

Proof. Circles and lines satisfy exactly the equations

$$az\bar{z} + bz + \bar{b}\bar{z} + c = 0 \tag{7.2}$$

with $a, c \in \mathbb{R}$, $b \in \mathbb{C}$, and $b\bar{b} - ac > 0$. It is enough to prove this for the elementary LFTs. The statement is clear for homotheties, rotations, translations and spiral similarities. We now consider the inversion $z \mapsto \frac{1}{z} = z'$. Then equation (7.2) becomes

$$cz'\bar{z'} + \bar{b}z' + b\bar{z'} + a = 0. \qquad \square$$

Remark 7.7. The lines and circles are the images of the circles on S^2 under the stereographic projection (see Chapter 5).

Definition 7.8. Let $f(z) = \frac{az+b}{cz+d}$ be an LFT. An element $z_0 \in \hat{\mathbb{C}}$ is called a *fixed point of f* if $f(z_0) = z_0$.

Lemma 7.9. *If $f(z) = \frac{az+b}{cz+d}$ is an LFT which is unequal to $\mathrm{id}_{\hat{\mathbb{C}}}$, then f has exactly one or two fixed points.*

Proof. Let first $c = 0$. If $\frac{a}{d} = 1$, then necessarily $a = d = \pm 1$ because $ad = 1$, and then ∞ is the only fixed point. If $\frac{a}{d} \neq 1$, then ∞ and $z = \frac{b}{d-a}$ are two fixed points.
Now let $c \neq 0$. Then

$$z = \frac{a - d}{2c} \pm \frac{1}{2c}\sqrt{(a + d)^2 - 4}$$

holds for a fixed point of f, and f has one fixed point if $(a+d)^2 = 4$ and two fixed points if $(a+d)^2 \neq 4$. □

Theorem 7.10. *An LFT is uniquely determined by its images at three distinct points.*

Proof. Let f, g be two LFTs with $f(z_i) = g(z_i)$, $i = 1, 2, 3$.

Then $g^{-1} \circ f$ has three fixed points if z_1, z_2, z_3 are pairwise distinct, which means then that $g^{-1} \circ f = \mathrm{id}_{\hat{\mathbb{C}}}$. □

Theorem 7.11. *Let f, g be two LFTs with $f \neq \mathrm{id}_{\hat{\mathbb{C}}} \neq g$. We have $f \circ g = g \circ f$ if and only if f and g have the same fixed points or $f^2 = g^2 = (f \circ g)^2 = \mathrm{id}_{\hat{\mathbb{C}}}$.*

Proof. The proof follows from a simple calculation. □

Theorem 7.12. *Let G be a finite subgroup of* $\mathrm{Aut}(\hat{\mathbb{C}})$. *Then one of the following cases occur:*

(1) *G is finite cyclic of order $m \geq 1$.*
(2) *$G = \langle f, g \mid f^2 = g^2 = (f \circ g)^n = 1 \rangle$, $n \geq 2$, that is, G is isomorphic to the dihedral group D_n, $n \geq 2$ (here D_2 is the Klein four group).*
(3) *$G = \langle f, g \mid f^2 = g^3 = (f \circ g)^3 = 1 \rangle$, that is, G is isomorphic to the alternating group A_4.*
(4) *$G = \langle f, g \mid f^2 = g^3 = (f \circ g)^4 = 1 \rangle$, that is, G is isomorphic to the symmetric group S_4.*
(5) *$G = \langle f, g \mid f^2 = g^3 = (f \circ g)^5 = 1 \rangle$, that is, G is isomorphic to the alternating group A_5.*

Proof. Let G be a finite subgroup of $\mathrm{Aut}(\hat{\mathbb{C}})$. Let $f \in G, f \neq \mathrm{id}_{\hat{\mathbb{C}}}$. Then f has two distinct fixed points. This can be seen as follows. It is an easy calculation that there exists an LFT g such that $g \circ f \circ g^{-1}(z) = \frac{az+b}{d}$, and f has two fixed points if and only if $g \circ f \circ g^{-1}$ has two fixed points. Now assume that f has only one fixed point. Then necessarily $a = d = \pm 1$ (see proof of Lemma 7.9), but this implies that f has infinite order which contradicts the fact that $f \in G$. So f has two fixed points.

We say that $v \in \hat{\mathbb{C}}$ is a vertex if v is fixed by some $g \in G, g \neq \mathrm{id}_{\hat{\mathbb{C}}}$. We denote the set of vertices by V. Now consider the number $|E|$ of elements of the finite set

$$E = \{(g, v) \mid g \in G, g \neq \mathrm{id}_{\hat{\mathbb{C}}}, v \in V, g(v) = v\}.$$

As seen before, g fixes exactly two vertices, and we have $|E| = 2(|G| - 1)$.

Let $G_v = \{g \in G \mid g(v) = v\}$ be the stabilizer of a vertex v.

Then we also have

$$|E| = \sum_{v \in V} (|G_v| - 1).$$

The set V is portioned by G into disjoint orbits $V_1, V_2, \ldots, V_s$. Recall that the orbit $G(v)$ of $v \in V$ is the subset of V defined by $G(v) = \{g(v) \in V \mid g \in G\}$.

As the stabilizer of each v in V_j has the same number, say n_j, of elements, we have

$$|E| = \sum_{j=1}^{s} \sum_{v \in V_j} (|G_v| - 1) = \sum_{j=1}^{s} |V_j|(n_j - 1).$$

Finally, each orbit $G(v)$ is in a 1–1 correspondence with the class of cosets gG_v, $g \in G$, so for v in V_j we have

$$|V_j| = \frac{|G|}{|G_v|} = \frac{|G|}{n_j}.$$

Eliminating $|V_j|$ we obtain

$$2\left(1 - \frac{1}{|G|}\right) = \sum_{j=1}^{s}\left(1 - \frac{1}{n_j}\right). \tag{7.3}$$

We shall exclude the trivial group, so $|G| \geq 2$ and $1 \leq 2(1 - \frac{1}{|G|}) < 2$. By definition, $n_j \geq 2$, so

$$\frac{1}{2}s \leq \sum_{j=1}^{s}\left(1 - \frac{1}{n_j}\right) < s.$$

These inequalities, together with (7.3), show that $s = 2$ or $s = 3$.

Case 1. $s = 2$.

In this case (7.3) becomes

$$2 = \frac{|G|}{n_1} + \frac{|G|}{n_2},$$

and hence

$$|G| = n_1 = n_2, \quad |V_1| = |V_2| = 1$$

because $|n_j| \leq G$.

In this case there are only two vertices and each is fixed by every element of G. By conjugation (as above), we may take the vertices to be 0 and ∞, and G is then a finite, cyclic group of rotations.

Case 2. $s = 3$.

In this case (7.3) becomes

$$\frac{1}{n_1} + \frac{1}{n_2} + \frac{1}{n_3} = 1 + \frac{2}{|G|},$$

and we may assume that $n_1 \le n_2 \le n_3$. Clearly, $n_1 \ge 3$ leads to a contradiction, thus $n_1 = 2$ and

$$\frac{1}{n_2} + \frac{1}{n_3} = \frac{1}{2} + \frac{2}{|G|}.$$

Now necessarily $n_2 = 2$ or 3.

The case $n_2 = 2$ leads to

$$(|G|, n_1, n_2, n_3) = (2n, 2, 2, n), \quad n \ge 2,$$

and G is isomorphic to the dihedral group D_n.

The remaining cases are those with $s = 3$, $n_1 = 2$, $n_2 = 3$ and

$$\frac{1}{n_3} = \frac{1}{6} + \frac{2}{|G|}, \quad n_3 \ge 3.$$

The integer solutions are
(1) $(|G|, n_1, n_2, n_3) = (12, 2, 3, 3)$;
(2) $(|G|, n_1, n_2, n_3) = (24, 2, 3, 4)$;
(3) $(|G|, n_1, n_2, n_3) = (60, 2, 3, 5)$.

These groups are isomorphic to A_4, S_4 and A_5, respectively. □

Remark 7.13.
(1) The finite groups in Theorem 7.12 may be realized as follows.
  (a) Consider an $f \in \mathrm{Aut}(\hat{\mathbb{C}})$ with

$$f(z) = -\frac{1}{z + 2\cos(\frac{\pi}{m})}, \quad m \ge 2.$$

  Then $\langle f \rangle$ is cyclic of order m.
  (b) Consider $f, g \in \mathrm{Aut}(\hat{\mathbb{C}})$ with

$$f(z) = -\frac{1}{z}, \quad g(z) = -\frac{\rho^{-1}}{\rho z}, \quad \rho + \rho^{-1} = 2\cos\left(\frac{\pi}{n}\right),$$

  $n \ge 2$. Then $\langle f, g \rangle$ is isomorphic to D_n.
  (c) Consider $f, g \in \mathrm{Aut}(\hat{\mathbb{C}})$ with

$$f(z) = -\frac{1}{z}, \quad g(z) = -\frac{\rho^{-1}}{\rho z + 1}, \quad \rho + \rho^{-1} = 1.$$

  Then $\langle f, g \rangle$ is isomorphic to A_4.

(d) Consider $f, g \in \mathrm{Aut}(\hat{\mathbb{C}})$ with

$$f(z) = -\frac{1}{z}, \quad g(z) = -\frac{\rho^{-1}}{\rho z + 1}, \quad \rho + \rho^{-1} = \sqrt{2}.$$

Then $\langle f, g \rangle$ is isomorphic to S_4.
(e) Consider $f, g \in \mathrm{Aut}(\hat{\mathbb{C}})$ with

$$f(z) = -\frac{1}{z}, \quad g(z) = -\frac{\rho^{-1}}{\rho z + 1}, \quad \rho + \rho^{-1} = 2\cos\left(\frac{\pi}{5}\right).$$

Then $\langle f, g \rangle$ is isomorphic to A_5.
We remark that $2\cos(\frac{\pi}{5}) = \frac{1}{2}(1+\sqrt{5})$, which is related to the golden section $\frac{1}{2}(\sqrt{5}-1)$, see [12].
(2) Via the stereographic projection we get that a finite subgroup of $SO(3) = O^+(3)$ is isomorphic to a finite subgroup of $\mathrm{Aut}(\hat{\mathbb{C}})$. On the other hand, each finite group which occurs in Theorem 7.12 can be realized as a subgroup of $SO(3)$ (see Chapter 5). Hence we get the following.

Corollary 7.14 (see Theorem 6.20). *Let G be a finite subgroup of $SO(3) = O^+(3)$. Then G is isomorphic to a finite cyclic group, a dihedral group D_n, $n \geq 2$, the alternating group A_4, the symmetric group S_4 or the alternating group A_5.*

Remark 7.15. We also get Corollary 7.14 directly by a modification of the proof of Theorem 7.12 for $SO(3)$ operating on S^2. We only have to remark the following. If G is a finite subgroup of $SO(3)$ then each nontrivial element of G fixes the center $\vec{O} = (0,0,0)$ of the ball $\{\vec{x} \in \mathbb{R}^3 \mid \|\vec{x}\| \leq 1\}$ and exactly one point of S^2, and G operates on $\{\vec{O}\} \cup S^2$.

We saw in Theorem 7.10 that an LFT is uniquely determined by the images of three distinct points. We now show the implication in the other direction that we get an LFT if we provide the images of three distinct points. We first give an LFT which maps three pairwise distinct complex numbers z_1, z_2, z_3 onto $0, 1, \infty$, respectively.
We define

$$f(z) = \frac{\frac{z-z_1}{z-z_3}}{\frac{z_2-z_1}{z_2-z_3}},$$

where $f(\infty) = \frac{z_2-z_3}{z_2-z_1}$. We have indeed that $f(z_1) = 0, f(z_2) = 1$ and $f(z_3) = \infty$. The term on the right side is called the *cross-ratio (or double ratio)* of the four points z, z_1, z_2, z_3 with z_1, z_2, z_3 pairwise distinct; written as

$$\mathrm{DV}(z, z_1, z_2, z_3) = \frac{\frac{z-z_1}{z-z_3}}{\frac{z_2-z_1}{z_2-z_3}}, \quad \text{if } z \neq \infty$$

and

$$DV(\infty, z_1, z_2, z_3) = \frac{z_2 - z_3}{z_2 - z_1}.$$

If now one of the z_i, $i = 1, 2, 3$, is ∞, then we write as usual $z_i = \frac{1}{t_i}$ and form the limit as $t_i \to 0$.

This gives the following cross-ratios:

$$DV(z, \infty, z_2, z_3) = \frac{z_2 - z_3}{z - z_3},$$

$$DV(z, z_1, \infty, z_3) = \frac{z - z_1}{z - z_3} \quad \text{and}$$

$$DV(z, z_1, z_2, \infty) = \frac{z - z_1}{z_2 - z_1}.$$

With this modification we get that the map $z \mapsto DV(z, z_1, z_2, z_3)$ in each case is an LFT which maps the three distinct points z_1, z_2, z_3 in $\hat{\mathbb{C}}$ onto $0, 1, \infty$, respectively.

Lemma 7.16. *If (z_1, z_2, z_3) and (w_1, w_2, w_3) are two triples of distinct points in $\hat{\mathbb{C}}$, then there exists exactly one LFT f with $f(z_i) = w_i$ for $i = 1, 2, 3$.*

This is the analog for hyperbolic geometry of the fact that Euclidean isometries are determined by their action on triangles.

Proof. Let $f_1(z) = DV(z, z_1, z_2, z_3)$ and $f_2(z) = DV(z, w_1, w_2, w_3)$. Then f_1 maps (z_1, z_2, z_3) onto $(0, 1, \infty)$, and f_2 maps (w_1, w_2, w_3) onto $(0, 1, \infty)$. Hence $f := f_2^{-1} \circ f_1$ is the desired LFT. $\square$

Theorem 7.17. *Let $z, z_1, z_2, z_3 \in \hat{\mathbb{C}}$ with z_1, z_2, z_3 pairwise distinct. Then*

$$DV(z, z_1, z_2, z_3) = DV(f(z), f(z_1), f(z_2), f(z_3))$$

for each $f \in \text{Aut}(\hat{\mathbb{C}})$.

Proof. We consider the map $g : z \mapsto DV(f(z), f(z_1), f(z_2), f(z_3))$; g is the composition of f and $h : z \mapsto DV(z, f(z_1), f(z_2), f(z_3))$, and hence an LFT. Further, $g(z_1) = 0$, $g(z_2) = 1$ and $g(z_3) = \infty$. Therefore g is the LFT $z \mapsto DV(z, z_1, z_2, z_3)$. $\square$

Remark 7.18. We saw that $f \in \text{Aut}(\hat{\mathbb{C}})$ is determined by the pairwise distinct elements $z_1, z_2, z_3 \in \hat{\mathbb{C}}$ and their images, and f can be realized by the relation $DV(z, z_1, z_2, z_3) = DV(f(z), f(z_1), f(z_2), f(z_3))$. We know that, given three distinct points z_1, z_2, z_3, there exists exactly one circle or line which passes through z_1, z_2, z_3. This also follows from the following construction for the case that z_1, z_2, z_3 are not collinear, see Figure 7.1.

Lemma 7.19. *Let $z_1, z_2, z_3 \in \hat{\mathbb{C}}$ be three distinct points. A point $z \in \hat{\mathbb{C}}$ is on the circle or line K determined by z_1, z_2, z_3 if and only if $DV(z, z_1, z_2, z_3) \in \mathbb{R} \cup \{\infty\}$.*

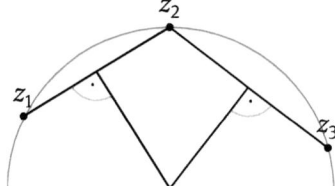

Figure 7.1: Circle through three distinct points.

Proof. Let f be the LFT with $(z_1, z_2, z_3) \mapsto (0, 1, \infty)$. By Theorem 7.6, we have that $z \in K$ if and only if $f(z) \in \mathbb{R} \cup \{\infty\}$. Since $f(z) = DV(f(z), 0, 1, \infty)$, the statement follows from Theorem 7.17. $\qquad\square$

7.2 A Model for a Planar Hyperbolic Geometry

In this section we use LFTs to present a model of planar hyperbolic geometry. To be precise, the points are the points of the Euclidean upper half-plane and the lines are the intersections with the upper half-plane of circles orthogonal to the real line as well as rays orthogonal to the real line.

Remark 7.20.
(1) From now on, we norm the LFT $f(z) = \frac{az+b}{cz+d}$ so that $ad - bc = 1$. This is possible by cancellation because

$$\ker(\varphi) = \left\{ \begin{pmatrix} a & 0 \\ 0 & a \end{pmatrix} \,\middle|\, a \in \mathbb{C}, a \neq 0 \right\}.$$

(2) Let $\mathbb{H} = \{z \in \mathbb{C} \mid \mathrm{Im}(z) > 0\}$ be the upper half-plane. Let $f(z) = \frac{az+b}{cz+d}$ with $a, b, c, d \in \mathbb{R}$ and $ad - bc = 1$.

Lemma 7.21. *We have*
(i) $f(r) \in \mathbb{R} \cup \{\infty\}$ *for* $r \in \mathbb{R} \cup \{\infty\}$.
(ii) $f(z) \in \mathbb{H}$ *for* $z \in \mathbb{H}$.

Proof. (i) is obvious because $a, b, c, d \in \mathbb{R}$.
We show (ii).
Let $z \in \mathbb{H}$ and $f(z) = \frac{az+b}{cz+d}$. Then

$$
\begin{aligned}
f(z) &= \frac{az+b}{cz+d} = \frac{(az+b)(c\bar{z}+d)}{(cz+d)(c\bar{z}+d)} \\
&= \frac{1}{|cz+d|^2}(acz\bar{z} + adz + bc\bar{z} + bd) \\
&= \frac{1}{|cz+d|^2}(acz\bar{z} + z + bc(z+\bar{z}) + bd)
\end{aligned}
$$

because $ad - bc = 1$.
Hence, $\mathrm{Im}(f(z)) = \frac{\mathrm{Im}(z)}{|cz+d|^2} > 0$. $\qquad\square$

Conclusion. f maps $\mathbb{R} \cup \{\infty\}$ onto itself and also $\mathbb{H}$ onto itself.

(3) The LFTs $f : \mathbb{H} \to \mathbb{H}, z \mapsto \frac{az+b}{cz+d}$, $a, b, c, d \in \mathbb{R}$ with $ad - bc = 1$, form a group which we denote by $\mathrm{Aut}(\mathbb{H})$, and we have

$$\mathrm{Aut}(\mathbb{H}) \cong {}^{\mathrm{SL}(2,\,\mathbb{R})}\!/_{\{\pm E_2\}} =: \mathrm{PSL}(2, \mathbb{R})$$

where again $E_2 = \left(\begin{smallmatrix} 1 & 0 \\ 0 & 1 \end{smallmatrix}\right)$.

We are now prepared to introduce a model of a planar hyperbolic geometry. For historical reasons one often calls this geometry the planar non-Euclidean geometry but we use here the term planar hyperbolic geometry. The planar hyperbolic geometry, which we consider, is the geometry on the upper half-plane $\mathbb{H}$.

Definition 7.22.
(1) A point z is an element of $\mathbb{H}$.
(2) A line g in $\mathbb{H}$ is either the intersection $\ell \cap \mathbb{H}$, with ℓ being a line parallel to the imaginary y-axis or the intersection $k \cap \mathbb{H}$ with k being a circle with center in $\mathbb{R}$ (ortho-circle on $\mathbb{R}$), see Figure 7.2.

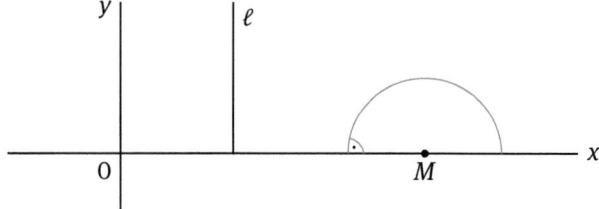

Figure 7.2: Hyperbolic lines.

Remark 7.23.
(1) If $z_1, z_2 \in \mathbb{H}$ then there exists exactly one line through z_1 and z_2.
    If $\mathrm{Re}(z_1) = \mathrm{Re}(z_2)$ then the line is parallel to the y-axis defined by $\mathrm{Re}(z_1)$.
    If $\mathrm{Re}(z_1) \neq \mathrm{Re}(z_2)$, then we see this from the construction in Figure 7.3.
    a and b are the intersection of the ortho-circle through z_1 and z_2 with the x-axis.
(2) Other geometric concepts are defined as in planar geometry but with straight lines replaced by lines in $\mathbb{H}$. The planar hyperbolic geometry has all the important axiomatic properties of planar geometry in $\mathbb{R}^2$ except for the parallel axiom which holds in $\mathbb{R}^2$.
    If g is a line in $\mathbb{H}$ and $z \in \mathbb{H}$ a point with $z \notin g$ then there exist (at least) two lines in $\mathbb{H}$ which have empty intersection with g.

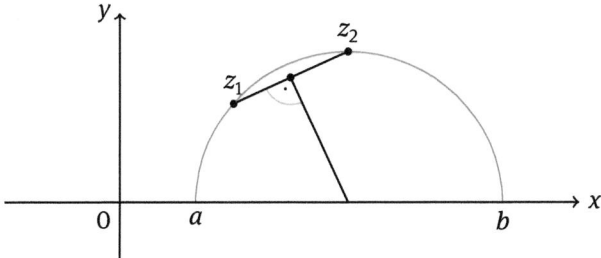

Figure 7.3: Hyperbolic line through z_1 and z_2.

This can be seen as follows. From the preliminary material we may assume that $g = \ell \cap \mathbb{H}$ where ℓ is the y-axis and $z = 2 + i$.

Then z is on the line g_1 parallel to g and given by $\mathrm{Re}(z) = 2$, and z is on the ortho-circle g_1 with center 2 and radius 1. We have $g \cap g_1 = \emptyset$ and $g \cap g_2 = \emptyset$.

We now define the *hyperbolic distance* $\delta(z_1, z_2)$ of two points $z_1, z_2 \in \mathbb{H}$.

Let first $z_1 \neq z_2$ and assume $\mathrm{Re}(z_1) \neq \mathrm{Re}(z_2)$. Then we consider, as in Figure 7.3, the ortho-circle through z_1 and z_2, and the intersections a and b with the x-axis. With the prescribed order, starting with a, we have $\mathrm{DV}(z_1, a, z_2, b) > 0$. The hyperbolic distance then is defined as $\delta(z_1, z_2) = |\ln(\mathrm{DV}(z_1, a, z_2, b))|$.

Now let $\mathrm{Re}(z_1) = \mathrm{Re}(z_2) = x$. Then $z_1 = x + iy_1$, $z_2 = x + iy_2$, $y_1 \neq y_2$. Let $y_1 < y_2$. Then take $a = x$ and $b = \infty$ so that

$$\mathrm{DV}(z_1, a, z_2, b) = \mathrm{DV}(z_1, x, z_2, \infty) > 0,$$

and again we may define the hyperbolic distance as

$$\delta(z_1, z_2) = \big|\ln(\mathrm{DV}(z_1, a, z_2, b))\big|.$$

If $z_1 = z_2$, then we define $\delta(z_1, z_2) = 0$.

Remark 7.24. We always must take the prescribed order between a, z_1, z_2 and b.

From the properties of the cross-ratio and Theorem 7.17, we get the following.

Lemma 7.25. *Let* $z_1, z_2, z_3 \in \mathbb{H}$. *Then*
(1) $\delta(z_1, z_2) > 0$ *if* $z_1 \neq z_2$, *and* $\delta(z_1, z_2) = 0$ *if* $z_1 = z_2$;
(2) $\delta(z_1, z_2) = \delta(z_2, z_1)$;
(3) $\delta(z_1, z_2) \leq \delta(z_1, z_3) + \delta(z_2, z_3)$;
(4) $\delta(z_1, z_2) = \delta(f(z_1), f(z_2))$ *for* $f \in \mathrm{Aut}(\mathbb{H})$.

Proof. (1) is clear. For (2) we only have to make sure that we also interchange a and b if we interchange z_1 and z_2.

(4) follows directly from Theorem 7.17.

To prove (3), we may map z_1, z_2, z_3 with an $f \in \text{Aut}(\mathbb{H})$ onto the y-axis; (3) now follows from the position of the image points under f to each other. $\qquad\square$

So far, we have constructed a model for a planar hyperbolic geometry with a suitable distance between two points.

From Lemma 7.25 and Corollary 7.5, we know the following.

Each $f \in \text{Aut}(\mathbb{H})$ is oriented, angle-preserving and length-preserving.

Hence it is reasonable to define $\text{Aut}(\mathbb{H})$ as the group of oriented isometries for the planar hyperbolic geometry on $\mathbb{H}$ (see [4] for a verification). As usual, we call a part of a hyperbolic line a *hyperbolic line segment*. We denote a hyperbolic line segment from z_1 to z_2 by $\overline{z_1 z_2}$. A *hyperbolic polygon* in $\mathbb{H}$ is a figure bounded by a finite set of hyperbolic line segments as edges.

We want to determine the hyperbolic area of a hyperbolic polygon in $\mathbb{H}$.

For this we need some preparations.

Let $f : D \to D$ be a function with $D \subset \mathbb{C}$. We say that f is (complex) differentiable in $z_0 \in D$ if

$$\lim_{\substack{z \to z_0 \\ z \in D \setminus \{z_0\}}} \frac{f(z) - f(z_0)}{z - z_0} = f'(z_0)$$

exists.

Let $f \in \text{Aut}(\mathbb{H})$, $f(z) = \frac{az+b}{cz+d}$, and $D = \mathbb{H}$ in the above definition.

Then f is differentiable in each $z \in \mathbb{H}$, and we get

$$\frac{|f'(z)|}{\text{Im}(f(z))} = \frac{1}{y} \quad \text{if } z = x + iy$$

for each $z \in \mathbb{H}$.

This leads to the following definition.

Definition 7.26. Let $\gamma : [0,1] \to \mathbb{H}$ be a curve such that γ is differentiable on $[0,1]$ and $\gamma' : [0,1] \to \mathbb{C}$ is continuous (see [12, Chapter 12]). We say that γ is *continuously differentiable*.

Then the *hyperbolic length of γ* is defined as

$$L_h(\gamma) = \int_0^1 \frac{|\gamma'(t)|}{\text{Im}(\gamma(t))} \, dt.$$

As usual, we write

$$L_h(\gamma) = \int_\gamma \frac{|dz|}{y}.$$

If $\gamma : [0,1] \to \mathbb{H}$ is continuously differentiable curve and $f \in \mathrm{Aut}(\mathbb{H})$. Then,

$$L_h(f \circ \gamma) = L_h(\gamma).$$

Now let $z_1, z_2 \in \mathbb{H}$, $z_1 \neq z_2$. Let γ be the hyperbolic line segment between z_1 and z_2, and let a and b be defined as above. We map a, z_1, z_2, b with a $f \in \mathrm{Aut}(\mathbb{H})$ onto $0, i, is$ and ∞, where $s > 1$.
   Then

$$\int\limits_\gamma \frac{|dz|}{y} = \int\limits_1^s \frac{dt}{t} = \ln(s) = \left|\ln(\mathrm{DV}(1,0,s,\infty))\right| = \delta(z_1, z_2).$$

Using this we may define the hyperbolic area F_h of a hyperbolic polygon P as $F_h = \iint \frac{dx\,dy}{y^2}$, where the integral is taken over the polygon.
   Before we calculate F_h in detail, we apply the above description for $\delta(z_1, z_2)$ to show the hyperbolic Theorem of Pythagoras.

7.3 The (Planar) Hyperbolic Theorem of Pythagoras in $\mathbb{H}$

Theorem 7.27. *A neutral geometry satisfying the (Euclidean) Pythagorean theorem is Euclidean.*

Proof. One of the equivalences for a Euclidean geometry is certainly the existence of non-congruent similar triangles (see Exercise 12 of Chapter 1). Here two triangles are similar if the corresponding angles have the same measure. Consider a right triangle in a neutral geometry where the Pythagorean theorem holds. Connect the midpoint of the hypotenuse to the midpoint of one of the sides. Using the Pythagorean theorem the resulting triangle is similar, but not congruent to the big triangle. □

We now want to explain a hyperbolic theorem of Pythagoras. For this we first consider the hyperbolic functions. Recall that the hyperbolic cosine is

$$\cosh(x) = \frac{1}{2}(e^x + e^{-x}),$$

and the hyperbolic sine is

$$\sinh(x) = \frac{1}{2}(e^x - e^{-x})$$

for $x \in \mathbb{R}$. We have in particular

$$\cosh(-x) = \cosh(x),$$
$$\sinh(-x) = -\sinh(x) \quad \text{and}$$
$$\cosh^2(x) - \sinh^2(x) = 1 \quad \text{for } x \in \mathbb{R}.$$

Further, $\cosh(x)$ is strictly monotonically increasing on $[0, \infty)$.

Theorem 7.28. *For all $z_1, z_2 \in \mathbb{H}$ we have*

$$\cosh(\delta(z_1, z_2)) = 1 + \frac{|z_1 - z_2|^2}{2 \operatorname{Im}(z_1) \cdot \operatorname{Im}(z_2)}.$$

Proof. As above, we may assume that $z_1 = i$ and $z_2 = i \cdot s$ with $s > 1$. Then $\delta(z_1, z_2) = \ln(s)$. Hence,

$$\cosh(\delta(z_1, z_2)) = \frac{1}{2}(e^{\ln(s)} + e^{-\ln(s)}) = \frac{1}{2}\left(s + \frac{1}{s}\right) = \frac{(s-1)^2 + 2s}{2s} = 1 + \frac{(s-1)^2}{2s},$$

which proves the statement. $\qquad\square$

Theorem 7.29 (Hyperbolic Theorem of Pythagoras). *Let Δ be a hyperbolic rectangular triangle with endpoints z_1, z_2, z_3, catheti $\overline{z_1 z_2}$, $\overline{z_1 z_3}$ and hypotenuse $\overline{z_2 z_3}$. Then*

$$\cosh(\delta(z_2, z_3)) = \cosh(\delta(z_1, z_2)) \cdot \cosh(\delta(z_1, z_3)).$$

Proof. Let Δ have the endpoints z_1, z_2, z_3 with $\delta(z_1, z_2) = a$, $\delta(z_1, z_3) = b$ and $\delta(z_2, z_3) = c$.

As above we may assume $z_1 = i$ and $z_2 = k \cdot i$ with $k > 1$.

Since Δ has a right angle at i, we must have that $z_3 = s + i \cdot t$ is on unit circle, that is, $s^2 + t^2 = 1$. We may assume that $s > 0$.

By Theorem 7.28 we have

$$\cosh(a) = 1 + \frac{(k-1)^2}{2k} = \frac{1 + k^2}{2k},$$

$$\cosh(b) = 1 + \frac{|-s + i(1-t)|^2}{2t} = \frac{2t + s^2 + (1-t)^2}{2t} = \frac{1}{t}$$

and

$$\cosh(c) = 1 + \frac{|s + i(t-k)|^2}{2tk} = \frac{2tk + s^2 + (t-k)^2}{2tk} = \frac{1 + k^2}{2tk}$$

because $s^2 + t^2 = 1$. This gives

$$\cosh(c) = \cosh(a) \cdot \cosh(b). \qquad\square$$

7.4 The Hyperbolic Area of a Hyperbolic Polygon

An important part of planar Euclidean geometry is the theory of area. An *area function* or measure on the Euclidean plane $\mathbb{R}^2$ is a function A from subsets of $\mathbb{R}^2$ to the reals $\mathbb{R}$ satisfying

(1) $A(S) \geq 0$ for all subsets of $\mathbb{R}$.

(2) $A(\cup A_i) = \sum_i A_i$ for any countable collection $\{A_i\}$ of disjoint sets.

We note that the set of functions to which the measure function applies is usually restricted to what are called measurable sets. We refer to [11] for a discussion of this. An area in the Euclidean plane is usually constructed using integration starting with the area of a rectangle. We now determine the hyperbolic area of hyperbolic polygons.

Theorem 7.30. *Let P be a hyperbolic polygon in $\mathbb{H}$. Let n be the number of vertices of P and let $\alpha_1, \alpha_2, \ldots, \alpha_n$ be the interior angles at the vertices. Then the hyperbolic area F_h of P is*

$$F_h = (n-2)\pi - (\alpha_1 + \alpha_2 + \cdots + \alpha_n).$$

Proof. It is enough to prove Theorem 7.30 for hyperbolic triangles.

Also the vertices of the triangle may in part lie on $\mathbb{R} \cup \{\infty\}$. This does not change the area F_h.

Let $\alpha_1 = \alpha$, $\alpha_2 = \beta$, $\alpha_3 = \gamma$.

Case 1.

The triangle may have the angles 0, $\frac{\pi}{2}$, α and the form as in Figure 7.4.

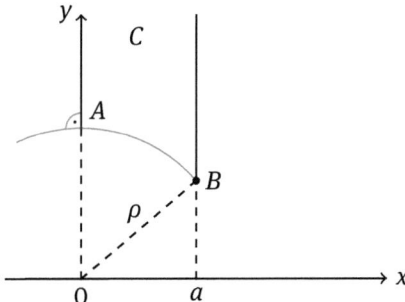

Figure 7.4: Triangle with angles 0, $\frac{\pi}{2}$, α.

Then

$$F_h = \int_0^a dx \int_{\sqrt{\rho^2 - x^2}} \frac{dy}{y^2} = \int_0^a \frac{dx}{\sqrt{\rho^2 - x^2}} = \arcsin\left(\frac{a}{\rho}\right) = \frac{\pi}{2} - \alpha = \pi - \left(0 + \frac{\pi}{2} + \alpha\right).$$

Case 2.

The triangle has the angles 0, α, β and the form in Figure 7.5.

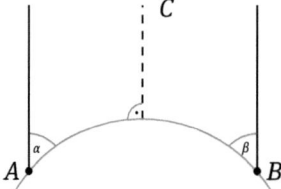

Figure 7.5: Triangle with angles 0, α, β.

The hyperbolic area of the triangle is the sum of the areas of the two triangles, given as in Case 1, and we get

$$F_h = \pi - \left(\alpha + \frac{\pi}{2}\right) + \pi - \left(\beta + \frac{\pi}{2}\right) = \pi - (0 + \alpha + \beta).$$

Case 3.

The triangle has an angle 0, and one of the vertices is on $\mathbb{R}$, that is, the triangle has a cusp on $\mathbb{R}$.

By an $f \in \mathrm{Aut}(\mathbb{H})$ we may move this cusp to ∞, and the result follows because P and $f(P)$ have the same hyperbolic area.

Case 4.

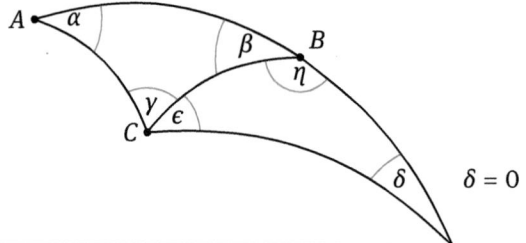

Figure 7.6: Triangle completely contained in $\mathbb{H}$.

The triangle in Figure 7.6 is completely contained in $\mathbb{H}$. We get

$$F_h = (\pi - (\alpha + \gamma + \epsilon)) - (\pi - (\epsilon + \eta)) = \eta - (\alpha + \gamma) = \pi - (\alpha + \beta + \gamma)$$

because $\beta + \eta = \pi$. $\qquad\qquad\qquad\qquad\qquad\qquad\qquad\qquad\qquad\qquad\qquad\Box$

Corollary 7.31.
(1) $\mathrm{Aut}(\mathbb{H})$ *is area-preserving with respect to the hyperbolic area.*
(2) *The sum of the interior angles $\alpha + \beta + \gamma$ is smaller than π for a hyperbolic triangle.*

Remark 7.32. Let $\widetilde{\mathrm{Aut}}(\mathbb{H})$ be the group which is generated by $\mathrm{Aut}(\mathbb{H})$ and the reflection $z \mapsto -\bar{z}$ at the y-axis. We define $\widetilde{\mathrm{Aut}}(\mathbb{H})$ to be the group of the hyperbolic isometries of planar hyperbolic geometry on $\mathbb{H}$ (see [4] for a verification).

$\widetilde{\mathrm{Aut}}(\mathbb{H})$ is composed of the maps

$$z \mapsto \frac{az+b}{cz+d}, \quad a, b, c, d \in \mathbb{R}, \ ad - bc = 1,$$

and

$$z \mapsto \frac{a\bar{z}+b}{c\bar{z}+d}, \quad a, b, c, d \in \mathbb{R}, \ ad - bc = -1.$$

All elements of $\widetilde{\mathrm{Aut}}(\mathbb{H})$ preserve angles, hyperbolic lengths and hyperbolic areas.

We remark that hereby we have explained what we understand under a reflection at hyperbolic lines in $\mathbb{H}$.

We now describe, analogously as we have done it for the Euclidean triangles in $\mathbb{R}^2$ and the spherical triangles in S^2, a tessellation of the hyperbolic plane $\mathbb{H}$ by hyperbolic triangles.

Theorem 7.33. *Let Δ be a hyperbolic triangle with interior angles $\frac{\pi}{p}, \frac{\pi}{q}, \frac{\pi}{r}; p, q, r \in \mathbb{N}\backslash\{1\}$, $\frac{1}{p} + \frac{1}{q} + \frac{1}{r} < 1$. Let G be the group generated by the reflections α, β, γ at the sides a, b, c of Δ, respectively.*

The images $g(\Delta), g \in G$, form a non-overlapping division of $\mathbb{H}$, a hyperbolic tessellation of $\mathbb{H}$, and G has a presentation

$$G = \langle \alpha, \beta, \gamma \mid \alpha^2 = \beta^2 = \gamma^2 = (\alpha \circ \beta)^p = (\beta \circ \gamma)^q = (\alpha \circ \gamma)^r = 1 \rangle.$$

Proof. The arguments are exactly as in the case of the Euclidean plane $\mathbb{R}^2$, and we get automatically the desired tessellation and the presentation for G (for more details in the case of hyperbolic triangles, see [4]). □

Certainly, G is infinite. The subgroup $G^+ = G \cap \mathrm{Aut}(\mathbb{H})$ has a presentation

$$G^+ = \langle x, y \mid x^p = y^q = (x \circ y)^r = 1 \rangle$$

with $x = \alpha \circ \beta$ and $y = \beta \circ \gamma$. The elements x and y are (hyperbolic) rotations with rotation angles $\frac{2\pi}{p}$ and $\frac{2\pi}{q}$, respectively (see Figure 7.7).

The subgroup G^+ is called a *hyperbolic triangle group* and can be realized as a group generated by $A, B \in \mathrm{Aut}(\mathbb{H})$ with

$$A : z \mapsto -\frac{1}{z + 2\cos(\frac{\pi}{p})} \quad \text{and} \quad B : z \mapsto -\frac{\rho^{-1}}{\rho z + 2\cos(\frac{\pi}{q})} \quad \text{with } \rho + \rho^{-1} = -2\cos\left(\frac{\pi}{r}\right).$$

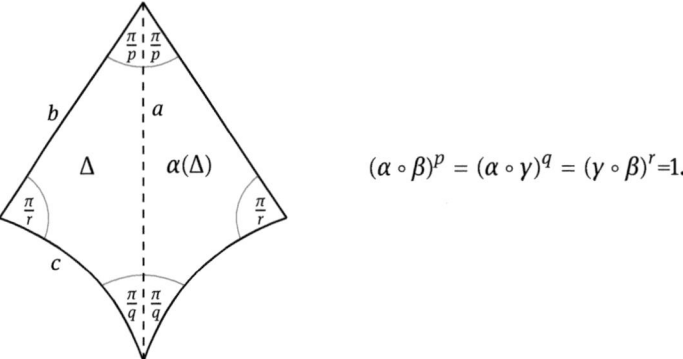

$$(\alpha \circ \beta)^p = (\alpha \circ \gamma)^q = (\gamma \circ \beta)^r = 1.$$

Figure 7.7: $\Delta' = \Delta \cup \alpha(\Delta)$ for a hyperbolic triangle group.

Remark 7.34. We may allow some of the interior angles to be 0, that is, we allow vertices at $\mathbb{R} \cup \{\infty\}$. For instance, the classical modular group

$$\Gamma = \left\{ S \in \mathrm{Aut}(\mathbb{H}) \ \middle| \ S : z \mapsto \frac{az + b}{cz + d}, a, b, c, d \in \mathbb{Z} \text{ and } ad - bc = 1 \right\}$$

is generated by the LFTs

$$T : z \mapsto -\frac{1}{z} \quad \text{and} \quad R : z \mapsto -\frac{1}{z + 1},$$

and has a presentation

$$\Gamma = \langle T, R \mid T^2 = R^3 = 1 \rangle.$$

This can be seen as follows. It is easy to see that Γ is generated by T and $U : z \to z + 1$. Then Γ is also generated by T and R because $U = T \circ R$.

Theorem 7.35. Γ is generated by T and U.

Proof. If $A : z \mapsto \frac{az+b}{cz+d}$ is in Γ and $k \in \mathbb{Z}$ then

$$T \circ A : z \mapsto \frac{-cz - d}{az + b} \quad \text{and} \quad U^k \circ A = \frac{(a + kc)z + b + kd}{cz + d}.$$

We assume that $|c| \leq |a|$ (for if this is not true for A, then we start with $T \circ A$).

If $c = 0$, then $A = U^{q_0}$, for some $q_0 \in \mathbb{Z}$.

If, however, $c \neq 0$, we apply the Euclidean algorithm (see [12]) to a and c (in modified form):

$$a = q_0 c + r_1, \quad -c = q_1 r_1 + r_2, \quad r_1 = q_2 r_2 + r_3, \ldots, (-1)^r r_{n-1} = q_n r_n + 0$$

which ends with $r_n = \pm 1$, since $\gcd(a, c) = 1$. Premultiplying

$$T \circ U^{-q_n} \circ T \circ \cdots \circ T \circ U^{-q_0} \quad \text{by } A,$$

we obtain $U^{q_{n+1}}$ with $q_{n+1} \in \mathbb{Z}$.

Thus we have, the case $|c| > |a|$ included:

$$A = T^m \circ U^{q_0} \circ T \circ U^{q_1} \circ \cdots \circ T \circ U^{q_n} \circ T \circ U^{q_{n+1}}$$

with $m = 0$ or 1, $q_0, q_1, \ldots, q_{n+1} \in \mathbb{Z}$ and $q_0, q_1, \ldots, q_n \neq 0$. $\qquad\square$

We now give a direct proof that Γ has a presentation

$$\Gamma = \langle T, R \mid T^2 = R^3 = 1 \rangle.$$

Theorem 7.36. *The modular group Γ has a presentation*

$$\Gamma = \langle T, R \mid T^2 = R^3 = 1 \rangle$$

with $T : z \mapsto -\frac{1}{z}$ and $R : z \mapsto -\frac{1}{z+1}$.

Proof. Γ is generated by T and R. We know that $R^3 = T^2 = 1 = \text{id}$. We show that these relations are defining relations for Γ. Let $\mathbb{R}^- = \{x \in \mathbb{R} \mid x < 0\}$ and $\mathbb{R}^+ = \{x \in \mathbb{R} \mid x > 0\}$. Then $T(\mathbb{R}^-) \subset \mathbb{R}^+$ and $R^\alpha(\mathbb{R}^+) \subset \mathbb{R}^-$ for $\alpha = 1, 2$.

Let $S \in \Gamma$. Applying the relations $T^2 = R^3 = 1$, we get that $S = 1$ is a consequence of $T^2 = R^3 = 1$, or that $S = R^{\alpha_1} \circ T \circ \cdots \circ R^{\alpha_n} \circ T \circ R^{\alpha_{n+1}}$ with $1 \leq \alpha_i \leq 2$ (eventually after a suitable conjugation). In the latter case, let $x \in \mathbb{R}^+$. Then $S(x) \in \mathbb{R}^-$, so in particular $S \neq 1$. Therefore $\Gamma = \langle T, R \mid T^2 = R^3 = 1 \rangle$. $\qquad\square$

From the proof of Theorem 7.36, we automatically get the following.

Corollary 7.37. *An element of finite order in Γ is either conjugate to T or to a power of R.*

We remark that a suitable triangle Δ which corresponds to Γ has a vertex at ∞, which comes from the fact that we get $z \mapsto z + 1$ for the product $U = T \circ R$, a vertex at i and a vertex at $\frac{1}{2} + \frac{i}{3}\sqrt{3}$. The hyperbolic polygon $\Delta' = \Delta \cup \alpha(\Delta)$ where α is the reflection $z \mapsto -\bar{z}$ at the y-axis looks like that in Figure 7.8.

We have $z_0 = e^{\frac{2\pi i}{3}} = -\frac{1}{2} + \frac{1}{2}\sqrt{3}\,i$, a cubic root of 1.

$\Delta' = \Delta \cup \alpha(\Delta)$ forms a fundamental domain for Γ, that is, if $z_1, z_2 \in \mathbb{H}$ with $|\text{Re}(z_i)| < \frac{1}{2}$ and $|z_i| > 1$ for $i = 1, 2$, then there is no $S \in \Gamma$, $S \neq \text{id}_{\mathbb{H}}$, with $S(z_1) = z_2$.

We complete this chapter with a nice number-theoretical application of the modular group.

Let H be the subgroup of Γ generated by

$$A = T \circ R \circ T \circ R^{-1} \quad \text{and} \quad B = T \circ R^{-1} \circ T \circ R.$$

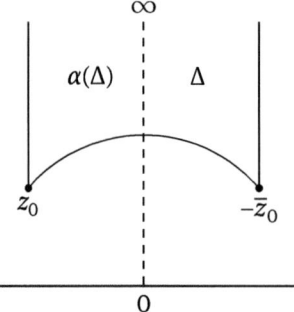

Figure 7.8: Hyperbolic polygon $\Delta' = \Delta \cup \alpha(\Delta)$ with α a reflection $z \mapsto -\bar{z}$ at the y-axis.

Now,

$$T \circ A \circ T = A^{-1}, \quad T \circ B \circ T = B^{-1}, \quad R \circ A \circ R^{-1} = A^{-1} \circ B, \quad R^{-1} \circ A \circ R = B^{-1},$$
$$R \circ B \circ R^{-1} = A^{-1} \quad \text{and} \quad R^{-1} \circ B \circ R = B^{-1} A.$$

Hence H is a normal subgroup of Γ with the factor group $\Gamma/H \cong C_6$, the cyclic group of order 6, and by Theorem 7.36 we get that H has a presentation

$$H = \langle A, B \mid \ \rangle,$$

that is, H is a (so-called) free group of rank 2, freely generated by A and B (see, for instance, [22] or [7]). In fact, H is the commutator subgroup Γ' of Γ, which is generated by all commutators $P \circ S \circ P^{-1} \circ S^{-1}$, $P, S \in \Gamma$. An automorphism of H is an isomorphism $\varphi : H \to H$. The set of automorphisms of H forms a group under composition, the automorphism group $\mathrm{Aut}(H)$ of H; $\mathrm{Aut}(H)$ is generated by the *elementary Nielsen transformations*

$$\alpha : (U, V) \mapsto (V, U),$$
$$\beta : (U, V) \mapsto (V \circ U, U^{-1}) \quad \text{and}$$
$$\gamma : (U, V) \mapsto (V \circ U \circ V, V^{-1}),$$

$\{U, V\}$ a (free) generating pair of H (see, for instance, [22] or [7]). Certainly, if $\{U, V\}$ is a generating pair of H, then the map $A \mapsto U$, $B \mapsto V$ defines an automorphism of H because $H = \langle A, B \mid \ \rangle$. Hence a pair $\{U, V\} \subset H$ is a generating pair of H if and only if there is a sequence of elementary Nielsen transformations from $\{A, B\}$ to $\{U, V\}$. We then say that $\{U, V\}$ is *Nielsen equivalent* to $\{A, B\}$.

If $P, S \in H$, let $[P, S] := P \circ S \circ P^{-1} \circ S^{-1}$ be the commutator of P and S. Since

$$[S, P] = [P, S]^{-1}, \quad [S \circ P, P^{-1}] = P^{-1} \circ [P, S] \circ P, \quad \text{and} \quad [S \circ P \circ S, S^{-1}] = [P, S]^{-1},$$

we get the following.

If $\{U, V\}$ is a generating pair of H then $[U, V] = W \circ [A, B] \circ W^{-1}$ for some $W \in H$. In fact, a pair $\{U, V\} \subset H$ is a generating pair of H if and only if

$$[U, V] = W \circ [A, B]^{\pm 1} \circ W^{-1}$$

for some $W \in H$ (see, for instance, [22] or [7]).

We now come back to our realization of $H = \langle A, B \mid \ \rangle$ as a subgroup of $\mathrm{Aut}(\mathbb{H})$. As mentioned above, $\mathrm{Aut}(\mathbb{H}) \cong \mathrm{PSL}(2, \mathbb{R})$. In this sense $\Gamma \cong \mathrm{PSL}(2, \mathbb{Z})$, and we identify Γ with $\mathrm{PSL}(2, \mathbb{Z})$.

Then $T = \pm \left(\begin{smallmatrix} 0 & 1 \\ -1 & 0 \end{smallmatrix} \right)$ and $R = \pm \left(\begin{smallmatrix} 0 & -1 \\ 1 & 1 \end{smallmatrix} \right)$, and we get $A = \pm \left(\begin{smallmatrix} 2 & 1 \\ 1 & 1 \end{smallmatrix} \right)$, $B = \pm \left(\begin{smallmatrix} 1 & 1 \\ 1 & 2 \end{smallmatrix} \right)$ and $AB^{-1} = \pm \left(\begin{smallmatrix} 3 & -1 \\ 1 & 0 \end{smallmatrix} \right)$.

We consider the subgroup $\tilde{H}$ of $\mathrm{SL}(2, \mathbb{Z})$, generated by $\left(\begin{smallmatrix} 2 & 1 \\ 1 & 1 \end{smallmatrix} \right)$ and $\left(\begin{smallmatrix} 1 & 1 \\ 1 & 2 \end{smallmatrix} \right)$. Without any misunderstanding, we also just write

$$A = \begin{pmatrix} 2 & 1 \\ 1 & 1 \end{pmatrix} \quad \text{and} \quad B = \begin{pmatrix} 1 & 1 \\ 1 & 2 \end{pmatrix}$$

for the corresponding matrices.

Since the subgroup H of Γ contains only products of commutators of Γ, we also have that $\tilde{H}$ has a presentation $\tilde{H} = \langle A, B \mid \ \rangle$, that is, also $\tilde{H}$ is a free group of rank 2. As constructed, $A = \left(\begin{smallmatrix} 2 & 1 \\ 1 & 1 \end{smallmatrix} \right)$, $B = \left(\begin{smallmatrix} 1 & 1 \\ 1 & 2 \end{smallmatrix} \right)$ and $AB^{-1} = \left(\begin{smallmatrix} 3 & -1 \\ 1 & 0 \end{smallmatrix} \right)$, and $\mathrm{tr}(A) = \mathrm{tr}(B) = \mathrm{tr}(AB^{-1}) = 3$. This means especially that $(\mathrm{tr}(A), \mathrm{tr}(B), \mathrm{tr}(AB^{-1}))$ is a solution of the diophantine equation $x^2 + y^2 + z^2 - xyz = 0$ because $9 + 9 + 9 - 27 = 0$.

Now let $\{U, V\}$ be a generating pair of $\tilde{H}$. Then $[U, V] = W[A, B]^{\pm 1} W^{-1}$ for some $W \in \tilde{H}$. Hence, $\mathrm{tr}([U, V]) = \mathrm{tr}([A, B]) = -2$. From the trace formula

$$\mathrm{tr}([U, V]) = \left(\mathrm{tr}(U) \right)^2 + \left(\mathrm{tr}(V) \right)^2 + \left(\mathrm{tr}(UV) \right)^2 - \mathrm{tr}(U) \cdot \mathrm{tr}(V) \cdot \mathrm{tr}(UV) - 2,$$

which is easy to compute, we get that also $(\mathrm{tr}(U), \mathrm{tr}(V), \mathrm{tr}(UV^{-1}))$ is a solution of the diophantine equation $x^2 + y^2 + z^2 - xyz = 0$.

We define

$$E = \{\{U, V\} \mid \{U, V\} \text{ is Nielsen equivalent to } \{A, B\}\}$$

and

$$L = \{(\mathrm{tr}(U), \mathrm{tr}(V), \mathrm{tr}(UV)) \mid \{U, V\} \in E\}.$$

Since $\tilde{H} = \langle U, V \rangle$ and $\mathrm{tr}([U, V]) = \mathrm{tr}([A, B])$ for all $\{U, V\} \in E$, the ternary form

$$F(x, y, z) = x^2 + y^2 + z^2 - xyz$$

is invariant under the automorphism of $\tilde{H}$. As mentioned, the automorphism group of $\tilde{H}$ is generated by the elementary Nielsen transformations

$$\alpha : (U, V) \mapsto (V, U),$$
$$\beta : (U, V) \mapsto (VU, U^{-1}) \quad \text{and}$$
$$\gamma : (U, V) \mapsto (VUV, V^{-1}).$$

These induce birational transformations of L by

$$\varphi : (x, y, z) \mapsto (y, x, z),$$
$$\omega : (x, y, z) \mapsto (z, x, y) \quad \text{and}$$
$$\psi : (x, y, z) \mapsto (x', y, z) \quad \text{with } x' = yz - x.$$

Let M be the permutation group of L generated by φ, ω and ψ.

Theorem 7.38. *Group M has a presentation*

$$M = \langle \varphi, \omega, \psi \mid \varphi^2 = \omega^3 = \psi^2 = (\varphi \circ \omega)^2 = (\psi \circ \varphi \circ \omega)^2 = 1 \rangle.$$

Proof. All relations in the presentation are evident by definition of φ, ω and ψ. We now show that these relations form a complete set of relations. We let $\psi_0 = \psi$, $\psi_1 = \omega \circ \psi \circ \omega^{-1}$ and $\psi_2 = \omega^{-1} \circ \psi \circ \omega$. We get

$$\psi_1(x, y, z) = (x, y', z) \quad \text{with } y' = xz - y$$

and

$$\psi_2(x, y, z) = (x, y, z') \quad \text{with } z' = xy - z.$$

Assume that $r = r(\varphi, \omega, \psi) = 1$ is an additional relation which is independent of the given relations for φ, ω and ψ, and in which ψ occurs. If we apply the given relations for φ, ω and ψ, we may write r as $r = y \circ \psi_{r_m} \circ \cdots \circ \psi_{r_1} = 1$ with $y \in \langle \varphi, \omega \rangle$, which is isomorphic to the permutation group S_3, $m \geq 1$, $r_j \in \{0, 1, 2\}$ for $j = 1, 2, \ldots, m$ and $r_i \neq r_{i-1}$ if $m > 1$, for $i = 2, 3, \ldots, m$ (otherwise there is a cancellation possible).

For $(x, y, z) \in L$ we define the height $h(x, y, z) = x + y + z$. Since $x^2 + y^2 + z^2 - xyz = 0$, we have $2 < x, y, z$ for $(x, y, z) \in L$. We choose $(x, y, z) \in L$ so that the components of (x, y, z) are pairwise distinct and that the $(r_1 + 1)$th component of (x, y, z) is not the biggest component (this is certainly possible after an application of φ, ω, ψ).

Then we get

$$h(\psi_{r_1}(x, y, z)) > h(x, y, z),$$

the components of $\psi_{r_1}(x, y, z)$ are pairwise distinct, and the $(r_1 + 1)$th component of $\psi_{r_1}(x, y, z)$ is here the biggest component. This is clear because if, for instance, $x < y$ then $x' = yz - x > x, y, z$.

If $m > 1$ then we get inductively that

$$h(y \circ \psi_{r_m} \circ \cdots \circ \psi_{r_1}(x, y, z)) > h(x, y, z),$$

which contradicts $y \circ \psi_{r_m} \circ \cdots \circ \psi_{r_1} = 1$.

This proves Theorem 7.38. $\qquad\square$

Corollary 7.39.

$$M \cong \mathrm{PGL}(2, \mathbb{Z}) = \left\{ \pm \begin{pmatrix} a & b \\ c & d \end{pmatrix} \,\middle|\, \begin{pmatrix} a & b \\ c & d \end{pmatrix} \in \mathrm{GL}(2, \mathbb{Z}) \right\}$$

and

$$M^0 \cong \mathrm{PSL}(2, \mathbb{Z}) \cong \Gamma,$$

where M^0 is the subgroup of M generated by ω and $\rho = \psi \circ \varphi \circ \omega$.

Proof. If we extend $\mathrm{PSL}(2, \mathbb{Z})$ by the element $Z = \pm \begin{pmatrix} -1 & 0 \\ 0 & 1 \end{pmatrix}$, we get easily the presentation

$$\langle X, Y, Z \mid X^2 = Y^3 = Z^2 = (XY)^2 = (ZXY)^2 = 1 \rangle$$

for $\mathrm{PGL}(2, \mathbb{Z})$, where

$$X = \pm \begin{pmatrix} -1 & -1 \\ 0 & 1 \end{pmatrix}, \quad Y = \pm \begin{pmatrix} -1 & -1 \\ 1 & 0 \end{pmatrix}, \quad Z = \pm \begin{pmatrix} -1 & 0 \\ 0 & 1 \end{pmatrix}.$$

The mapping

$$\varphi \mapsto X, \quad \omega \mapsto Y, \quad \psi \mapsto Z$$

defines an isomorphism between M and $\mathrm{PGL}(2, \mathbb{Z})$. Then certainly

$$M^0 \cong \mathrm{PSL}(2, \mathbb{Z}) \cong \Gamma. \qquad\square$$

We now describe the announced number-theoretical application of the modular group Γ.

Theorem 7.40. *The natural numbers x, y, z are solutions of the equation $x^2 + y^2 + z^2 - xyz = 0$ if and only if there exists a generating pair $\{U, V\}$ of $\tilde{H}$ with $\mathrm{tr}(U) = x$, $\mathrm{tr}(V) = y$ and $\mathrm{tr}(UV) = z$.*

Proof. Let $x, y, z \in \mathbb{N}$ with $x^2 + y^2 + z^2 - xyz = 0$. Certainly, $2 < x, y, z$. Let

$$L' = \{(x, y, z) \in \mathbb{N}^3 \mid x^2 + y^2 + z^2 - xyz = 0\}.$$

We have to show that $L = L'$. Starting with $(x, y, z) \in L'$, we may apply the above defined transformations φ, ω and ψ also to L', and we get ongoing new triples from L'. We apply φ, ω and ψ in a minimizing manner. With this we get a triple $(x, y, z) \in L'$ with $2 < x \le y \le z$ and $h(x, y, z) = x + y + z$ minimal for all triples from L'.

With this we get $z \le xy - z$. Hence altogether we get

$$2 < x \le y \le z \le \frac{xy}{2}.$$

Since $x \le y \le z$, we also have $\frac{1}{3} \le z$. Hence $\frac{1}{3}xy \le z \le \frac{1}{2}xy$, that is, also $\frac{1}{2}xy - z \le \frac{1}{6}xy$. From this we get

$$0 = x^2 + y^2 + z^2 - xyz = x^2 + y^2 + \left(\frac{1}{2}xy - z\right)^2 - \left(\frac{1}{2}xy\right)^2$$

$$\le 2y^2\left(1 - \frac{1}{9}x^2\right).$$

This gives $x = 3$, and further $y \le z = \frac{3}{2}y - \frac{1}{2}\sqrt{5y^2 - 36}$, that is, $y^2 \le 9$, and therefore also $y = 3$, and finally, $z = 3$. But for this solution we have the above matrices $A, B^{-1} \in \tilde{H}$ with $\mathrm{tr}(A) = \mathrm{tr}(B^{-1}) = \mathrm{tr}(AB^{-1}) = 3$. This gives $L = L'$ from the preparing considerations. $\square$

Exercises

1. Prove Theorem 7.11.
2. Verify statement (1) in Remark 7.13.
3. Give a proof of Corollary 7.14 along the lines in Remark 7.15.
4. Let $f \in \mathrm{Aut}(\mathbb{H})$, $f(z) = \frac{az+b}{cz+d}$, $z \in \mathbb{H}$.
   (a) Show that

   $$\frac{|f'(z)|}{\mathrm{Im}(f(z))} = \frac{1}{\mathrm{Im}(z)}.$$

   (b) Let $\gamma : [0, 1] \to \mathbb{H}$ be a continuously differentiable curve. Show that

   $$L_h(f \circ \gamma) = L_h(\gamma).$$

5. Let $f, g \in \mathrm{Aut}(\mathbb{H})$, $f \ne \mathrm{id}_{\mathbb{H}} \ne g$. Show that $f \circ g = g \circ f$ if and only if f and g have the same fixed points.

6. Show that
   (a) $\cosh(x_1 + x_2) = \cosh(x_1) \cdot \cosh(x_2) + \sinh(x_1) \cdot \sinh(x_2)$ for all $x_1, x_2 \in \mathbb{R}$.
   (b) $\cosh(x_1 - x_2) = \cosh(x_1) \cdot \cosh(x_2) - \sinh(x_1) \cdot \sinh(x_2)$ for all $x_1, x_2 \in \mathbb{R}$.

7. (a) Consider $z_1, z_2 \in \mathbb{H}$, $z_1 \neq z_2$. Describe the construction of the hyperbolic perpendicular bisector $\{z \in \mathbb{H} \mid \delta(z_1, z) = \delta(z_2, z)\}$ of the line segment $\overline{z_1 z_2}$.
   (b) Let $z_0 \in \mathbb{H}$. The hyperbolic circle with center z_0 and hyperbolic radius $r \in \mathbb{R}$, $r \geq 0$, is defined as

$$C_h = \{z \in \mathbb{H} \mid \delta(z_0, z) = r\}.$$

   Show that the hyperbolic circles are Euclidean circles, possibly with a different center. Give an example where the centers are equal and an example where the centers are different.

8. Consider two hyperbolic lines ℓ_1, ℓ_2 and complex numbers $z, z_1, z_2, s_1, s_2, s_1', s_2'$ as in Figure 7.9.

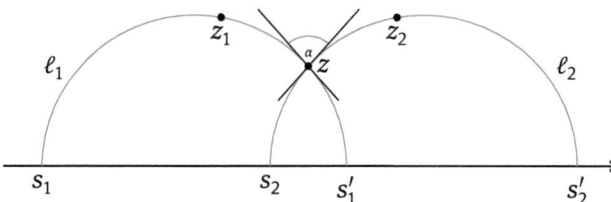

Figure 7.9: Hyperbolic lines ℓ_1, ℓ_2 and complex numbers $z, z_1, z_2, s_1, s_2, s_1', s_2'$.

Let α be the angle between the hyperbolic line segments $\overline{z z_1}$ and $\overline{z z_2}$. Show that

$$\cos(\alpha) = \frac{(s_1 - s_2')(s_1' - s_2) + (s_1 - s_2)(s_1' - s_2')}{(s_1 - s_1')(s_2 - s_2')}.$$

(*Hint*: Apply the Euclidean Cosine rule in a suitable manner.)

9. Given a hyperbolic triangle with endpoints z_1, z_2, z_3, line segments $\overline{z_1 z_2}$, $\overline{z_1 z_3}$, $\overline{z_2 z_3}$ and opposite angles α, β, γ, respectively.
   Let $a = \delta(z_1, z_2)$, $b = \delta(z_2, z_3)$ and $c = \delta(z_2, z_3)$.
   Show the
   (a) rule of sine:

$$\frac{\sinh(a)}{\sin(\alpha)} = \frac{\sinh(b)}{\sin(\beta)} = \frac{\sinh(c)}{\sin(\gamma)},$$

   (b) first cosine rule:

$$\cosh(c) = \cosh(a) \cdot \cosh(b) - \sinh(a) \cdot \sinh(b) \cos(\gamma),$$

(c) second cosine rule:

$$\cos(\gamma) = -\cos(\alpha) \cdot \cos(\beta) + \sin(\alpha) \cdot \sin(\beta) \cdot \cosh(c).$$

(*Hint*: Let the triangle be given with vertices as in Figure 7.10.

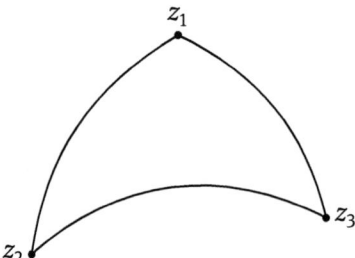

Figure 7.10: Hyperbolic triangle with endpoints z_1, z_2, z_3.

Consider the two rectangular triangles which we get if we draw the line through z_1 orthogonal to the line $z_2 z_3$.)

10. The hyperbolic congruence axiom 4 says that two hyperbolic triangles $z_1 z_2 z_3$ and $w_1 w_2 w_3$ are congruent if

$$\delta(z_1, z_2) = \delta(w_1, w_2), \quad \delta(z_1, z_3) = \delta(w_1, w_3) \quad \text{and} \quad \sphericalangle(\overline{z_1 z_2}, \overline{z_1 z_3}) = \sphericalangle(\overline{w_1 w_2}, \overline{w_1 w_3}).$$

Show that two hyperbolic triangles are congruent if they coincide in all three angles.

11. (a) Let $\alpha : \mathbb{H} \to \mathbb{H}$, $\alpha(z) = -\bar{z} = -x + iy$, if $z = x + iy$, be the reflection at the y-axis. Show that

$$\delta(\alpha(z_1), \alpha(z_2)) = \delta(z_1, z_2) \quad \text{for all } z_1, z_2 \in \mathbb{H}.$$

(b) Show that $\widetilde{\text{Aut}}(\mathbb{H})$ is composed of the maps

$$z \mapsto \frac{az + b}{cz + d}, \quad a, b, c, d \in \mathbb{R}, \, ad - bc = 1$$

and

$$z \mapsto \frac{a\bar{z} + b}{c\bar{z} + d}, \quad a, b, c, d \in \mathbb{R}, \, ad - bc = -1.$$

12. The Hecke group $G(q)$, $q \in \mathbb{N}$ with $q > 2$, is a subgroup of $\text{Aut}(\mathbb{H})$, which is generated by the two linear fractional transformations

$$T : z \mapsto -\frac{1}{z} \quad \text{and} \quad U(q) : z \mapsto z + 2\cos\left(\frac{\pi}{q}\right).$$

It is named after E. Hecke (1887–1947).

Show that $G(q)$ has a presentation

$$G(q) = \langle T, R(q) \mid T^2 = R(q)^q = 1 \rangle,$$

with $R(q) = TU(q)$.

(*Hint*: Proceed as in the proof of Theorem 7.36.)

13. Let K be a field and $X, Y \in \mathrm{SL}(2, K)$. Show the trace identities:

(a) $\mathrm{tr}(XY^{-1}) = \mathrm{tr}(X) \cdot \mathrm{tr}(Y) - \mathrm{tr}(XY)$,

(b) $\mathrm{tr}([X, Y]) = (\mathrm{tr}(X))^2 + (\mathrm{tr}(Y))^2 + (\mathrm{tr}(XY))^2 - \mathrm{tr}(X) \cdot \mathrm{tr}(Y) \cdot \mathrm{tr}(XY) - 2$.

14. Let $X = \pm \left(\begin{smallmatrix} -1 & -1 \\ 0 & 1 \end{smallmatrix} \right)$, $Y = \pm \left(\begin{smallmatrix} -1 & -1 \\ 1 & 0 \end{smallmatrix} \right)$ and $Z = \pm \left(\begin{smallmatrix} -1 & 0 \\ 0 & 1 \end{smallmatrix} \right)$.

Show that the PGL$(2, \mathbb{Z})$ is generated by X, Y and Z and has the presentation

$$\langle X, Y, Z \mid X^2 = Y^3 = Z^2 = (XY)^2 = (ZXY)^2 = 1 \rangle.$$

(*Hint*: Use that PGL$(2, \mathbb{Z}) = \mathrm{PSL}(2, \mathbb{Z}) \cup X\,\mathrm{PSL}(2, \mathbb{Z})$ and the known presentation for PSL$(2, \mathbb{Z}) = \langle X, Y \mid X^2 = Y^3 = 1 \rangle \cong \Gamma$.)

8 Simplicial Complexes and Topological Data Analysis

In Chapter 5 we have discussed a principle to model data in the form of objects and relations between them as the vertices and edges of a graph, respectively. In this chapter we aim at higher-dimensional data, that is, for instance, a point cloud in some $\mathbb{R}^n$. By nature of things, this data is very complex. Sometimes one is only interested in the 'shape' of the data. This boils down the task to a purely topological question. Topology is a rather young branch of mathematics with rich applications in pure and applied mathematics. Roughly speaking, in topology we only investigate features of geometric figures that do not change under continuous deformation, that is, no gluing and cutting is allowed. However, we will not give an introduction to topology here; instead our exposition will be laid out on a purely combinatorial level. We use methods from Combinatorial Algebraic Topology and we apply them to give a little insight in the emerging field of Topological Data Analysis.

8.1 Simplicial Complexes

For standard references concerning the following sections we refer to the classical [23] as well as to the recent [20]. We will start with the combinatorial definition of a simplicial complex that can be understood as a higher-dimensional analogue of a graph.

Definition 8.1. An *abstract simplicial complex* on a finite set V is a collection K of subsets of V such that whenever $\sigma \in K$ and $\tau \subset \sigma$, then also $\tau \in K$.

We sometimes leave the ground set V implicit. We call the elements $\sigma \in K$ with $|\sigma| = n + 1$ the *n-simplices*. If $\tau \subset \sigma$ and $|\tau| = k + 1$ we call τ a *k-face* of σ. We denote the set of n-simplices of K by $K(n)$. In particular, if $|\sigma| = 4$, $|\sigma| = 3$, $|\sigma| = 2$, or $|\sigma| = 1$, we call σ a tetrahedron, a triangle, an edge or a vertex, respectively. We will see in the following that there is a geometric intuition for these notions but first we will give some examples of abstract simplicial complexes.

Example 8.2. The following collections of subsets are simplicial complexes on $[n] = \{0, 1, \ldots, n\}$ for a suitable n:
1. $K_1 = \{\{0\}, \{1\}, \{2\}, \emptyset\}$,
2. $K_2 = \{\{0, 1, 2\}, \{0, 1\}, \{1, 2\}, \{0, 2\}, \{1, 3\}, \{0, 3\}, \{0\}, \{1\}, \{2\}, \{3\}, \emptyset\}$, and
3. $\Delta^n = \{A \subset [n]\}$, the standard n-simplex.

Recall that we have implicitly visualized graphs by assigning a point in some $\mathbb{R}^n$ to each node and connecting two adjacent points by a line segment between them. This procedure is called *geometric realization*. We will describe a standard procedure to do this in higher dimensions.

https://doi.org/10.1515/9783110740783-008

To this end, we discuss the concepts of affine independence and convex hulls. Recall that vectors $v_1, v_2, \ldots, v_n \in \mathbb{R}^n$ are linear independent if for $\lambda_1, \lambda_2, \ldots, \lambda_n \in \mathbb{R}$ the equation $\lambda_1 v_1 + \cdots + \lambda_n v_n = 0$ implies $\lambda_1 = \lambda_2 = \cdots = \lambda_n = 0$. We now define a stronger concept.

Definition 8.3. Vectors $v_0, \ldots, v_n \in \mathbb{R}^n$ are *affinely independent* if for $\lambda_0, \ldots, \lambda_n \in \mathbb{R}$ and $\sum_{i=0}^n \lambda_i = 0$ the equation $\lambda_0 v_0 + \cdots + \lambda_n v_n = 0$ implies $\lambda_0 = \cdots = \lambda_n = 0$.

We give the following characterization of affinely independent vectors.

Lemma 8.4. *Let $v_0, v_1, \ldots, v_n \in \mathbb{R}^n$. Then $v_0, v_1, \ldots, v_n$ are affinely independent if and only if $v_1 - v_0, \ldots, v_n - v_0$ are linearly independent.*

Proof. First assume that $v_0, \ldots, v_n$ are affinely independent and that $\sum_{i=1}^n \lambda_i (v_i - v_0) = 0$ for $\lambda_1, \ldots, \lambda_n \in \mathbb{R}$. Then $\sum_{i=1}^n \lambda_i =: -\lambda_0 \in \mathbb{R}$ and $\sum_{i=0}^n \lambda_i = 0$. We have

$$\sum_{i=0}^n \lambda_i v_i = \sum_{i=1}^n \lambda_i (v_i - v_0) + \left(\sum_{i=0}^n \lambda_i \right) v_0 = 0$$

and hence, as $v_0, \ldots, v_n$ are affinely independent, it follows that $\lambda_i = 0$ for all $0 \leq i \leq n$.

Conversely, assume that $v_1 - v_0, \ldots, v_n - v_0$ are linearly independent and that $\sum_{i=0}^n \lambda_i v_i = 0$ with $\lambda_0, \ldots, \lambda_n \in \mathbb{R}$ and $\sum_{i=0}^n \lambda_i = 0$. Then also $(\sum_{i=0}^n \lambda_i) v_0 = 0$ and $\sum_{i=1}^n \lambda_i (v_i - v_0) = 0$ by the above equation. Hence $\lambda_i = 0$ for all $1 \leq i \leq n$ by linear independence and thus also $\lambda_0 = 0$. $\square$

We now consider the *convex hull* of vectors.

Definition 8.5. Let $S = \{v_0, \ldots, v_n\} \subset \mathbb{R}^n$. Then their convex hull is defined as

$$conv(S) := conv(v_0, \ldots, v_n) = \left\{ \sum_{i=0}^n \lambda_i v_i \,\middle|\, \lambda_i \in \mathbb{R}, \sum_{i=0}^n \lambda_i = 1, \lambda_i \geq 0 \right\}.$$

This, together with the above characterization gives us the following geometric intuition: If vectors $v_0, \ldots, v_n$ are not affinely independent but $v_1, \ldots, v_n$ are, then we have for their convex hulls $conv(v_0, \ldots, v_n) = conv(v_1, \ldots, v_n)$. Hence, we see that in $\mathbb{R}^N$, N large enough, one vector is affinely independent, two distinct vectors are affinely independent, the convex hull of three vectors needs to constitute a triangle and the convex hull of four vectors needs to give a tetrahedron.

This is why we work with affinely independent vectors in the following in order to have convex hulls uniquely defined by their spanning vectors. We now give a canonical recipe to geometrically realize abstract simplicial complexes.

Definition 8.6 (Canonical geometric realization). Given an abstract simplicial complex K on some set V with $K(0) = \{v_1, \ldots v_n\}$. Then its *canonical geometric realization* is the union of $\{e_1, \ldots e_n\}$, where e_i denotes the ith standard vector in $\mathbb{R}^{|K(0)|}$, and $\{conv(e_{i_0}, \ldots, e_{i_k}) \mid \{v_{i_0}, \ldots, v_{i_k}\} \in K\}$.

By abuse of notation, for an abstract simplicial complex K we will denote its geometric realization also by K. It will always be clear from the context if we consider the abstract or geometric versions. We get the pictures in Figure 8.1 for the geometric realization of the low-dimensional standard simplices Δ^0, Δ^1, and Δ^2.

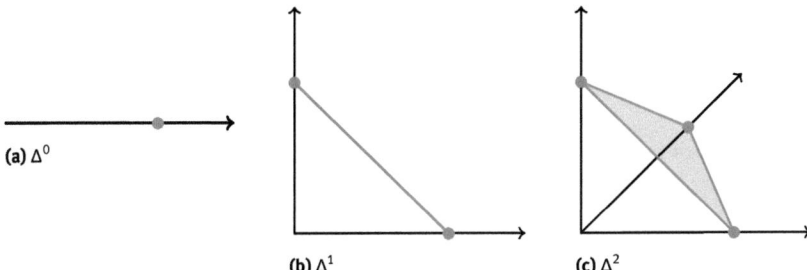

(a) Δ^0

(b) Δ^1 (c) Δ^2

Figure 8.1: Canonical geometric realizations of the first three standard simplices Δ^0, Δ^1, and Δ^2.

We already see that this canonical construction does not yield a minimal embedding dimension as the examples could be visualized in lower dimensions. We will address this issue below but we note that the canonical geometric realization of the standard simplices and of a general complex K are instances of geometric simplices and geometric simplicial complexes, respectively.

Definition 8.7.
1. A *geometric n-simplex* σ in $\mathbb{R}^N$ is the convex hull $conv(v_0, \ldots, v_n)$ of affinely independent vectors $v_0, \ldots, v_n \in \mathbb{R}^N$ with $n \leq N$.
2. A *k-face* τ of a geometric n-simplex $\sigma = conv(v_0, \ldots, v_n)$ is the convex hull $\tau = conv(S)$ with $S \subset \{v_0, \ldots, v_n\}$ and $|S| = k + 1$.
3. A *geometric simplicial complex L* in $\mathbb{R}^N$ is a collection of geometric simplices in $\mathbb{R}^N$ such that
    (a) every face of a simplex is also a simplex and
    (b) the intersection of two simplices is a face of each of them.
4. For a geometric simplicial complex L we define its *polyhedron* as $|L| = \cup_{\sigma \in L} \sigma$.

To get some intuition we give non-examples of simplicial complexes in Figure 8.2.

We have seen how to obtain a geometric simplicial complex from an abstract one. We would now like to reverse this process and define the vertex scheme of a geometric simplicial complex.

Definition 8.8. Given a geometric simplicial complex $L = \{\sigma_1, \ldots, \sigma_n\}$, we then define its *vertex scheme* as $\{\{v_0, v_1, \ldots, v_k\} \mid conv(v_0, v_1, \ldots, v_k) \in L\}$.

With this notion we can give the general definition of geometric realizations.

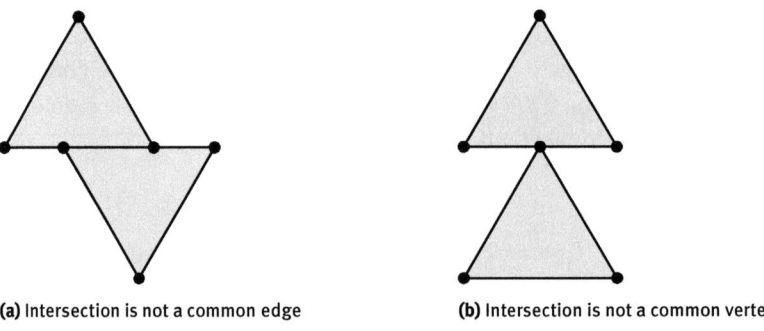

(a) Intersection is not a common edge   **(b)** Intersection is not a common vertex

Figure 8.2: Non-examples of geometric simplicial complexes.

Definition 8.9. Let L be a geometric simplicial complex and K be an abstract simplicial complex. Then L is a *geometric realization* of K if K is the vertex scheme of L.

We would like to recall Example 8.2.1 and 8.2.2. In this case, the canonical geometric realization would yield geometric realizations in $\mathbb{R}^3$ and $\mathbb{R}^4$. However, in Figure 8.3 we see that both can be realized in $\mathbb{R}^2$.

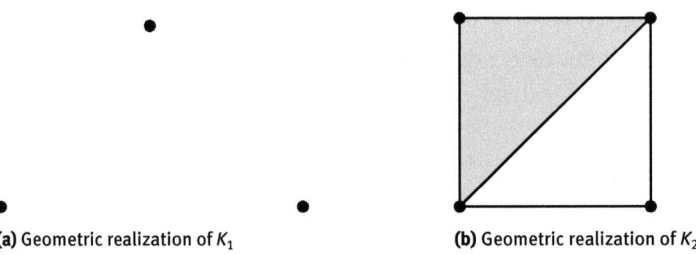

(a) Geometric realization of K_1 **(b)** Geometric realization of K_2

Figure 8.3: Geometric realizations of Examples 8.2.1 and 8.2.2.

We end this section with the following remark.

Remark 8.10. Any abstract simplicial complex K of dimension d, that is,

$$\max\{|\sigma| \mid \sigma \in K\} = d + 1,$$

can be embedded into $\mathbb{R}^{2d+1}$, see [22]. This bound is sharp in the sense that there exist abstract simplicial complexes of dimension d that cannot be geometrically realized in $\mathbb{R}^{2d}$. More explicitly, the latter assertion is known as the van Kampen–Flores theorem (first version proved 1932 by E. Kampen and independently 1933 by A. Flores) which states that the d-skeleton of a $(2d + 2)$-simplex, that is,

$$\mathrm{sk}_d \Delta^n = \{\sigma \subset [n] \mid |\sigma| \leq d + 1\},$$

cannot be geometrically realized in $\mathbb{R}^{2d}$.

8.2 Sperner's Lemma

We do a little detour and present a beautiful lemma that is broadly used in important proofs in Combinatorial Topology. In order to formulate the lemma we need some additional concepts concerning simplicial complexes.

We introduce the subdivision of the (geometric) standard simplex.

Definition 8.11. Let Δ^n denote a geometric realization of the standard n-simplex and let L be a geometric simplicial complex. Then L is called a *subdivision* of Δ^n if $|L| = \Delta^n$.

In Figure 8.4 we present two-dimensional instances of the barycentric subdivision that plays an important role in the simplicial approximation theorem in topology, and its chromatic analogue, the standard chromatic subdivision, that occurs within the context of distributed computing as a protocol complex and that is a subject of current research. In order to give a general definition of the barycentric subdivision, and for further reference, we would like to take the opportunity to introduce the face poset of a simplicial complex and the notion of a flag complex.

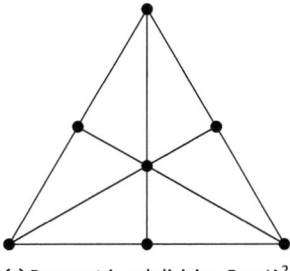
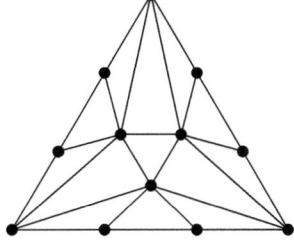

(a) Barycentric subdivision *Bary*(Δ^2) **(b)** Chromatic subdivision $\chi(\Delta^2)$

Figure 8.4: Examples of subdivisions.

Definition 8.12.
1. Let K be an abstract simplicial complex. Then we define $P(K)$ to be the partially ordered set with elements being the simplices and the order is induced by the inclusion relation. We call $P(K)$ the poset of K.
2. If P is a partially ordered set, we define $F(P)$ to be the abstract simplicial complex with vertex set P, and a set of vertices $\{v_0, \ldots, v_n\}$ constitutes an n-simplex if and only if $v_0 < \cdots < v_n$ with respect to the partial order.

Observe that $P(K)$ always contains $\emptyset$ if $K \neq \emptyset$. We will later see examples of face posets of abstract simplicial complexes when we take a look at discrete Morse theory, named after M. Morse (1892–1977). For now we will just give the following general definition of the barycentric subdivision of a simplex.

Definition 8.13. The barycentric subdivision $Bary(\Delta^n)$ of the standard simplex Δ^n is a geometric realization of the abstract simplicial complex $F(P(\Delta^n))$.

For a general combinatorial definition and a proof of the subdivision property of the chromatic subdivision we refer the reader to [19].

We now discuss a Sperner labeling for a subdivision of the standard simplex, named after E. Sperner (1905–1980).

Definition 8.14. Let L be a subdivision of Δ^n. A *Sperner labeling* is a function $f: L(0) \to [n]$ that assigns to each $x \in |L|$ with $x \in conv(S)$, $S \subset [n]$, a value in S. We call $f(x)$ the color of x.

In particular, the vertices of Δ^n have distinct labels and any vertex in a face gets labelled with the spanning vertices of that face. *Sperner's lemma* is the following:

Lemma 8.15 (Sperner's lemma). *Let L be a subdivision of Δ^n and $f: L(0) \to [n]$ be a Sperner labeling. Then there exists an odd number of simplices $\sigma \in L$, and at least one, with $f(\sigma) = [n]$.*

Proof. We give a direct combinatorial proof of the lemma in the general case by induction over n. The case $n = 1$ is immediate: If we consider a subdivided interval with the endpoints labeled 0 and 1 we must switch the labeling at least once and obtain an odd number of colored edges.

We now assume that the lemma holds for $n - 1$ and consider a Sperner labeling f of L. Let C denote the set of colored simplices and D denote the set of the simplices that admit all colors but n. We let A be the set of $(n - 1)$-dimensional colored faces in the interior of L and B their analogue in the boundary. We set $a = |A|, b = |B|, c = |C|$, and $d = |D|$. Observe that each $\sigma \in D$ contributes two elements to $A \cup B$ and each $\sigma \in C$ exactly one. However, elements $\sigma \in A$ are the faces of either two simplices $\tau, \tau' \in D$ or of simplices $\tau \in C$ and $\tau' \in D$. Either way they get counted twice. Hence we obtain

$$c + 2d = b + 2a.$$

But b is an odd number by the induction hypothesis and so is c. This shows the assertion for all dimensions. □

In Figure 8.5 we demonstrate the lemma with respect to different Sperner labelings on our subdivisions. For the Sperner labeling f_3 we see that all triangles obtain all colors. Observe that this is not possible for $Bary(\Delta^2)$.

We have already mentioned that Sperner's lemma has rich applications in mathematics. We end this section by presenting two of them. First, we state that Sperner's lemma is equivalent to a famous theorem by L. E. J. Brouwer (1881–1966):

Theorem 8.16 (Brouwer fixed point theorem). *Any continuous map $f: D^n \to D^n$ has a fixed point. Here $D^n = \{x \in \mathbb{R}^n \mid \|x\| \leq 1\}$.*

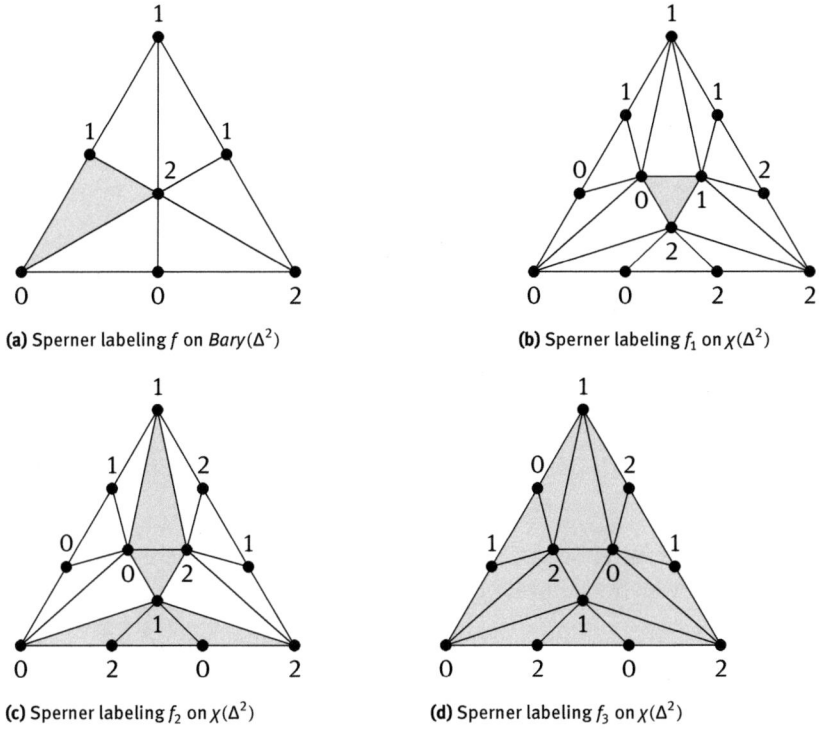

(a) Sperner labeling f on $Bary(\Delta^2)$

(b) Sperner labeling f_1 on $\chi(\Delta^2)$

(c) Sperner labeling f_2 on $\chi(\Delta^2)$

(d) Sperner labeling f_3 on $\chi(\Delta^2)$

Figure 8.5: Examples of Sperner labelings f, f_1, f_2, f_3.

The second application we would like to give here is the board game Hex, see Figure 8.6, for two players. Here, in each turn a player places a stone in his or her color on an arbitrary hexagon upon the board. The game ends if all stones are placed and/or a player wins if he or she obtains a connected path from his or her side to the opposite side. The hexagons of the four corners belong to both players.

Theorem 8.17. *The game of Hex always has a winner.*

We would also like to hint at [15] that shows how one can prove Hall's Marriage Theorem 5.35 using Sperner's lemma. See also [6] for the other direction.

8.3 Simplicial Homology

We now discuss an algebraic tool to determine the shape of a simplicial complex that is called simplicial homology. Again, we will work completely combinatorially and define our theory for abstract simplicial complexes. However, we first need to enhance them with a notion of orientation.

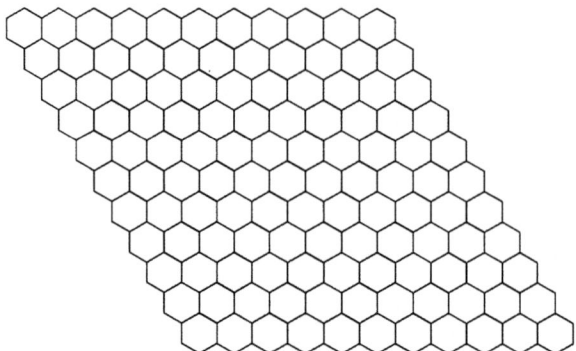

Figure 8.6: The game of Hex.

Definition 8.18. An *ordered abstract simplicial complex* is an abstract simplicial complex K with a total order on its vertex set $K(0)$.

We remark that if the vertex set of an abstract simplicial complex K is $[n]$, we have a total order inherited from the natural numbers.

Definition 8.19. Let K be an ordered abstract simplicial complex and $\sigma = (v_0, \ldots, v_n)$ be an ordered n-simplex, that is, $v_0 < \cdots < v_n$. We define the nth boundary $\partial_n(\sigma)$ as follows:

$$\partial_n(\sigma) = \sum_{i=0}^{n} (-1)^i (v_0, \ldots, \hat{v}_i, \ldots, v_n)$$

where $\hat{v}_i$ says that v_i should be omitted.

We next define the group of n-chains which are instances of free Abelian groups. We recall Remark 4.2 where we described some fundamental definitions and facts in group theory. In addition to those we need some more notations in the following. A group G is called *free Abelian* of rank n, $n \in \mathbb{N} \cup \{0\}$, if G is isomorphic to $\mathbb{Z}^n$. Recall that the Cartesian product $\mathbb{Z}^n$ is an Abelian group with componentwise addition, that is,

$$(z_1, \ldots, z_n) + (u_1, \ldots, u_n) = (z_1 + u_1, \ldots, z_n + u_n).$$

In general, the *rank* of a finitely generated group G is the minimal cardinality of a generating set of G and is denoted by $rk(G)$.

In Chapter 4 we have already seen how to present a group G with respect to a set of generators X and defining relations R, that is, $G = \langle X \mid R \rangle$. If our group G is Abelian we use $G = \langle X \mid R \rangle_{ab}$ to denote that G has generators X and its set of defining relations is the union of R and the commutators. As a prototypical example we consider

$$\mathbb{Z} \times \mathbb{Z} = \langle x, y \mid xy = yx \rangle = \langle x, y \mid \ \rangle_{ab}.$$

Definition 8.20. Given an ordered abstract simplicial complex K. Then its nth *chain group* $C_n(K)$ is the free Abelian group generated by the ordered n-simplices.

We look at the boundary map again. As we apply it to n-simplices we can linearly extend it to the nth chain group of a simplicial complex K and thus obtain a map:

$$\partial_n \colon C_n(K) \to C_{n-1}(K).$$

We check the following assertion that will be of use when we take the quotient of certain subgroups of $C_n(K)$.

Lemma 8.21. *For an n-simplex $\sigma = (v_0, \dots, v_n) \in K$ we have $(\partial_{n-1} \circ \partial_n)(\sigma) = 0$.*

Proof. The proof is a direct calculation:

$$
\begin{aligned}
\partial_{n-1}(\partial_n(\sigma)) &= \partial_{n-1}\left(\sum_{i=0}^{n} (-1)^i (v_0, \dots, \hat{v}_i, \dots, v_n) \right) \\
&= \sum_{i=0}^{n} (-1)^i \partial_{n-1}(v_0, \dots, \hat{v}_i, \dots, v_n) \\
&= \sum_{i=0}^{n} (-1)^i \left(\sum_{j=0}^{i-1} (-1)^j (v_0, \dots, \hat{v}_j, \dots, \hat{v}_i, \dots, v_n) \right. \\
&\qquad \left. + \sum_{j=i+1}^{n} (-1)^{j-1} (v_0, \dots, \hat{v}_i, \dots, \hat{v}_j, \dots, v_n) \right) \\
&= \sum_{0 \le k < i \le n} \left((-1)^{k+l} + (-1)^{k+l-1} \right) (v_0, \dots, \hat{v}_k, \dots, \hat{v}_l, \dots, v_n) \\
&= 0. \qquad\qquad\qquad\qquad\qquad\qquad\qquad\qquad\qquad\qquad \square
\end{aligned}
$$

We define two more groups associated to $C_n(K)$ as well as their quotient.

Definition 8.22. Let K be an ordered abstract simplicial complex. Then we define
1. the nth *cycle group* $Z_n(K)$ as $\ker \partial_n$,
2. the nth *boundary group* $B_n(K)$ as $\operatorname{im} \partial_{n+1}$, and
3. the nth *homology group* $H_n(K)$ as $Z_n(K)/B_n(K)$.

Note that a factor group is not necessarily well-defined, also see Remark 4.2. However, as $Z_n(K)$ and $B_n(K)$ are subgroups of Abelian groups we just need to check that $B_n(K)$ is a subset of $Z_n(K)$ but this is just the assertion of Lemma 8.21.

We will now study a collection of examples.

Example 8.23.

1. Consider the simplicial complex $K_1 = \{\{0\}, \{1\}, \{2\}, \emptyset\}$ of three isolated vertices. We immediately see that we have no simplices in dimensions 1 or higher. Hence we just calculate $\ker \partial_0$ and this trivially consists of linear combinations of the vertices (0), (1) and (2). Hence $H_n(K) = \{0\}$ for all $n \neq 0$ and

    $$H_0(K) = \langle (0), (1), (2) \mid \quad \rangle_{ab} \cong \mathbb{Z}^3.$$

2. Now we consider the abstract simplicial complex where the three vertices of K_1 are connected, that is, a hollow triangle. We set $\partial \Delta^2 = \{\{0, 1\}, \{1, 2\}, \{0, 2\}, \{0\}, \{1\}, \{2\}, \emptyset\}$ and write (vw) for the ordered simplex $(v, w), v < w$. This time we need to calculate which 1-chains are cycles. We have

    $$\partial_1(a(01) + b(12) + c(02)) = a(1) - a(0) + b(2) - b(1) + c(2) - c(0)$$
    $$= -(a + c)(0) + (a - b)(1) + (b + c)(2)$$
    $$= 0$$

    if and only if $a = -c$, $a = b$, and $b = -c$ because (0), (1), (2) cannot cancel each other. That is, we have $H_1(\partial \Delta^2) = \langle (01) + (12) - (02) \mid \quad \rangle_{ab}$ as all cycles are multiples of $(01) + (12) - (02)$ and there are no relations (no 2-simplices). We immediately see that $Z_0(\partial \Delta^2) = \langle (0), (1), (2) \mid \quad \rangle_{ab}$, but this time we have boundaries $\partial(01) = (1) - (0)$, $\partial(12) = (2) - (1)$ and $\partial(02) = (2) - (0)$, and thus $H_0(\partial \Delta^2) = \langle (0), (1), (2) \mid (0) - (1), (1) - (2), (0) - (2) \rangle_{ab} \cong \langle (0) \mid \quad \rangle \cong \mathbb{Z}$.

3. We now consider the solid triangle Δ^2. We see that this calculation is almost verbatim but we have to pay attention to 2-cycles and 1-boundaries. The calculation $\partial_2(012) = (12) - (02) + (01) \neq 0$ shows that we have no 2-cycles, but that the 1-cycle from the previous example is now 'bounded'. That is, $H_1(\Delta^2) = \langle (01) + (12) - (02) \mid (01) + (12) - (02) \rangle_{ab} \cong \{0\}$ and also all other homology groups but $H_0(\Delta^2) \cong \mathbb{Z}$ vanish.

We give geometric realizations of the complexes in Example 8.23 in Figure 8.7.

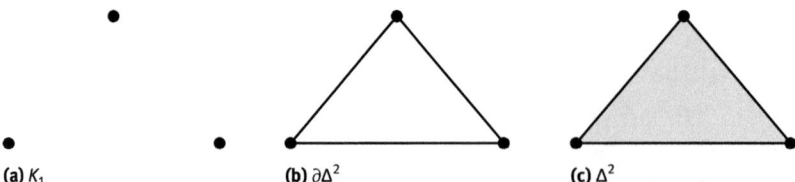

(a) K_1 **(b)** $\partial \Delta^2$ **(c)** Δ^2

Figure 8.7: Geometric realizations of the complexes in Example 8.23.

We just discuss the intuition we should have gained from the sample computation. In the first example, the rank of $H_0(K_1)$ was equal to the number of vertices. In the second example we have computed the homology of a connected complex, that is, there is a path from one point to another. The rank of $H_0(\partial\Delta^2)$ reflected this fact. Also $H_1(\partial\Delta^2)$ counted the number of 1-dimensional 'holes' the complex has. In the third example there was no homology at all expect for $H_0(\Delta^2)$ which again reflected the connectedness of the complex.

This observation is best phrased with the *Betti numbers* $\beta_i(K) = rk(H_i(K))$, named after B. Betti (1823–1892). In the above sense, β_0 counts the number of connected components, β_1 the number of one-dimensional or circular holes, and β_2 the number of two-dimensional holes or voids. We will summarize, and make precise, these observations in the following theorem.

Theorem 8.24. *Let $k \leq n$ and $\mathrm{sk}_k \Delta^n = \{\sigma \subset [n] \mid |\sigma| \leq k + 1\}$ be the k-skeleton of Δ^n. Then*

1. *for $0 < d = n$ we have $H_d(\Delta^n) = \begin{cases} \{0\} & \text{for } d \neq 0, \\ \mathbb{Z} & \text{for } d = 0, \end{cases}$*

2. *for $0 < k < n$ we have $H_d(\mathrm{sk}_k(\Delta^n)) = \begin{cases} \{0\} & \text{for } d \neq k, d \neq 0, \\ \mathbb{Z} & \text{for } d = 0, \\ \binom{n}{k+1} & \text{for } d = k, \end{cases}$ and*

3. *for $k = 0$ we have $H_d(\mathrm{sk}_0 \Delta^n) = \begin{cases} \mathbb{Z}^{n+1} & \text{for } d = 0, \\ \{0\} & \text{for } d \neq 0. \end{cases}$*

Theorem 8.24.1 follows from the fact that the standard simplex is an instance of a *cone*, that is a simplicial complex K with a vertex $v \in K(0)$ such that for $\sigma \in K$ we also have $\{v\} \cup \sigma \in K$, and these have trivial homology but in degree 0. Theorem 8.24.3 is immediate and Theorem 8.24.2 follows from the beautiful *Euler–Poincaré formula*, named after L. Euler (1707–1783) and J. H. Poincaré (1854–1912):

Theorem 8.25 (Euler–Poincaré). *For a simplicial complex K of dimension n we have*

$$\sum_{i=0}^{n} (-1)^i K(i) = \sum_{i=0}^{n} (-1)^i \beta_i(K).$$

We leave the proof as an exercise. We are now ready to prove Theorem 8.24.2.

Proof. We assume $k = d$ and derive from Theorem 8.25 that

$$1 + (-1)^k \beta_k(K) = \sum_{i=0}^{k} (-1)^i \binom{n+1}{i+1}$$

$$= 1 + (-1)^k \binom{n}{k+1},$$

where the last equality follows from induction, and hence $\beta_k(K) = \binom{n}{k+1}$, that is, $H_k(\mathrm{sk}_k(\Delta^n)) = \mathbb{Z}^{\binom{n}{k+1}}$. □

We close this section with two more examples that we will compute in a different way and this way gives an outlook to the recent field of the discrete Morse theory.

Consider again the simplicial complex K_2 of Example 8.2.2. Just by looking at its geometric realization we see that the complex is connected and has a 1-dimensional hole. We already might feel that the solid triangle does not add to homology. Hence, we conjecture that the homology groups will be $H_1(K) \cong H_0(K) \cong \mathbb{Z}$ and all other homology groups will vanish. A topological way to justify this observation is the concept of deformation retracts that we will not discuss. However, we will quote a main definition of discrete Morse theory that will shorten our computation and we follow [20] for the following definition and theorem, though in a special case.

Definition 8.26. Let $P := P(K)$ be the poset of an abstract simplicial complex K. Then we write $x \prec y$ if $x < y$ and there is no $z \in P$ with $x < z < y$. A *matching* on P is a matching $M \subset E$ of the underlying graph $G = (V, E)$. We call a matching on P *acyclic* if there are no edges $e_1 = \{b_1, b_1'\}, e_2 = \{b_2, b_2'\}, \ldots, e_n = \{b_n, b_n'\}$, $n \geq 2$, and no chain

$$b_1 \succ b_1' \prec b_2 \succ b_2' \prec \cdots \prec b_n \succ b_n' \prec b_1.$$

If for P such an acyclic matching exists, we call the unmatched simplices *critical*. We give the following special case of a main theorem of discrete Morse theory, see for instance Theorem B in [20], in order to keep our exposition at a purely combinatorial and simplicial setting.

Theorem 8.27. *Let K be an abstract simplicial complex.*
1. *If K has a complete acyclic matching, then $H_n(K) \cong \{0\}$ for all $n \neq 0$ and $H_0(K) = \mathbb{Z}$.*
2. *If K has an acyclic matching with all critical cells in dimension d, $d \geq 1$, and their number is l, then $H_d(K) = \mathbb{Z}^l$, $H_0(K) \cong \mathbb{Z}$ and $H_n(K) \cong \{0\}$ for all $n \neq 0, d$.*
3. *If K has an acyclic matching with all critical cells in even dimensions, then the homology is also concentrated in even dimensions.*

We consider the following applications.

Example 8.28.
1. We consider the complex K_2 and the acyclic matching on its face poset given in Figure 8.8.
   We conclude by Theorem 8.27.2 that $H_1(K_2) \cong \mathbb{Z} \cong H_0(K_2)$ and $H_n(K_2) \cong \{0\}$ for all $n \neq 0, 1$.
2. We consider Δ^3 and the acyclic matching in Figure 8.9. We conclude by Theorem 8.27.1 that $H_n(\Delta^3) \cong \{0\}$ for all $n \neq 0$ and $H_0(\Delta^3) \cong \mathbb{Z}$.

The face poset $P(\Delta^3)$ and the partially ordered subset of $P(K_2)$ that corresponds to the copy of Δ^2 are instances of Boolean algebras that we will consider in Chapter 11.

This section just gave a little insight into simplicial homology theory and a special case of an important theorem in discrete Morse theory. In particular we have only considered the so-called homology with integer coefficients and just applied discrete Morse theory to the purely simplicial setting. Just going beyond these restrictions gives a much richer setting to investigate the shape of data. Furthermore, many algebraic, combinatorial or topological theories that are in connection with homology theories

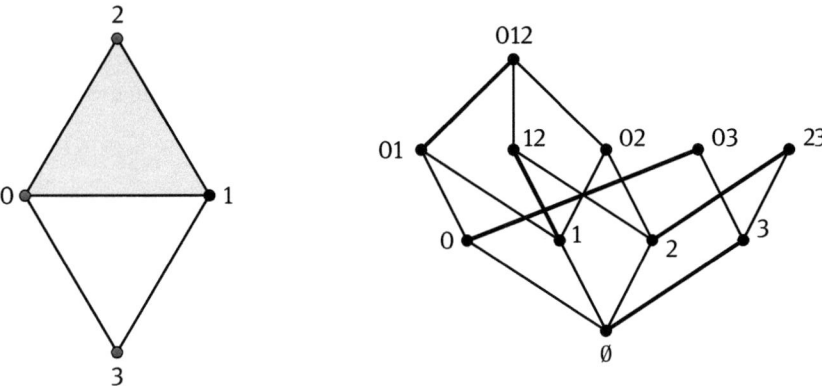

Figure 8.8: Acyclic matching for the complex K_2.

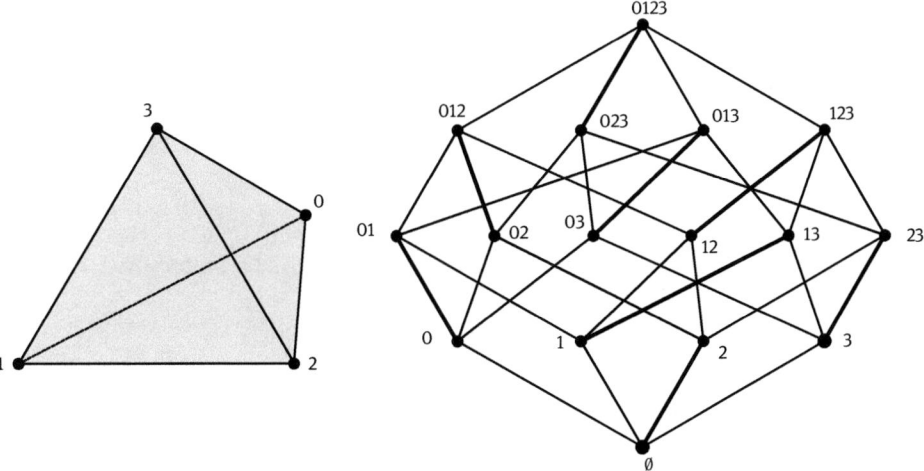

Figure 8.9: Acyclic matching for the complex Δ^3.

are out of the scope of our book. Our aim here is just to equip the reader with some intuition that will be helpful to understand the following section.

8.4 Persistent Homology

In the last section we have settled the theory and intuition to study and understand first examples and applications in the emerging field of topological data analysis. Here, we aim to understand data in the form of a point cloud in a metric space, associate to it a family of simplicial complexes and study its homology. These results then are visualized in a suitable way such that we read off so-called persistent features of the data and give a conjecture about its shape. This procedure is sometimes referred to as the 'TDA pipeline'.

We first gather the remaining notions and then discuss an instructive example.

Definition 8.29. A (finite) *filtration* of an abstract simplicial complex K is a family $\{K_0, \ldots, K_n\}$ of simplicial complexes with the property

$$\emptyset = K_0 \subset K_1 \subset \cdots \subset K_{n-1} \subset K_n = K.$$

We consider an example in Figure 8.10.

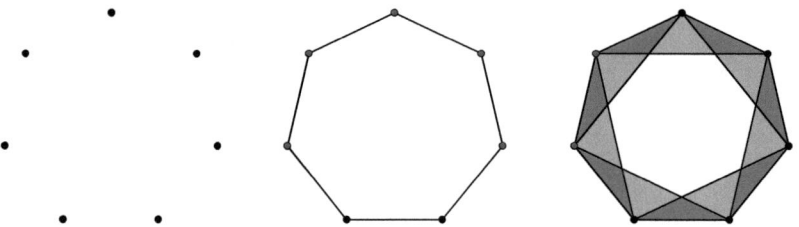

Figure 8.10: Filtration of a complex.

We recall the notion of a metric space.

Definition 8.30. Let X be set. A map $d\colon X \times X \to \mathbb{R}$ is called a *metric* on X if for all $x, y, z \in X$ we have:
1. $d(x,y) \geq 0$ and $d(x,y) = 0$ if and only if $x = y$,
2. $d(x,y) = d(y,x)$,
3. $d(x,y) \leq d(x,z) + d(z,y)$, the *triangle inequality*.

The pair (X, d) is called a *metric space*.

We will discuss metric spaces in the exercises. Now we define an abstract simplicial complex associated to a point cloud that we consider as a finite metric space (X, d).

Definition 8.31. Let (X, d) be a finite metric space and $\varepsilon > 0$. Then we define the abstract simplicial complex V_ε, the *Vietoris–Rips complex* after L. Vietoris (1891–2002) and E. Rips (born 1948), with respect to ε as follows: The vertex set of V_ε is X and the simplex set is $\{\sigma \subset X \mid d(x,y) < \varepsilon \text{ for all } x, y \in \sigma\}$.

We will show that for a finite sequence $\varepsilon_1, \ldots, \varepsilon_n$ the associated Vietoris–Rips complexes yield a filtration.

Lemma 8.32. *For $\varepsilon_1 < \varepsilon_2$ we have $V_{\varepsilon_1} \subset V_{\varepsilon_2}$.*

Proof. If $\sigma \in V_{\varepsilon_1}$, then $\sigma \in V_{\varepsilon_2}$, because for $x, y \in \sigma$ we have $d(x,y) < \varepsilon_1 < \varepsilon_2$. $\qquad\square$

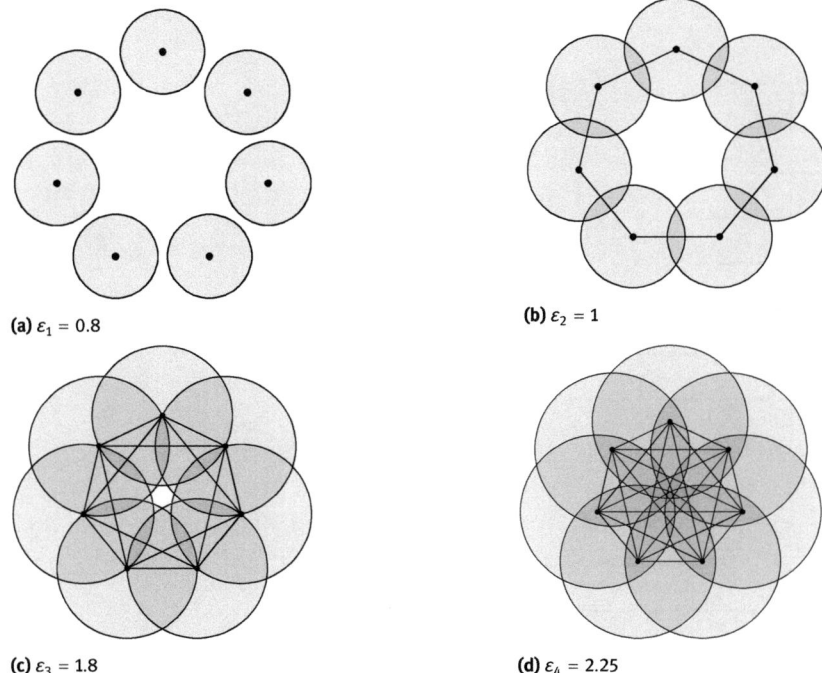

(a) $\varepsilon_1 = 0.8$

(b) $\varepsilon_2 = 1$

(c) $\varepsilon_3 = 1.8$

(d) $\varepsilon_4 = 2.25$

Figure 8.11: Examples of (underlying graphs) of Vietoris–Rips complexes.

We give an example of such a filtration in Figure 8.11.

Now, a naive approach would be to compute homology of each V_r. However, persistant homology is more subtle and makes use of the filtration property. The inclusion of the complexes gives rise to maps $f_p^{i,j}: H_p(K_i) \to H_p(K_j)$ for $i \leq j$.

Definition 8.33. Given a filtration $\{K_i\}_{0 \leq i \leq n}$ of a simplicial complex K, we define its pth persistant homology group as $H_p^{i,j}(K) = Z_p(K_i)/(B_p(K_j) \cap Z_p(K_i))$ for $i \leq j$.

We can interpret $H_p^{i,j}(K)$ as the group of all homology classes of K_i that are still present in K_j. This enables us to record which topological features, in terms of homology generators, persist.

We use the following terms:

Definition 8.34. We say that a homology class $[c] \in H_p(K_i)$
1. is *born* in K_i if $[c] \notin H_p^{i-1,i}(K)$, and
2. *dies* in K_j if $f_p^{i,j-1}([c]) \notin H_p^{i-1,j-1}(K)$ and $f_p^{i,j}([c]) \in H_p^{i,j}(K)$.

In the case of the Vietoris–Rips complexes we say that a class $[c]$ that is born in V_{ε_1} and that dies in V_{ε_2} has *persistence* $p([c]) = \varepsilon_2 - \varepsilon_1$. The persistence of homology generators is often visualized in so-called *barcode*, compare Figure 8.12.

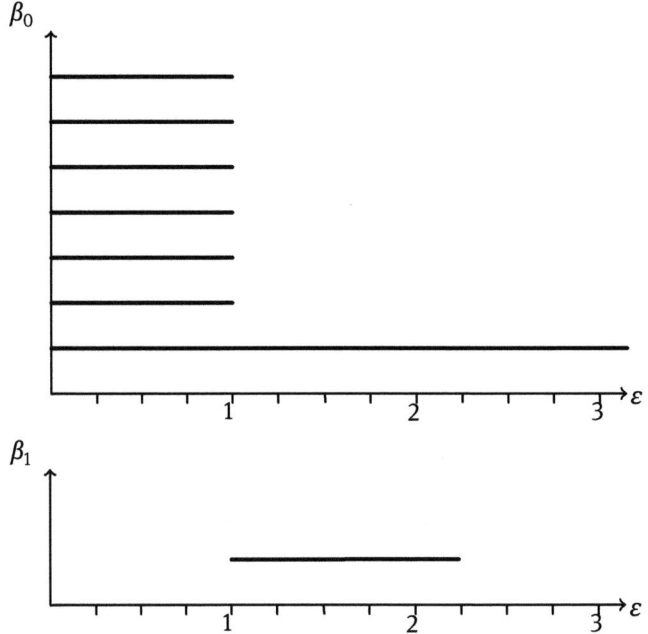

Figure 8.12: Barcode for the running example.

Example 8.35. We discuss the barcode of our running example. We see that from the beginning ($V_0 = \emptyset$) we have seven generators for the zeroth homology group. But for $\varepsilon_1 = 1$ they all get identified, so six of them die at K_2.

However, at K_2 a generator for the first homology group is born that persists until K_4 where it becomes bounded. In this situation one would hence conjecture that the point cloud comes from a circle as its features persist.

We have seen a glimpse of topological data analysis that is a very active field of research and also developed outside academia with a plethora of real world applications. From the mathematical point of view we would just like to mention that there are many variants (alternative complexes to construct, different metrics to use, or even a manipulation of the initial point cloud itself) that are suited for certain situations.

A big topic within TDA is the computability of persistant homology within a reasonable amount of time. Here, the project *Ripser* seems to be one of the most widely used packages for the computation of barcodes using the Vietoris–Rips complexes.

Exercises

1. The *mesh* of a geometric simplicial complex K is the length of its longest edge. Let $Bary^N(\Delta^n)$ be the Nth iterated barycentric subdivision of Δ^n and let m_N denote its mesh. Prove that the barycentric subdivision is mesh-shrinking, that is, $\lim_{N \to \infty} m_N = 0$.
2. Show that $H_0(K)$ is always free Abelian for a simplicial complex K.
3. Find a simplicial complex K with $H_1(K) = \mathbb{Z}/2\mathbb{Z} \oplus \mathbb{Z}$.
4. Show the induction in the proof of Theorem 8.24.
5. Prove Theorem 8.25.
6. Recall that a simplicial complex K is a cone if there exists a vertex $v \in K(0)$ with the following property: If $\sigma \in K$, then $\sigma \cup v \in K$. Show that for a cone K we have $H_0(K) \cong \mathbb{Z}$ and $H_n \cong \{0\}$ for all $n \neq 0$.
7. Show that in the definition of a metric space, the property $d(x,y) \geq 0$ follows from the other axioms.
8. Show that, if $d: X \times X \to \mathbb{R}$ is a metric, then also
    (a) $k \cdot d$ with $k > 0$ is a metric,
    (b) $\frac{d}{1+d}$ is a metric.
9. Prove that $d: \mathbb{R} \times \mathbb{R} \to \mathbb{R}$ with $d(x,y) = \ln(1 + |x + y|)$ is a metric on $\mathbb{R}$.

9 Combinatorics and Combinatorial Problems

9.1 Combinatorics

Combinatorics is the area of mathematics that is primarily concerned with counting and the structure of finite sets. Finite probability theory that we examine in the next chapter is heavily dependent on combinatorics and combinatorial techniques. A *combinatorial problem* is one that involves combinatorics. Combinatorial problems arise in many areas of both pure and applied mathematics. The portion of combinatorics that deals with counting and finding the sizes of finite sets is usually referred to as *enumerative combinatorics*.

In this chapter we give some overview about enumerative combinatorics. We start with elementary counting problems, consider the sieve method, partitions of sets and numbers, recursion and generating functions. This is just an overview of some principles and material a mathematics teacher and working mathematicians just should know. Some people consider graph theory as part of combinatorics but we already considered some of the most important principles in graph theory in Chapter 5.

9.2 Basic Techniques and the Multiplication Principle

For the most part of this chapter we will consider finding the sizes of finite sets and combinations of finite sets. We make the following notational conventions.

If $\{a_1, \ldots, a_n\}$ is a finite indexed set of real numbers then

$$\sum_{i=1}^{n} a_i = a_1 + a_2 + \cdots + a_n$$

is the sum of these elements and

$$\prod_{i=1}^{n} a_i = a_1 \cdot a_2 \cdots a_n$$

is the product of these elements. Then we note that

$$\sum_{k=m}^{n} a_k = 0 \quad \text{and} \quad \prod_{k=m}^{n} a_k = 1$$

if $m, n \in \mathbb{N}_0$ and $m > n$.

If M is a set then $|M|$ denotes its cardinality or the number of elements in M. Now let $M_1, \ldots, M_r$ be finite sets. We have the following straightforward results.

https://doi.org/10.1515/9783110740783-009

(1)

$$\left| \bigcup_{k=1}^{r} M_k \right| = \sum_{k=1}^{r} |M_k|$$

if the sets M_i are pairwise disjoint, that is, $M_i \cap M_j = \emptyset$ for $i \neq j$.

(2)

$$|M_1 \times M_2 \times \cdots \times M_r| = \prod_{k=1}^{r} |M_k|.$$

The first of these is called the *basic rule of summation* or *summation rule*. The contraposition of the rule of summation is the *rule of difference* or *difference rule*. It is the idea that if we have n ways of doing something where exactly m of the n ways have an additional property, then $n - m$ ways do not have this additional property.

The second is called the *multiplication principle* or *multiplication rule* and is the basic idea in most enumerative techniques. It is usually described in terms of choices.

Suppose a choice is to be made in two steps. If the first step has m choices and the second step has n choices then there is a total of mn choices, that is, we multiply the numbers of choices at each step.

The idea in the multiplication principle can be extended to more than two steps. We give some illustrations.

Example 9.1. There are 8 applicants for one job and 12 applicants for a second job. In how many ways can these people be hired?

Here the final choice is in two steps, fill the first job and then fill the second. In the first case there are 8 choices while in the second we have 12. Therefore by the multiplication principle there are a total of $8 \cdot 12 = 96$ choices.

Example 9.2. A hospital cross-classifies patients in three ways: gender, age and type of payment. There are 2 gender classifications, 6 age classifications and 4 payment type classifications. In how many ways can the patients be cross-classified?

Here a cross-classification consists of a choice for gender, followed by a choice for age, followed by a choice for type of payment. There are 2 choices in the first step, 6 in the second and 4 in the third. Thus by the multiplication principle there are $2 \cdot 6 \cdot 4 = 48$ cross-classifications.

Example 9.3. The elements of the set $\{A, B\}$ can be combined with the elements of the set $\{1, 2, 3\}$ in six different ways, see Figure 9.1.

The multiplication principle is a special case of a more general counting technique known as a *tree diagram*. In this type of diagram, the decision maker or choice maker is represented as the trunk of a tree. The initial choices are represented by *branches*. A branch terminates at a *node* where further branching is possible. If there

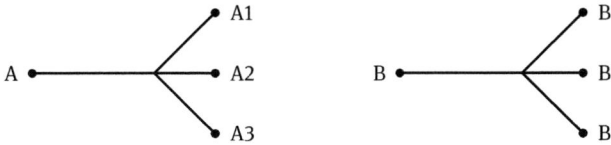

Figure 9.1: Combination of two sets.

is no branching at a node it is called a *leaf*. The total number of leaves gives the total number of choices. We illustrate this in the example below.

Example 9.4. A production item has 4 weight classifications, W_1, W_2, W_3, W_4, 3 length classifications, L_1, L_2, L_3, and 2 hardness classifications, H_1, H_2. The heaviest weights W_1, W_2 are made with all lengths and hardnesses while the lighter weights are only made in L_3; W_3 is made with both hardness classifications but W_4 is only made with H_2. How many cross-classifications are there of this item?

The tree diagram for this situation is given in Figure 9.2 below. At the first stage we can make all 4 weight choices but at the first set of nodes the possible branchings are different. From the W_1 and W_2 nodes all 3 length choices are possible but at the W_3 and W_4 nodes only the L_3 choice is possible. Finally, at branches which begin with W_1 or W_2 both hardness classifications are possible while at branches beginning with W_4 only H_2 is possible.

From the tree we see that there are 15 cross-classifications. This is found by counting the number of leaves on the tree. Each path – such as $W_1L_1H_1$ – corresponds to a cross-classification. The example given by $W_1L_1H_1$ corresponds to the cross-classification of weight factor W_1, length factor L_1 and hardness factor H_1.

There are some basic results that follow directly from the above.

Theorem 9.5. *Let R be a relation between two finite sets M and N, that is, $R \subset M \times N$. For $x \in M$ let $r(x)$ be the number of $y \in N$ which are related to x, that is,*

$$r(x) = |\{y \in N \mid (x, y) \in R\}|.$$

Analogously, let $s(y), y \in N$, be the number of $x \in M$ which are related to y. Then

$$\sum_{x \in M} r(x) = \sum_{y \in N} s(y).$$

Proof. For $(x, y) \in M \times N$ we define

$$a(x, y) = \begin{cases} 1, & \text{if } (x, y) \in R, \\ 0, & \text{if } (x, y) \notin R. \end{cases}$$

Then

$$r(x_0) = \sum_{y \in N} a(x_0, y), \quad x_0 \in M,$$

and

$$s(y_0) = \sum_{x \in M} a(x, y_0), \quad y_0 \in N.$$

Hence we get

$$\sum_{x \in M} r(x) = \sum_{(x,y) \in M \times N} a(x, y) = \sum_{y \in N} s(y). \qquad \square$$

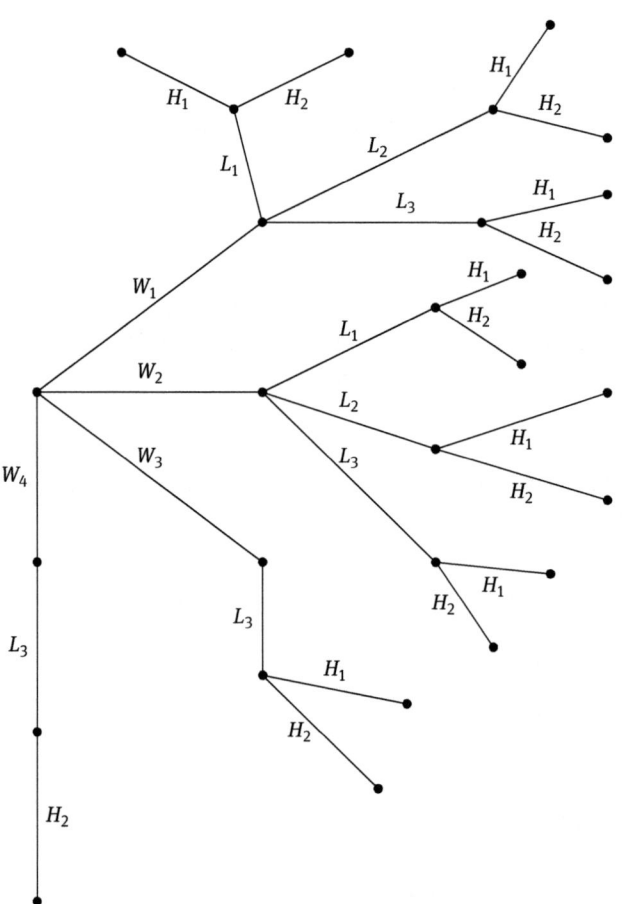

Figure 9.2: Tree diagram.

This double counting, or counting in two ways, is a combinatorial proof technique for showing that two expressions are equal by demonstrating that they are two ways of counting the size of one set. In this method one describes a finite set X from two perspectives, leading to two expressions for the size of X. Since both expressions equal the size of the same set X, they are equal to each other.

Remarks 9.6.
(1) Let $M = \{x_1, x_2, \ldots, x_m\}$ and $N = \{y_1, y_2, \ldots, y_n\}$ be given by a concrete listing. The characteristic function $a : M \times N \to \{0, 1\}$ for the subset $R \subset M \times N$ corresponds to an $(m \times n)$-matrix over $\{0, 1\}$, the *incidence matrix* of the relation R.
   The proof of Theorem 9.5 implies that one goes once column-by-column and the other time row-by-row through the matrix.
(2) Counting arguments are often used to give existential (non-constructive) proofs. The basic principle here is often the *drawer principle of Dirichlet* (*pigeonhole principle*):
   If one distributes n objects into k drawers and if $n > k$, then there exists one drawer which contains at least two objects.

Let us state this principle in a mathematical way.

Theorem 9.7. *Let $f : X \to Y$ be a map between two finite sets, and let $|X| > |Y|$. Then there exists a $y \in Y$ with $|f^{-1}(y)| \geq 2$.*

Example 9.8.
(1) A library has more than 4000 books. None of the books has more than 4000 pages. Then there are (at least) two books with the same number of pages.
(2) **The friendship problem.** In each set of $n \geq 2$ persons there are at least two persons which have the same number of friends (here we assume the relation of friendship is symmetric).

   Proof. If a person has $n - 1$ friends, then each person is his/her friend, hence nobody is friendless. This means, that 0 and $n - 1$ do not occur simultaneously as numbers of friends. Hence we have to distribute only $n - 1$ possible numbers of friends to n persons. By the principle of Dirichlet, at least two persons have the same number of friends.   □

(3) In each set $A \subset \{1, 2, \ldots, 2m\}$ with at least $m + 1$ elements there are two numbers a, b such that $a \mid b$, that is, a is a divisor of b.

   Proof. Let $\{a_1, a_2, \ldots, a_{m+1}\} \subset M$. We write each a_i in the form $a_i = 2^{r_i} q_i$ with q_i odd. There are only m odd numbers in M. Hence, by the principle of Dirichlet, at least one of the q_i occurs for two distinct a_i and a_j, that is, $a_i = 2^{r_i} q$ and $a_j = 2^{r_j} q$ with $q = q_i = q_j$. If $a_i < a_j$, then $a_i \mid a_j$.   □

9.3 Sizes of Finite Sets and the Sampling Problem

Dealing with counting in finite sets is related to two basic problems: the *sampling problem* and the *occupancy problem*. They turn out to be equivalent. We describe the sampling problem first.

The multiplication principle can be used to determine the number of ways one set of items can be sampled out of a larger set of items. Specifically, we suppose that we have a total of n items and k items are to be sampled from them. The question of interest is in how many ways this can be done. The answer to this depends on the manner in which the sampling is done. In particular, the sampling can be done *with replacement*, meaning the items are put back after they are sampled, or *without replacement*, meaning they are not put back. In addition, we can have *ordered samples*, that is, samples in which the ordering of the sample is relevant, or *unordered samples* where the ordering is irrelevant. Therefore to answer the question of how many ways x items can be sampled from n items, we must consider four situations:

(1) *Case 1.* Ordered Samples Without Replacement.
(2) *Case 2.* Unordered Samples Without Replacement.
(3) *Case 3.* Ordered Samples With Replacement.
(4) *Case 4.* Unordered Samples With Replacement.

We concentrate first on the case of sampling without replacement. The number of ordered samples of k items that can be taken from n items without replacement is called the number of *permutations* of k items out of n. This is denoted by $_nP_k$ or $P(n, k)$. An ordered sample of k items taken from n items is called a k-permutation.

The number of unordered samples of k items that can be taken from n items without replacement is called the number of *combinations* of k items out of n. This is denoted by $_nC_k$, or most commonly by $\binom{n}{k}$, which is called a *binomial coefficient* because these numbers appear in the expansion of a binomial expression $(a + b)^n$. An unordered sample of k items taken from n items is called a k-combination. We summarize these definitions:

$$_nP_k = \text{number of permutations of } k \text{ objects out of } n$$
$$= \text{number of ways to sample } k \text{ objects out of } n \text{ objects}$$
$$\text{without replacement but with order.}$$

$$_nC_k = \binom{n}{k} = \text{number of combinations of } k \text{ objects out of } n$$
$$= \text{number of ways to sample } k \text{ objects out of } n \text{ objects}$$
$$\text{without replacement and without order.}$$

The multiplication principle can be applied to determine formulas for both permutations and combinations. First of all,

$$_nP_k = n(n-1)(n-2)\cdots(n-k+1).$$

This expression $_nP_k = (n)(n-1)\cdots(n-k+1)$ is also referred to as the falling factorial and denoted $(n)_k$.

Example 9.9. Compute $_7P_3$.

From the formula for $_nP_k$ we have $_7P_3 = 7\cdot 6\cdot 5 = 210$. Thus there are 210 different ways to choose ordered samples of size 3 from a collection of 7 objects without replacement.

Example 9.10. How many 1–2–3 finishes are possible in a horse race with 11 entries?

A 1–2–3 finish consists of a choice of 3 objects from the total of 11 objects. The order is certainly relevant and therefore the number of these choices is

$$_{11}P_3 = 11\cdot 10\cdot 9 = 990.$$

Example 9.11. How many unordered samples of size 6 are there out of 49 numbers?

Here the result is $_{49}C_6 = \binom{49}{6} = 1398316$.

If all n objects are chosen from the total of n objects what is really being done is that the items are being rearranged. Thus $_nP_n$ counts the total possible rearrangements of n objects. We see that

$$_nP_n = n(n-1)(n-2)\cdots 1.$$

This expression is given the name n *factorial* and is denoted by $n!$. Then

$$n! = n(n-1)(n-2)\cdots 1 = \text{number of ways to rearrange } n \text{ objects.}$$

Example 9.12.
(a) $6! = 6\cdot 5\cdot 4\cdot 3\cdot 2\cdot 1 = 720$, so there are 720 different ways to rearrange 6 objects.
(b) $3! = 6$ so there are 6 different ways to rearrange 3 objects. For example, suppose the 3 objects were A, B, C. Then the six rearrangements are ABC, ACB, BAC, BCA, CAB and CBA.

To see how the formula for permutations is derived, consider sampling k objects from n objects. At the first step there are n choices. At the second there are only $(n-1)$ choices since the sampling is done without replacement. At the third step there are $(n-2)$, and so on down for k choices, the final number being $(n-k+1)$. Finally, by the multiplication principle, these numbers would be multiplied, yielding the result.

By algebraic manipulation it is clear that the formula for permutations can also be expressed in factorial notation as

$$_nP_k = \frac{n!}{(n-k)!} = (n)_k.$$

To determine a method for computing combinations we reason in the following manner. Each choice of an unordered sample of k objects from n objects can be rearranged in $k!$ different ways. Therefore the total number of rearrangements multiplied by the total number of unordered samples will give the total number of ordered samples. Writing this in terms of permutations and combinations we have $_nC_k \cdot k! = {_nP_k}$. Solving this for $_nC_k$ gives us the following formula:

$$_nC_k = \binom{n}{k} = \frac{_nP_k}{k!} = \frac{n(n-1)\cdots(n-k+1)}{k!} = \frac{n!}{k!(n-k)!}.$$

Example 9.13. Compute $\binom{9}{4}$. From the formula for $_nC_k$,

$$\binom{9}{4} = \frac{_9P_4}{4!} = \frac{9\cdot 8\cdot 7\cdot 6}{4\cdot 3\cdot 2\cdot 1} = 126.$$

Example 9.14. How many 3-person committees can be chosen from among 8 people?
In choosing a committee the order of choice is irrelevant so that we are choosing an unordered sample of size 3 from 8 objects without replacement. The number of ways to do this is then $_8C_3$, and therefore the number of possible committees is

$$_8C_3 = \binom{8}{3} = \frac{8\cdot 7\cdot 6}{3\cdot 2\cdot 1} = 56.$$

We now turn our attention to sampling with replacement, that is, where the objects are replaced after each pick. If order counts, then at every stage there are n choices and there are a total of k stages. Therefore by the multiplication principle the number of ordered samples of size k chosen from n objects with replacement is given by

$$n^k = \text{number of ordered samples of size } k$$
$$\text{chosen from } n \text{ objects with replacement.}$$

We call an ordered sample of size k chosen with replacement from n objects a k-variation.

In a somewhat more complicated manner, the number of unordered samples with replacement is determined by

$$\binom{n+k-1}{k} = \text{number of unordered samples of size } k$$
$$\text{chosen from } n \text{ objects with replacement.}$$

We call an unordered sample of size k chosen with replacement from n objects a k-repetition.

To see how this last formula is derived, consider, without loss of generality, $M = \{1, 2, \ldots, n\}$. Then M is ordered in a natural manner. We write respectively each k-repetition, which is unordered, as a monotonically decreasing sequence

$$a_1 \leq a_2 \leq \cdots \leq a_k, \quad \text{where } a_1, a_2, \ldots, a_k \in M.$$

The assignment

$$(a_1, a_2, \ldots, a_k) \mapsto (a_1 + 1, a_2 + 2, \ldots, a_k + k)$$

defines a map of the set N_1 of the monotonically increasing k-sequences of elements of M onto the set N_2 of strictly monotonically increasing k-sequences of elements from $\{2, 3, \ldots, n, n + 1, \ldots, n + k\}$. This map is bijective, and N_2 corresponds to the set of all k-subsets of $\{2, 3, \ldots, n + k\}$.

We now give a short recapitulation of the different sampling problems in the form of the following theorem.

Theorem 9.15. *Let M be a set with $|M| = n$ and $k \in \mathbb{N}$ with $1 \leq k \leq n$.*
(1) *A k-combination of M is a k-subset of M. Their number is $\binom{n}{k} = {}_nC_k$.*
(2) *A k-permutation of M is an ordered k-tuple of distinct elements from M. Their number is*

$$\frac{n!}{(n-k)!} = n(n-1)\cdots(n-k+1) = (n)_k = \binom{n}{k}k! = {}_nP_k.$$

(3) *A k-repetition of M is an unordered choice of k not necessarily distinct elements from M. Their number is $\binom{n+k-1}{k}$.*
(4) *A k-variation of M is an ordered k-tuple of elements from M which are not necessarily distinct. Their number is n^k.*

This is summarized in Table 9.1.

We close the section sizes of finite sets with three further examples.

Table 9.1: Summary of Theorem 9.15.

Choice of k elements from n elements	unordered	ordered
without replacement	k-combination $\binom{n}{k} = {}_nC_k$	k-permutation $(n)_k = {}_nP_k$
with replacement	k-repetition $\binom{n+k-1}{k}$	k-variation n^k

Example 9.16. Let $M = \{a, b, c\}$, $n = |M| = 3$ and $k = 2$. Then we get
(1) k-combination

$$
\begin{array}{c}
ab \\
ac \\
bc
\end{array}
$$

(2) k-permutation

$$
\begin{array}{cc}
ab & ac \\
ba & bc \\
ca & cb
\end{array}
$$

(3) k-repetition

$$
\begin{array}{cc}
aa & bb \\
ab & bc \\
ac & cc
\end{array}
$$

(4) k-variation

$$
\begin{array}{ccc}
aa & ba & ca \\
ab & bb & cb \\
ac & bc & cc
\end{array}
$$

Example 9.17.
(1) Suppose 5 members of a group of 12 are to be chosen to work as a team. The number of distinct 5-person teams is $\binom{12}{5} = 792$.
(2) Now suppose two members A and B of the group of 12 refuse to work together on a team. The number of distinct 5-person teams that do not contain both A and B is $\binom{12}{5} - \binom{10}{3} = 792 - 120 = 672$, where $\binom{10}{3}$ is the number of distinct 5-person teams that contain both A and B (difference rule).
(3) Now suppose the group of 12 consists of 5 men and 7 woman.
   The number of 5-person teams that contain 3 men and 2 women is

$$
\binom{5}{3} \cdot \binom{7}{2} = 210
$$

(multiplication rule).
The number of 5-person teams that contain at least one men is $\binom{12}{5} - \binom{7}{5} = 792 - 21 = 771$, where $\binom{7}{5}$ is the number of 5-person teams that do not contain any man (difference rule).
Hence, the number of 5-person teams that contain at most one man is $\binom{7}{5} + \binom{5}{1}\binom{7}{4} = 21 + 175 = 196$. Here $\binom{5}{1}\binom{7}{4}$ is the number of 5-person teams with one man (multiplication rule).

Example 9.18. Given a dice with 6 numbers, the number t of possible occurrences in case of five times rolling is as follows:

(1) Without chronological order

$$t = \binom{6 + 5 - 1}{5} = \binom{10}{5} = 252;$$

(2) With chronological order

$$t = 6^5 = 7776.$$

9.3.1 The Binomial Coefficients

The number of combinations of k objects out of n is given by the binomial coefficients $\binom{n}{k}$. These have many important properties and applications that we explore in this subsection.

The basic properties of the binomial coefficients are:

(1) $\binom{n}{k} = \frac{n!}{k!(n-k)!}$ for $k, n \in \mathbb{N}_0, n \geq k$;
(2) $\binom{n}{k} = 0$ for $k, n \in \mathbb{N}_0, n < k$;
(3) $\binom{n}{0} = \binom{n}{n} = 1$ for $n \in \mathbb{N}_0$;
(4) $\binom{n}{k} = \binom{n}{n-k}$ for $k, n \in \mathbb{N}_0, k \leq n$.

We have the following recursion formula.

Theorem 9.19 (Recursion formula).

$$\binom{n+1}{k+1} = \binom{n}{k+1} + \binom{n}{k}.$$

A proof is given in [12, Chapter 12.2].

This formula can be visualized through Pascal's triangle which is named after B. Pascal (1623–1662).

$$\binom{0}{0}$$

$$\binom{1}{0} \quad \binom{1}{1}$$

$$\binom{2}{0} \quad \binom{2}{1} \quad \binom{2}{2}$$

$$\binom{3}{0} \quad \binom{3}{1} \quad \binom{3}{2} \quad \binom{3}{3}$$

$$\binom{4}{0} \quad \binom{4}{1} \quad \binom{4}{2} \quad \binom{4}{3} \quad \binom{4}{4}$$

$$\binom{5}{0} \quad \binom{5}{1} \quad \binom{5}{2} \quad \binom{5}{3} \quad \binom{5}{4} \quad \binom{5}{5}$$

and so on; or in concrete numbers

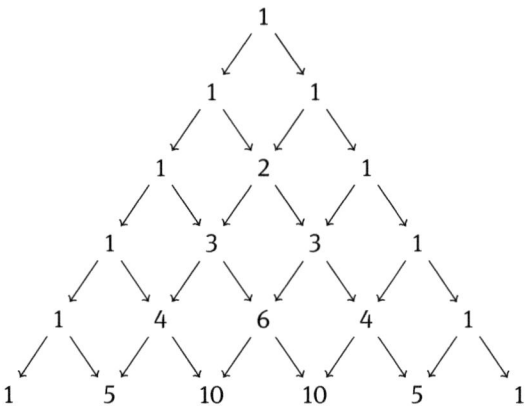

and so on.

The binomial coefficients can be used to count the number of subsets of size k in a set with n elements.

Theorem 9.20. *Let M be a finite set with $|M| = n$, $n \in \mathbb{N}_0$, elements. A subset of M with exactly k elements is called a k-subset of M. The number of k-subsets of M is $\binom{n}{k}$.*

Proof. Let $a(n, k)$ be the number of k-subsets of M.

Certainly, $a(0, k) = \binom{0}{k}$ for $k > 0$ and $a(n, 0) = 1 = \binom{n}{0}$ for all n. In particular, the statement holds for $n = 0$. Now if $|M| = n + 1$ and $x \in M$, then for an arbitrary $(k + 1)$-subset N exactly one of the following holds:
(1) $x \notin N$;
(2) $x \in N$.

If $x \notin N$, then N is a $(k + 1)$-subset of $M \setminus \{x\}$.

There is a bijection between the $(k + 1)$-subsets of M which contain x and the k-subsets of $M \setminus \{x\}$.

Hence we have

$$a(n + 1, k + 1) = a(n, k + 1) + a(n, k).$$

Therefore, the $a(n, k)$ satisfy the same recursion formula and the same initial conditions as the binomial coefficients $\binom{n}{k}$. Hence $a(n, k) = \binom{n}{k}$. □

Remark 9.21. The idea to get a suitable recursion formula for numbers a_n by case analysis is typical for counting problems.

The numbers $\binom{n}{k}$ are called binomial coefficients because of their role in the *binomial theorem*, or *binomial expansion*.

Theorem 9.22 (Binomial formula). *Let R be an integral domain with unity 1 and $n \in \mathbb{N}_0$. Then*

$$(a + b)^n = \sum_{k=0}^{n} \binom{n}{k} a^k b^{n-k} \quad \text{for } a, b \in R.$$

Recall that an integral domain with unity 1 is a commutative ring R with a unity $1 \neq 0$ and the property that if $ab = 0 \in R$ then $a = 0$ or $b = 0$.

Proof. This is correct for $n = 0$ because $x^0 = 1$ for all $x \in R$ by definition.

Let now $n > 0$. By the expansion of

$$(a + b)^n = \underbrace{(a + b)(a + b) \cdots (a + b)}_{n\text{-times } (a+b)}$$

we have to decide either for a or b in each factor $(a + b)$. If we decide in the factors with the numbers $1 \leq i_1 < i_2 < \cdots < i_k \leq n$ for a, then each such choice corresponds uniquely to a k-subset of $\{1, 2, \ldots, n\}$. For this we have by Theorem 9.20 exactly $\binom{n}{k}$ possibilities. If we decide k times for a, then b occurs exactly $n - k$ times. Hence, we have exactly $\binom{n}{k}$ summands $a^k b^{n-k}$. This means for the whole sum that

$$(a + b)^n = \sum_{k=0}^{n} \binom{n}{k} a^k b^{n-k}.$$

$\square$

From the binomial theorem we can prove that the total number of subsets of a set with n elements is 2^n.

Corollary 9.23. *A set M with n elements, $n \in \mathbb{N}_0$, has 2^n subsets, that is,*

$$|\mathcal{P}(M)| = 2^n \quad \text{if } |M| = n.$$

Here $\mathcal{P}(M)$ is the power set of M.

Proof. By Theorem 9.20, we get

$$|\mathcal{P}(M)| = \sum_{k=0}^{n} \binom{n}{k}$$

and

$$\sum_{k=0}^{n} \binom{n}{k} = (1 + 1)^n = 2^n$$

by Theorem 9.22.

$\square$

Corollary 9.24. *For all $n \in \mathbb{N}$ we have*

$$\sum_{k=0}^{n} \binom{n}{k}(-1)^k = 0.$$

Proof. Take $a = -1$, $b = 1$ in Theorem 9.22. □

Corollary 9.25.

$$\binom{m+n}{m} = \sum_{k=0}^{m} \binom{m}{k}\binom{n}{m-k}$$

for all $n, m \in \mathbb{N}_0$.

Proof. We apply Theorem 9.22 for both sides of the equation

$$(1+t)^{m+n} = (1+t)^m (1+t)^n$$

and compare the coefficients of t^m. □

Remark 9.26. We get

$$\sum_{k=0}^{m} \binom{m}{k}\binom{n}{m-k} \underset{\substack{\uparrow \\ \text{let } j=m-k}}{=} \sum_{j=0}^{m} \binom{m}{m-j}\binom{n}{j} = \sum_{k=0}^{m} \binom{m}{k}\binom{n}{k}$$

because $\binom{m}{m-j} = \binom{m}{j}$.

9.3.2 The Occupancy Problem

The sampling problem is equivalent to the *occupancy problem*. Here we have n distinguishable cells and k particles. In how many ways can the particles be distributed in the cells?

The solution to this is equivalent to sampling k objects out of n objects with the following translations:

$$\text{distinguishable particles} \equiv \text{ordered sample}$$
$$\text{with exclusion} \equiv \text{without replacement}$$

This is illustrated in the following Table 9.2.

The occupancy problem plays a role in atomic structure, in particular, in finding how electrons are distributed among the atomic orbitals. The names in the table indicate the physical model for this: F. C. Maxwell (1831–1879), C. Boltzmann

Table 9.2: Occupancy Problems.

Distribute k objects in n cells	distinguishable particles	indistinguishable particles	
without exclusion	n^k Maxwell–Boltzmann statistics	$\binom{n+k-1}{k}$ Bose–Einstein statistics	with replacement
with exclusion	$_nP_k = (n)_k$	$_nC_k = \binom{n}{k}$ Fermi–Dirac statistics	without replacement
	ordered sample	unordered sample	sample k objects out of n objects

(1849–1906), S. Bose (1894–1974), A. Einstein (1879–1974), E. Fermi (1901–1954), and P. Dirac (1902–1984).

9.3.3 Some Further Comments

We have some further interpretations of the basic sampling formulas in terms of maps between sets.

Let $k, n \in \mathbb{N}$.

(1) $\frac{n!}{(n-k)!} = \,_nP_k$ is the number of all injective maps of k-set A into an n-set B. To see this, consider, without loss of generality, $A = \{1, 2, \ldots, k\}$ and $B = \{a_1, a_2, \ldots, a_n\}$. For each $a \in A$ there is an arrow starting at a, and for each $b \in B$, which is affected, there is exactly one arrow ending at b. We have necessarily $1 \le k \le n$.

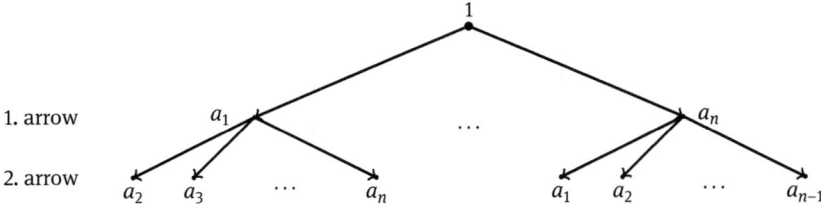

Figure 9.3: Injective maps.

There are n possibilities for the first arrow. There are $n - 1$ possibilities for the second arrow. There are $n - k + 1$ possibilities for the kth arrow, see Figure 9.3. Hence altogether

$$_nP_k = n(n-1)\cdots(n-k+1) = (n)_k.$$

Remark 9.27. If $n = k$, then each injective map $f : A \to B$ is already bijective, and there are $n!$ bijective maps.

If $A = B$, then these form a group, isomorphic to the permutation group S_n. If $n > k$, then $(n)_k$ may be considered as the number of k-tuples of B, in which all elements are pairwise distinct (k-permutations of B).

Application. *Pigeonhole principle*
If $|A|$ balls are put into $|B|$ drawers, and if $m \cdot |B| < |A| < \infty$, $m \in \mathbb{N}$, then there is one drawer with $m + 1$ or more balls.

(2) n^k is the number of all maps from a k-set to an n-set. Let, without loss of generality, $A = \{1, 2, \ldots, k\}$ and $B = \{a_1, a_2, \ldots, a_n\}$.

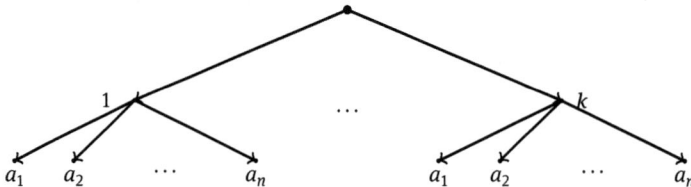

Figure 9.4: All maps.

For each $a \in A$ there are n possibilities to choose the image of a, and there are k elements in A, hence we have

$$\underbrace{n \cdot n \cdots n}_{k \text{ times}} = n^k \quad \text{possibilities, see Figure 9.4.}$$

Application. Let $|M| = n$. For each subset $A \subset M$ there exists a unique characteristic function

$$\chi_A : M \to \{0, 1\},$$

$$\chi_A(a) = \begin{cases} 1, & \text{if } a \in A, \\ 0, & \text{if } a \notin A. \end{cases}$$

This means that there exists a bijection between $\mathcal{P}(M)$ and the set $\mathcal{A}$ of maps from M into $\{0, 1\}$. In particular,

$$|\mathcal{P}(M)| = |\mathcal{A}| = 2^n.$$

(3) $\binom{n}{k} = {}_nC_k$ is the number of injective, monotonically increasing maps from $\{1, 2, \ldots, k\}$ to $\{1, 2, \ldots, n\}$.

We necessarily have $1 \leq k \leq n$. This we see as follows. We may order the images $f(i) = a_i$ by size as

$$a_1 < a_2 < \cdots < a_k.$$

(4) $\binom{n+k-1}{k}$ is the number of monotonically increasing maps from $\{1, 2, \ldots, k\}$ to $\{1, 2, \ldots, n\}$. This we see as follows.

As in part 3, we may order the images $f(i) = a_i$ by size, but the images this time are not necessarily distinct so that

$$a_1 \leq a_2 \leq \cdots \leq a_k.$$

Hence the number of monotonically increasing maps from $\{1, 2, \ldots, k\}$ to $\{1, 2, \ldots, n\}$ is equal to the number of k-repetitions of $\{1, 2, \ldots, n\}$, which is $\binom{n+k-1}{k}$.

9.4 Multinomial Coefficients

An extension of the binomial coefficients are the *multinomial coefficients* which count the number of ways to partition a set with n elements into k subsets of respective sizes $n_1, \ldots, n_k$ with $n_1 + \cdots + n_k = n$.

Definition 9.28. Let $n_1, n_2, \ldots, n_k \in \mathbb{N}_0$, $k \in \mathbb{N}$ and $n = n_1 + n_2 + \cdots + n_k$. Then

$$\binom{n}{n_1, n_2, \ldots, n_k} = \frac{n!}{n_1! n_2! \cdots n_k!}$$

is called the *multinomial coefficient*. If $n = k + m$, then

$$\binom{n}{k, m} = \binom{n}{k} = \binom{n}{m}.$$

The following extension of the binomial theorem is called the *multinomial theorem*.

Theorem 9.29 (Multinomial Theorem). *Let $x_1, x_2, \ldots, x_k$, $k \geq 1$, be elements of an integral domain with unity 1, and let $n \in \mathbb{N}_0$. Then*

$$(x_1 + x_2 + \cdots + x_k)^n = \sum_{\substack{n_1, n_2, \ldots, n_k \geq 0 \\ n_1 + n_2 + \cdots + n_k = n}} \binom{n}{n_1, n_2, \ldots, n_k} x_1^{n_1} x_2^{n_2} \cdots x_k^{n_k}.$$

Proof. The proof uses the binomial theorem (Theorem 9.22) and induction on k.

First, for $k = 1$, both sides are equal x_1^n since there is only one term $n_1 = n$. For the induction step, suppose the multinomial theorem holds for k. Then

$$(x_1 + x_2 + \cdots + x_k + x_{k+1})^n = (x_1 + x_2 + \cdots + x_{k-1} + (x_k + x_{k+1}))^n$$

$$= \sum_{n_1+n_2+\cdots+n_{k-1}+m=n} \binom{n}{n_1, n_2, \ldots, n_{k-1}, m} x_1^{n_1} x_2^{n_1} \cdots x_{k-1}^{n_{k-1}} (x_k + x_{k+1})^m$$

by the induction hypothesis. Applying the binomial theorem to the last factor, we get

$$(x_1 + x_2 + \cdots + x_k + x_{k+1})^n$$

$$= \sum_{n_1+n_2+\cdots+n_{k-1}+m=n} \binom{n}{n_1, n_2, \ldots, n_{k-1}, m} x_1^{n_1} x_2^{n_2} \cdots x_{k-1}^{n_{k-1}} \sum_{n_k+n_{k+1}=m} \binom{m}{n_k, n_{k+1}} x_k^{n_k} x_{k+1}^{n_{k+1}}$$

$$= \sum_{n_1+n_2+\cdots+n_{k+1}=n} \binom{n}{n_1, n_2, \ldots, n_k} x_1^{n_1} x_2^{n_2} \cdots x_{k+1}^{n_{k+1}},$$

which completes the induction. The last step follows because

$$\binom{n}{n_1, n_2, \ldots, n_{k-1}, m} \binom{m}{n_k, n_{k+1}} = \binom{n}{n_1, n_2, \ldots, n_{k+1}}. \qquad \square$$

Theorem 9.30. *Consider n objects of k varieties, where n_j (not distinguishable) elements from the jth variety exist ($j = 1, 2, \ldots, k$). We call $(n_1, n_2, \ldots, n_k)$ the specification of the n objects. Then the number of the possible arrangements of the n objects as a n-sequence is given by*

$$\binom{n}{n_1, n_2, \ldots, n_k}.$$

Proof. Consider a fixed specification $(n_1, n_2, \ldots, n_k)$. Let a be the wanted number. If we replace the n_1 elements of the first variety by n_1 distinguishable elements, then the number of the possible arrangements increases by the factor $n_1!$. The same holds if we do that for all varieties, and we finally get $a \cdot n_1! \cdot n_2! \cdots n_k!$ possibilities. Since now all elements are distinct, we get

$$a \cdot n_1! \cdot n_2! \cdots n_k! = n!,$$

which gives

$$a = \binom{n}{n_1, n_2, \ldots, n_k}. \qquad \square$$

Example 9.31. For four objects with specification $(1, 1, 2)$, we have the following 12 arrangements:

a	b	c	c		b	a	c	c		c	b	c	a		c	a	b	c
a	c	b	c		b	c	a	c		c	c	a	b		c	a	c	b
a	c	c	b		b	c	c	a		c	c	b	a		c	b	a	c.

9.5 Sizes of Finite Sets and the Inclusion–Exclusion Principle

The inclusion–exclusion principle is a counting technique which generalizes the familiar method of obtaining the numbers of elements in the union of two finite sets. Let A, B be two finite sets. If $A \cap B = \emptyset$, then

$$|A \cup B| = |A| + |B|.$$

But, if $A \cap B \neq \emptyset$, then

$$|A \cup B| = |A| + |B| - |A \cap B|.$$

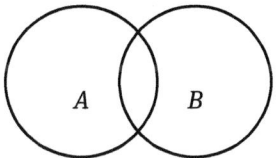

Figure 9.5: Intersection of two sets.

This expresses the fact that the sum of $|A|$ and $|B|$ may be too large since some elements may be counted twice. The double-counted elements are those in the intersection $A \cap B$, and the count is corrected by subtracting the size of $A \cap B$, see Figure 9.5.

The principle is more clearly seen in the case of three finite sets A, B and C. We then get

$$|A \cup B \cup C| = |A| + |B| + |C| - |A \cap B| - |A \cap C| - |B \cap C| + |A \cap B \cap C|.$$

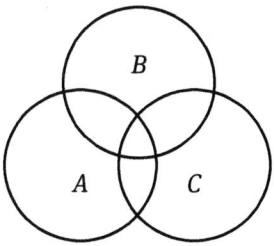

Figure 9.6: Intersection of three sets.

When removing the contributions of the over-counted elements in the pairwise intersections, the number of elements in the mutual intersection of the three sets has been subtracted too often, so must be added back in to get the correct total, see Figure 9.6.

Example 9.32. How many integers in $\{1, 2, \ldots, 1000\}$ are divisible by $3, 5$ or 8. Define A and B and C to be the sets of numbers in $\{1, 2, \ldots, 1000\}$ divisible by 3 and 5 and 8, respectively. Then

$$|A| = \left\lfloor \frac{1000}{3} \right\rfloor = 333,$$

$$|B| = \left\lfloor \frac{1000}{5} \right\rfloor = 200,$$

$$|C| = \left\lfloor \frac{1000}{8} \right\rfloor = 125,$$

$$|A \cap B| = \left\lfloor \frac{1000}{15} \right\rfloor = 66,$$

$$|A \cap C| = \left\lfloor \frac{1000}{24} \right\rfloor = 41,$$

$$|B \cap C| = \left\lfloor \frac{1000}{40} \right\rfloor = 25 \quad \text{and}$$

$$|A \cap B \cap C| = \left\lfloor \frac{1000}{120} \right\rfloor = 8.$$

Here $\lfloor x \rfloor$ means the biggest integer less than or equal to x (Gauss bracket). Hence

$$|A \cup B \cup C| = 333 + 200 + 125 - 66 - 41 - 25 + 8 = 534.$$

Consequently, the number of integers in $\{1, 2, \ldots, 1000\}$, which are not divisible by $3, 5$ or 8, is then

$$1000 - 534 = 466.$$

Generalizing the result of these examples gives the principle of inclusion–exclusion. To find the number of elements of the union of r finite sets $M_1, M_2, \ldots, M_r$:
(1) Include the numbers $|M_i|$,
(2) Exclude the numbers $|M_i \cap M_j|$ for $i \neq j$,
(3) Include the numbers $|M_i \cap M_j \cap M_k|$ for $i \neq j \neq k \neq i$,
(4) Exclude the numbers of elements of the quadruple-wise intersections,
(5) Include the numbers of elements of the quintuple-wise intersections,
(6) Continue, until the number of elements of the r-tuple-wise intersections is included (if r is odd) or excluded (if r is even).

The name comes from the idea that the principle is based on over-generous inclusion, followed by compensating exclusion. The concept is attributed to A. de Moivre (1667–1754), but it first appears in a paper of D. da Silva (1814–1878) and later in a paper by J. J. Sylvester (1814–1897). Often the principle is referred to as the formula of Sylvester.

Theorem 9.33. *Let $M_1, M_2, \ldots, M_r$ be finite subsets of the set M. Then*

$$|M_1 \cup M_2 \cup \cdots \cup M_r| = \sum_{k=1}^{r} (-1)^{k+1} \sum_{1 \le i_1 < \cdots < i_k \le r} |M_{i_1} \cap \cdots \cap M_{i_k}|.$$

Proof. The proof uses the formula for two sets and induction on r.
The statement holds for $r = 1$ and $r = 2$. Now, let $r \ge 3$. Then

$$|M_1 \cup M_2 \cup \cdots \cup M_r| = \left|(M_1 \cup M_2 \cup \cdots \cup M_{r-1}) \cup M_r\right|$$

$$= |M_1 \cup M_2 \cup \cdots \cup M_{r-1}| + |M_r| - \left|(M_1 \cap M_2 \cap \cdots \cap M_{r-1}) \cap M_r\right|$$

$$= |M_1 \cup M_2 \cup \cdots \cup M_{r-1}| + |M_r| - \left|(M_1 \cap M_r) \cup (M_2 \cap M_r) \cup \cdots \cup (M_{r-1} \cap M_r)\right|.$$

Now we use the induction hypothesis for the first and third term in the last sum. The third term gives the summands for which $k \ge 2$ and M_r is present. This finally gives the result. $\qquad\square$

Remark 9.34.
(1) In applications it is common to see the principle in its complementary form. That is, let M be a finite set and $M_1, M_2, \ldots, M_r$ be subsets of M. Then

$$\left|M \setminus (M_1 \cup M_2 \cup \cdots \cup M_r)\right|$$

$$= |M| - |M_1 \cup M_2 \cup \cdots \cup M_r|$$

$$= |M| + \sum_{k=1}^{r} (-1)^k \sum_{1 \le i_1 < i_2 < \cdots < i_k \le r} |M_{i_1} \cap M_{i_2} \cap \cdots \cap M_{i_k}|.$$

More precisely, let M be a finite set. We ask for the number of all elements of M which do not have certain properties $E_1, E_2, \ldots, E_r$. We then define

$$M_i = \{x \mid x \in M \mid x \text{ has property } E_i\}$$

and ask for $|M \setminus (M_1 \cup M_2 \cup \cdots \cup M_r)|$.

(2) Let M be a finite set and $M_1, M_2, \ldots, M_r$ be subsets of M. Let G be an Abelian group with addition $+$. Suppose there exists a weight function $\omega : M \to G$ with $\omega(A) = \sum_{x \in A} \omega(x)$ for $A \subset M$. Then

$$\omega(M_1 \cup M_2 \cup \cdots \cup M_r) = \sum_{k=1}^{r} (-1)^{k+1} \sum_{1 \le i_1 < i_2 < \cdots < i_k \le r} \omega(M_{i_1} \cap M_{i_2} \cap \cdots \cap M_{i_k}).$$

In Theorem 9.33 we have $G = \mathbb{Z}$ and $\omega(x) = 1$ for each $x \in M$.

Applications.

(1) Let $n = p_1^{\alpha_1} p_2^{\alpha_2} \cdots p_r^{\alpha_r}$, $r \in \mathbb{N}_0$, $\alpha_i \in \mathbb{N}$ for $i = 1, 2, \ldots, r$, and $p_1 < p_2 < \cdots < p_r$ be prime numbers, the prime factorization of $n \in \mathbb{N}$, where $n = 1$ if $r = 0$ (see [12]). Let $\varphi(n)$ be the number of elements a in the set $\{1, 2, \ldots, n\}$ which are coprime to n, that is,

$$\varphi(n) = |\{a \mid a \in \{1, 2, \ldots, n\} \text{ and } \gcd(a, n) = 1\}|.$$

The function $\varphi : \mathbb{N} \to \mathbb{N}$ is called *Euler's φ-function* (see [12]). Then

$$\varphi(n) = n\left(1 - \frac{1}{p_1}\right)\left(1 - \frac{1}{p_2}\right)\cdots\left(1 - \frac{1}{p_r}\right).$$

Proof. This is clear for $n = 1$. Now let $n \geq 2$. We first remark that, if $d \in \mathbb{N}$ is a divisor of n, then there are exactly $\frac{n}{d}$ numbers in $\{1, 2, \ldots, n\}$ which are divisible by d. This holds because from $d \leq md \leq n$ we get $1 \leq m \leq \frac{n}{d}$.
Now define

$$M = \{1, 2, \ldots, n\} \quad \text{and} \quad M_i = \{x \mid x \in M \text{ and } p_i \mid x\} \quad \text{for } i = 1, 2, \ldots, r.$$

We ask for $|M \setminus (M_1 \cup M_2 \cup \cdots \cup M_r)|$, which we get from

$$|M \setminus (M_1 \cup M_2 \cup \cdots \cup M_r)| = |M| + \sum_{k=1}^{r}(-1)^k \sum_{1 \leq i_1 < i_2 < \cdots < i_k \leq r} |M_{i_1} \cap M_{i_2} \cap \cdots \cap M_{i_k}|$$

$$= n + \sum_{k=1}^{n}(-1)^k \sum_{1 \leq i_1 < i_2 < \cdots < i_k \leq r} \frac{n}{p_{i_1} p_{i_2} \cdots p_{i_k}}$$

$$= n\left(1 - \frac{1}{p_1}\right)\left(1 - \frac{1}{p_2}\right)\cdots\left(1 - \frac{1}{p_r}\right). \qquad \square$$

(2) Let $M_1, M_2, \ldots, M_r$ be finite subsets of the set M. The number of elements of M which belong to exactly m of the subsets ($1 \leq m \leq r$) is given by

$$\sum_{k=m}^{r}(-1)^{k+m}\binom{k}{m} \sum_{1 \leq i_1 < i_2 < \cdots < i_k \leq r} |M_{i_1} \cap M_{i_2} \cap \cdots \cap M_{i_k}|.$$

Proof. Let $x \in M$ be an element which belongs to exactly s of the subsets ($0 \leq s \leq r$). We count what x contributes to the above sum.
If $s < m$ then x does not contribute anything to the sum because $k \geq m$ and, hence, x does not belong to a k-tuple-wise intersection.
If $s \geq m$ then the contribution is

$$\sum_{k=m}^{s}(-1)^{k+m}\binom{k}{m}\binom{s}{k}.$$

This can be seen as follows: x belongs to exactly $\binom{s}{k}$ of the k-tuple-wise intersections of those sets, in which x is an element. If $s = m$ then this sum is equal to 1. If $s > m$ we get 0 (which proves the statement):

$$\sum_{k=m}^{s} (-1)^{k+m} \binom{k}{m}\binom{s}{k} = \sum_{\ell=0}^{s-m} (-1)^\ell \binom{\ell + m}{m}\binom{s}{\ell + m}$$

$$= \binom{s}{m} \sum_{\ell=0}^{s-m} (-1)^\ell \frac{(s-m)!}{\ell!(s-\ell-m)!} = 0,$$

by Corollary 9.24. □

(3) Here we start with a classical problem, the "Problème de recontres" of R. de Montmart (1678–1719):

n married couples meet together for a dancing party. How many possibilities exist to form dancing couples so that no married couple dances together? This problem will be solved by an interpretation through the number of permutations of $\{1, 2, \ldots, n\}$ with no fixed point.

We repeat some of the definitions from [12].

(a) Let M be a set and $f : M \to M$ be a bijection. We call f a *permutation of M*.

(b) An element $x \in M$ is called a *fixed point of the permutation* $f : M \to M$ if $f(x) = x$.

(c) A permutation $f : M \to M$ is called a *derangement* if f has no fixed point, that is, $f(x) \neq x$ for all $x \in M$.

(d) The set of all permutations of $\{1, 2, \ldots, n\}$ forms a group S_n with respect to the function composition, the symmetric group S_n. The order of S_n is $n!$.

The number of derangements in S_n is

$$D_n = n! \sum_{k=0}^{n} \frac{(-1)^k}{k!}.$$

The numbers D_n are called *Recontres numbers*.

Proof. Let $M = \{1, 2, \ldots, n\}$, and let P_k be the set of all permutations which fix k; P_k is a subgroup of S_n. Then

$$D_n = |S_n \setminus (P_1 \cup P_2 \cup \cdots \cup P_n)|$$

$$= n! + \sum_{k=1}^{n} (-1)^k \sum_{1 \le i_1 < i_2 < \cdots < i_k \le n} (n-k)!$$

$$= n! + \sum_{k=1}^{n} (-1)^k \binom{n}{k}(n-k)!$$

$$= n! \sum_{k=0}^{n} (-1)^k \frac{1}{k!}.$$

□

We give a different interpretation. A person writes n letters and the corresponding envelops. Then he randomly puts the letters into the envelopes. What is the probability that no recipient gets the letter which is meant for him? This probability is given by

$$p_n = \frac{D_n}{n!} = \sum_{k=0}^{n} \frac{(-1)^k}{k!} \to \frac{1}{e}$$

as $n \to \infty$ (see [12]).

For a formal definition of a finite probability, we refer to the following Chapter 10. The number of permutations of S_n with exactly m fixed points is

$$D_{n,m} = \frac{n!}{m!} \sum_{k=0}^{n-m} (-1)^k \frac{1}{k!}.$$

Proof. For the choice of m fixed points we have $\binom{n}{m}$ possibilities, the remaining elements have to be mapped then without fixed points.

Hence, $D_{n,m} = \binom{n}{m} D_{n-m}$ which gives the statement. $\qquad\square$

(4) Let M, N be sets with $|M| = m$ and $|N| = n$ where $1 \leq n \leq m$. The number of surjective maps $f : M \to N$ is given by

$$\sum_{k=0}^{n} (-1)^k \binom{n}{k} (n-k)^m.$$

Proof. Without loss of generality, let $N = \{1, 2, \ldots, n\}$. Let A be the set of all mappings $M \to N$ and A_k the set of mappings $f \in A$ with $k \notin f(M)$.

Then $A \setminus (A_1 \cup A_2 \cup \cdots \cup A_n)$ is the set of all surjective maps and their number is

$$S_{m,n} = n^m + \sum_{k=1}^{n} (-1)^k \sum_{1 \leq i_1 < i_2 < \cdots < i_k \leq n} |A_{i_1} \cap A_{i_2} \cap \cdots \cap A_{i_k}|.$$

Here $A_{i_1} \cap A_{i_2} \cap \cdots \cap A_{i_k}$ is the set of all mappings $M \to N \setminus \{i_1, i_2, \ldots, i_k\}$; their number is $(n-k)^m$. Hence

$$S_{m,n} = n^m + \sum_{k=1}^{n} (-1)^k \binom{n}{k} (n-k)^m = \sum_{k=0}^{n} (-1)^k \binom{n}{k} (n-k)^m.$$

$\qquad\square$

Theorem 9.35 (Inversion formula). *Let R be an integral domain with unity 1 and $\varphi :$ $\mathbb{N}_0 \to R$. Define $\psi(k) := \sum_{\ell=0}^{k} \binom{k}{\ell} \varphi(\ell)$ for $k \in \mathbb{N}_0$. Then*

$$\varphi(k) = \sum_{\ell=0}^{k} (-1)^\ell \binom{k}{\ell} \psi(k - \ell).$$

Proof.

$$\sum_{\ell=0}^{k}(-1)^{\ell}\binom{k}{\ell}\psi(k-\ell)=\sum_{\ell=0}^{k}\sum_{r=0}^{k-\ell}(-1)^{\ell}\binom{k}{\ell}\binom{k-\ell}{r}\varphi(r)$$

$$=\sum_{\substack{\ell,r\\0\le\ell,r\le k}}(-1)^{\ell}\binom{k}{\ell}\binom{k-\ell}{r}\varphi(r)$$

$$=\sum_{r=0}^{k}\varphi(r)\left(\sum_{\ell=0}^{k}(-1)^{\ell}\binom{k}{\ell}\binom{k-\ell}{r}\right).$$

Now

$$\binom{k}{\ell}\binom{k-\ell}{r}=\begin{cases}0 & \text{if } k-\ell<r, \text{ that is, } \ell>k-r,\\ \binom{k}{r}\binom{k-r}{\ell} & \text{if } \ell\le k-r.\end{cases}$$

If we plug this into the above equation, we get

$$\sum_{\ell=0}^{k}(-1)^{\ell}\binom{k}{\ell}\psi(k-\ell)=\sum_{r=0}^{k}\varphi(r)\underbrace{\left(\sum_{\ell=0}^{k-r}(-1)^{\ell}\binom{k-r}{\ell}\right)}_{=\begin{cases}1 & \text{if } k=r,\\ 0 & \text{if } k>r.\end{cases}\\ \text{by Corollary 9.24}}\binom{k}{r}=\varphi(k). \qquad \square$$

Applications.

(1) Let $k,n\in\mathbb{N}$ with $k\le n$. We denote the number of surjective maps

$$f:\{1,2,\ldots,n\}\rightarrow\{1,2,\ldots,k\}$$

by $S_{n,k}$ and define numbers $S(n,k)$ by $S(n,k)=\frac{1}{k!}S_{n,k}$. We will see in the next section that the $S(n,k)$ are the Stirling numbers (of the second type). We have the relation

$$k^{n}=\sum_{v=0}^{k}\frac{k!}{(k-v)!}S(n,v).$$

Proof. We consider the maps

$$f:\{1,2,\ldots,n\}\rightarrow\{1,2,\ldots,k\}.$$

There are k^{n} of them. Now we split these as follows:
For a v-subset of $\{1,2,\ldots,k\}$ there are $S_{n,v}=v!S(n,v)$ surjective maps from $\{1,2,\ldots,n\}$ onto this v-subset, and there are $\binom{n}{v}$ such v-subsets of $\{1,2,\ldots,k\}$. Hence

$$k^{n}=\sum_{v=0}^{k}\frac{k!}{(k-v)}S(n,v). \qquad \square$$

Now $\frac{k!}{(k-v)!} = \binom{k}{v}v!$, and from the inversion formula we get

$$k!S(n,k) = \sum_{v=0}^{k}(-1)^v \binom{k}{v}(k-v)^n = S_{n,k}$$

for the number of surjective maps $\{1,2,\ldots,n\} \to \{1,2,\ldots,k\}$. This we got also with the inclusion–exclusion principle.

(2) Let D_n be the number of $\sigma \in S_n$ without fixed points, that is, the Recontres numbers. We know that

$$S_n = \bigcup_{k=0}^{n} \underbrace{\{\sigma \mid \sigma \text{ as exactly } k \text{ fixed points}\}}_{=:F_k}.$$

We have from above that

$$|F_k| = D_{n,k} = \binom{n}{k}D_{n-k},$$

where $D_{n,k}$ is the number of permutations in S_n with exactly k fixed points. Hence

$$n! = \sum_{k=0}^{n}\binom{n}{k}D_{n-k} = \sum_{k=0}^{n}\binom{n}{k}D_k.$$

From the inversion formula we now get

$$D_n = \sum_{k=0}^{n}(-1)^k\binom{n}{k}(n-k)! = n!\sum_{k=0}^{n}(-1)^k\frac{1}{k!}.$$

This we also got with the inclusion–exclusion principle.

9.6 Partitions and Recurrence Relations

In this section we consider the number of possible partitions of a finite set. We recall some definitions.

Let $M \neq \emptyset$ be a set and $S \subset \mathcal{P}(M)$. If

(1) $\emptyset \notin S$, $M = \bigcup_{X \in S} X$ and
(2) $X \cap Y = \emptyset$ for $X, Y \in S$, $X \neq Y$,

then S is called a *partition of M*.

Remark 9.36. A partition S defines an equivalence relation on M by

$$a \sim b \quad \Leftrightarrow \quad a, b \in X \text{ for some } X \in S \quad (a, b \in M).$$

Conversely, if ~ is an equivalence relation on M, then the set of the equivalence classes defines a partition of M.

Hence the partitions of a nonempty set M correspond uniquely to the equivalence relations on M.

Definition 9.37. Let $M \neq \emptyset$ be a set and $k \in \mathbb{N}$.

(1) A k-partition of M is a partition which is composed of k nonempty subsets of M, that is,

$$M = \bigcup_{i=1}^{k} M_i \quad \text{with } M_i \subset M, M_i \neq \emptyset$$

for $i = 1, 2, \ldots, k$ and $M_i \cap M_j = \emptyset$ for $i \neq j$.

(2) Let $|M| = n \in \mathbb{N}$. We denote the number of distinct k-partitions of M as $S(n, k)$. Further, we set $S(0, 0) := 1$, $S(n, 0) = 0$ for $n > 0$ and $S(n, k) = 0$ for $0 \leq n < k$. The numbers $S(n, k)$ are called *Stirling numbers* (of the second kind).

From the discussion above we get the following.

Theorem 9.38. *Let M be a nonempty set and $|M| = n$. The number of equivalence relations with exactly k equivalence classes of M is $S(n, k)$.*

We now give a recursion formula for the Stirling numbers. They are named after J. Stirling (1692–1770).

Theorem 9.39 (Recursion formula). *Let $k, n \in \mathbb{N}$. Then*

$$S(n, k) = S(n - 1, k - 1) + k \cdot S(n - 1, k).$$

Proof. The formula is correct for $k = 1$ because $S(n, 1) = 1$ for $n = 1$ and

$$1 = S(1, 1) = S(0, 0) + S(0, 1) = 1 + 0$$

and

$$1 = S(n, 1) = S(n - 1, 0) + S(n - 1, 1) = 0 + 1$$

for $n \geq 2$.

If $k > n$, then the formula is correct because

$$0 = S(n, k) = S(n - 1, k - 1) = S(n - 1, k).$$

Hence, let now $|M| = n \geq 2$ and $1 < k \leq n$.

Let $x \in M$. We divide the k-partitions of M into two distinct types:

(1) $\{x\}$ is an element of the k-partition. Then the remaining elements of the partition form a $(k-1)$-partition of $M \setminus \{x\}$. Hence, the number of k-partitions of this type is $S(n-1, k-1)$.

(2) $x \in A$ where $|A| \geq 2$ and A is an element of the k-partition. If we remove x from M, then we get a k-partition of $M \setminus \{x\}$. On the other hand, if

$$M \setminus \{x\} = \bigcup_{i=1}^{k} N_i$$

is a k-partition of $M \setminus \{x\}$, then we may extend each of the N_i by x to get a k-partition of M of the second type.

Therefore, the number of k-partitions of this second type is $kS(n-1, k)$. Hence,

$$S(n, k) = S(n-1, k-1) + kS(n-1, k). \qquad \square$$

Example 9.40.
(1) $S(n, n) = 1$ for $n \in \mathbb{N}_0$.
(2) $S(4, 2) = 7$.
(3) $S(4, 3) = 6$.
(4) $S(n, n-1) = \binom{n}{2}$ for $n \geq 2$.

Proof. We prove this by induction on n. If $n = 2$, then $S(2, 1) = 1 = \binom{2}{2}$. Now, let $n \geq 3$ and suppose the statement is correct for $n-1$. Then

$$S(n, n-1) = S(n-1, n-2) + (n-1)S(n-1, n-1)$$
$$= \binom{n-1}{2} + (n-1) = \frac{(n-1)!}{2!(n-3)!} + (n-1)$$
$$= \frac{(n-1)!}{2!(n-3)!}\left(1 + \frac{2}{n-2}\right)$$
$$= \frac{n!}{2!(n-2)!} = \binom{n}{2}. \qquad \square$$

(5) $S(n, 2) = 2^{n-1} - 1$ for $n \geq 1$.

Proof. We prove this by induction. The statement is correct for $n = 1$. Now, let $n \geq 2$. Then

$$S(n, 2) = S(n-1, 1) + 2S(n-1, 2) = 1 + 2(2^{n-2} - 1) = 2^{n-1} - 1$$

using the induction hypothesis. $\qquad \square$

Theorem 9.41. *Let $n, k \in \mathbb{N}$. Then*
(1) $S(n, k) = \sum_{i=0}^{n-1} k^{n-i-1} S(i, k-1)$.
(2) $S(n, k) = \sum_{i=0}^{n-1} \binom{n-1}{i} S(i, k-1)$.

Proof.
(1) The statement follows directly by induction from the recursion formula.
(2) We prove this combinatorially by counting the k-partitions of an n-set in a different manner.

Let $x \in M$ be chosen fixed. For each $i \in \{0, 1, \ldots, n-1\}$ there are $\binom{n-1}{n-1-i} = \binom{n-1}{i}$ possibilities to choose a subset A of M with $x \in A$ and $|A| = n - i$.
For each of these subsets, the number of $(k-1)$-partitions of $M \setminus A$ is equal to $S(i, k-1)$. $\qquad\square$

Theorem 9.42. *Let $k, n \in \mathbb{N}$. The number $S_{n,k}$ of surjective maps from an n-set onto a k-set is $S_{n,k} = k! S(n, k)$.*

Proof. If $k > n$, then there does not exist any surjective maps from an n-set to a k-set, in other words, $S(n, k) = 0$.

Now let $1 \le k \le n$.

Let $f : \{1, 2, \ldots, n\} \to \{1, 2, \ldots, k\}$ be surjective; f defines a k-partition of

$$\{1, 2, \ldots, n\} = \overset{\cdot}{\bigcup_{1 \le i \le k}} f^{-1}(\{i\}).$$

Each k-partition of $\{1, 2, \ldots, n\}$ provides $k!$ surjective maps from $\{1, 2, \ldots, n\}$ to $\{1, 2, \ldots, k\}$ given by the $k!$ permutations of $\{1, 2, \ldots, k\}$.

Hence the number $S_{n,k}$ of surjective maps of an n-set onto a k-set is $S_{n,k} = k! S(n, k)$. $\qquad\square$

Remark 9.43. This shows how meaningful the definition of $S(n, k)$ was in the last section:

$$S(n, k) = \frac{1}{k!} S_{n,k}$$

for $k, n \in \mathbb{N}$, $1 \le k \le n$.

Definition 9.44. Let $n \in \mathbb{N}_0$. Then

$$B_n := \sum_{k=0}^{n} S(n, k)$$

is called the nth *Bell number*; we have $B_0 = S(0, 0) = 1$. It is named after E. T. Bell (1883–1960) who wrote science fiction books with the pseudonym John Taire.

Theorem 9.45.

$$B_{n+1} = \sum_{k=0}^{n} \binom{n}{k} B_k \quad \text{for } n \in \mathbb{N}_0.$$

Proof. Let $M = \{1, 2, \ldots, n + 1\}$. For each $k \in \{0, 1, \ldots, n\}$ there are $\binom{n}{k}$ possibilities to choose a subset A of M with $|A| = k + 1$ and $(n + 1) \in A$. Further, there are B_{n-k} possibilities to partition the remaining set $M \setminus A$. □

The following formula is due to G. Dobinski who found it in 1877.

Theorem 9.46.

$$B_n = \frac{1}{e} \sum_{k=0}^{\infty} \frac{k^n}{k!}.$$

Proof. We use the formula

$$k! S(n, k) = \sum_{v=0}^{k} (-1)^v \binom{k}{v} (k - v)^n,$$

$n, k \in \mathbb{N}$, $1 \leq k \leq n$, from the last section.
This formula certainly also holds for $n = k = 0$.
Now let $N \geq n$. Then

$$B_n = \sum_{m=0}^{n} S(n, m) = \sum_{0 \leq m \leq N} \left(\frac{1}{m!} \sum_{k=0}^{N} (-1)^{m-k} \binom{m}{k} k^n \right)$$

$$= \sum_{0 \leq m \leq N} \left(\frac{1}{m!} \sum_{0 \leq k \leq m} (-1)^{m-k} \frac{m!}{k!(m-k)!} k^n \right)$$

$$= \sum_{0 \leq k \leq N} \left(\frac{k^n}{k!} \sum_{k \leq m \leq N} (-1)^{m-k} \frac{1}{(m-k)!} \right)$$

$$= \sum_{0 \leq k \leq N} \left(\frac{k^n}{k!} \sum_{0 \leq \ell \leq N-k} \frac{(-1)^\ell}{\ell!} \right).$$

Since $e^{-1} = \sum_{\ell=0}^{\infty} \frac{(-1)^\ell}{\ell!}$, we get the statement with $N \to \infty$. □

Remark 9.47. So far, we introduced the Stirling numbers $S(n, k)$ of the second type. There exist also Stirling numbers $s(n, k)$ of the first type.
To introduce these, we have to recall some facts about permutations from S_n (see [12]).

Definition 9.48. Let $M = \{1, 2, \ldots, n\}$. A *cycle* is a permutation σ of M which maps the elements of some subset A of M to each other in a cyclic fashion, while fixing all other elements of M. If A has k elements, the cycle is called a *k-cycle*.
In other words, if $A = \{a_1, a_2, \ldots, a_k\} \subset M$, then for each $i \in \{1, 2, \ldots, k\}$ there exists a j with $\sigma^j(a_1) = a_i$, and $\sigma(v) = v$ for all $v \notin A$.

Example 9.49. Let $M = \{1, 2, \ldots, 7\}$ and $A = \{2, 4, 5, 6, 7\}$. Let σ be given as in Table 9.3.

Table 9.3: Values for σ.

i	1	2	3	4	5	6	7
$\sigma(i)$	1	5	3	7	4	2	6

Now

$$2 = \sigma^0(2), \quad 5 = \sigma(2), \quad 4 = \sigma(5) = \sigma^2(2),$$
$$7 = \sigma(4) = \sigma^3(2) \quad \text{and} \quad 6 = \sigma(7) = \sigma^4(2).$$

Hence σ is a cycle.

We just write $\sigma = (2, 5, 4, 7, 6)$ for the cycle σ (recall that, with this writing, we may cyclically permute in $(2, 5, 4, 7, 6)$).

The set A is called the *orbit* of the cycle. Every permutation of S_n can be decomposed into a collection of cycles on disjoint orbits (see [12]).

Example 9.50. Let $M = \{1, 2, \ldots, 8\}$ and $\sigma \in S_8$ be given as in Table 9.4.

Table 9.4: Values for σ.

i	1	2	3	4	5	6	7	8
$\sigma(i)$	3	5	1	2	4	8	7	6

Then $\sigma = (1, 3)(2, 5, 4)(6, 8)(7)$.

Now we may define the Stirling numbers $s(n, k)$ of the first type.

Definition 9.51. The number of permutations from S_n, $n \in \mathbb{N}$, which are composed of k cycles, is called the *Stirling number $s(n, k)$ of the first kind*.

In addition, we define $s(0, 0) = 1$, $s(n, k) = 0$ for $k = 0$, $n \geq 1$, or $0 \leq n < k$.

Remarks 9.52.
(1) Since $n!$ is the number of all permutations in S_n, $n \in \mathbb{N}$, we get

$$\sum_{k=1}^{n} s(n, k) = n!.$$

This gives in particular that $s(n, k) \leq n!$.

(2) We have $s(n, n) = 1$ for $n \in \mathbb{N}_0$. This is clear by definition for $n = 0$.

Now let $n \geq 1$. Let $\sigma \in S_n$ be composed of n cycles. Then each cycle must have length 1, that is, each element of M is fixed. Then σ is the identity.

Theorem 9.53 (Recursion formula). *Let $n, k \in \mathbb{N}$. Then*

$$s(n, k) = s(n - 1, k - 1) + (n - 1)s(n - 1, k).$$

Proof. This is clear for $k > n$ by definition.

Now let $1 \leq k \leq n$. The proof is analogous to that of Theorem 9.38.

On the left side of the equation we have all permutations of $M = \{1, 2, \ldots, n\}$ with k cycles. There are two types of them:

(1) (n) is one of the cycles.

A permutation of this type arises by adding (n) to a permutation of the set $\{1, 2, \ldots, n - 1\}$ with $n - 1$ cycles. Here there are $s(n - 1, k - 1)$ possibilities.

(2) (n) is not one of the cycles.

All such permutations arise by inserting the element n to a permutation of $\{1, 2, \ldots, n - 1\}$ with k cycles. There are $s(n - 1, k)$ of such permutations, and for each of these we may insert in a cycle the element n after any element of the $n - 1$ remaining elements.

Hence there are $(n - 1)s(n - 1, k)$ possibilities. Altogether we get the recursion formula. □

Corollary 9.54. *Let $n \in \mathbb{N}$. Then*

$$s(n, 1) = (n - 1)!.$$

Proof. This is correct for $n = 1$. Now let $n \geq 2$. Then $s(n - 1, k - 1) = s(n - 1, 0) = 0$ for $k = 1$, and

$$s(n, 1) = (n - 1)(n - 2) \cdots 2 \cdot 1 = (n - 1)!,$$

by induction. □

Remarks 9.55.

(1) We consider the polynomial

$$f(x) = x(x - 1) \cdots (x - n + 1).$$

By expanding this polynomial we get

$$f(x) = \sum_{k=1}^{n} \tilde{s}(n, k)x^k.$$

We further define $\tilde{s}(0, 0) = 1$, $\tilde{s}(n, 0) = 0$ for $n \geq 1$ and $\tilde{s}(n, k) = 0$ for $n \in \mathbb{N}_0$, $k > n$. Via the recursion formula for $s(n, k)$ we get

$$\tilde{s}(n, k) = (-1)^{n-k} s(n, k) \quad \text{for } n, k \in \mathbb{N}_0.$$

(2) In many books one uses the *Karamata notation*

$$S(n,k) = \begin{Bmatrix} n \\ k \end{Bmatrix}, \quad s(n,k) = \begin{bmatrix} n \\ k \end{bmatrix},$$

which is named after J. Karamata (1902–1967), to demonstrate the analogy to the recursion formula for the binomial coefficients.

9.7 Decompositions of Naturals Numbers, Partition Function

So far we have considered decomposition of nonempty sets M with $|M| = n \in \mathbb{N}$. Let $M = M_1 \cup M_2 \cup \cdots \cup M_k$ be a k-partition of M with $k \geq 1$ nonempty, pairwise disjoint sets. This k-partition induces a decomposition of n into k positive summands:

$$n = |M_1| + |M_2| + \cdots + |M_k|.$$

We described the number of k-partitions of M by the Stirling numbers $S(n,k)$ (of second type). Nevertheless, distinct k-partitions may generate the same decomposition of n into the same k summands. In general, a partition of a natural number n is a way of writing n as a sum of positive integers where two sums that differ only in the order of their summands are considered the same partition. Hence we make the following definition.

Definition 9.56. A *partition* of $n \in \mathbb{N}$ into k, $1 \leq k \leq n$, positive summands is a sequence $(n_1, n_2, \ldots, n_k) \in \mathbb{N}^k$ with $n_1 \geq n_2 \geq \cdots \geq n_k \geq 1$ and $n = n_1 + n_2 + \cdots + n_k$.

Example 9.57. Number 4 can be partitioned in five distinct ways:

$$4, \quad 3+1, \quad 2+2, \quad 2+1+1, \quad 1+1+1+1.$$

Definition 9.58.
(1) A summand in a partition is called a *part*.
(2) Let $k, n \in \mathbb{N}_0$.
If $1 \leq k \leq n$ then $p(n,k)$ denotes the number of partitions of n with exactly k parts. Further we define $p(0,0) = 1$ and $p(n,k) = 0$ for $k = 0$, $n \geq 1$ or for $n < k$. The numbers $p(n,k)$ are called the (*arithmetic*) *partition numbers*.

Example 9.59.
(1) $p(4,2) = 2$ with $4 = 3+1 = 2+2$,
(2) $p(7,3) = 4$ with $7 = 5+1+1 = 4+2+1 = 3+3+1 = 3+2+2$.

Theorem 9.60 (Recursion formula). *If* $1 \leq k \leq n$ *then*

$$p(n,k) = p(n-1,k-1) + p(n-k,k).$$

Proof. The partitions of n into exactly k parts fall into two types:

(1) 1 is a part.

    If we cancel the part 1, then we get a partition of $n-1$ into exactly $k-1$ parts. There are $p(n-1, k-1)$ possibilities.

(2) 1 is not a part.

    Then we may subtract 1 from each part, and we get a partition of $n-k$ into exactly k parts. There are $p(n-k, k)$ possibilities. These types together give the recursion formula. □

There is a common diagram method to represent partitions as Ferrers diagrams, named after N. M. Ferrers (1829–1903), which is given by the points (i, j) with $i \in \{1, 2, \ldots, k\}, j = \{1, 2, \ldots, n_i\}$.

Example 9.61.

$$15 = 5 + 3 + 3 + 2 + 1 + 1$$

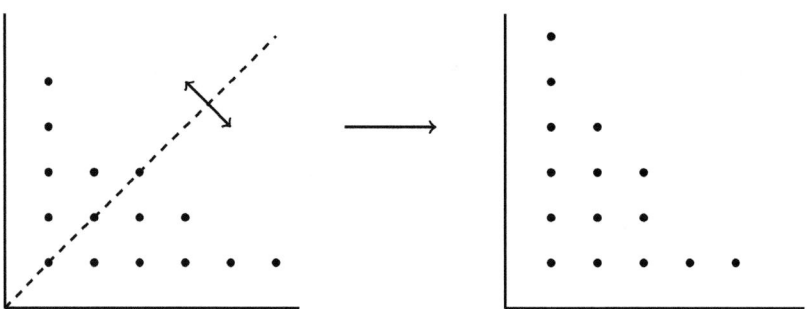

Figure 9.7: Ferrers diagram.

Reflection at the angle bisector of the quadrant translates a Ferrers diagram into a Ferrers diagram (for the same number n), see Figure 9.7. This way we get a bijection between the set of partitions of n with exactly k parts and the set of partitions of n with k as the biggest part.

Hence $p(n, k)$ are the partitions of n with k as the biggest part.

Finally, we define the summary partition function

$$p : \mathbb{N} \to \mathbb{N},$$

$$p(n) = \sum_{k=1}^{n} p(n, k).$$

Example 9.62. $p(4) = 5.$

9.8 Catalan Numbers

In this section, we finally just mention the Catalan numbers. The Catalan numbers form a sequence of natural numbers that occur in various counting problems, often involving recursively defined objects. They are named after E. C. Catalan (1814–1894). There are several ways to define the Catalan numbers.

Definition 9.63. Let $n \in \mathbb{N}_0$. Then C_n is the number of different ways $n + 1$ factors can be completely parenthesized, in pairs if $n \geq 1$, omitting the pair of a right-most and left-most bracket (or the number of ways of associating n applications of a binary operator).

The C_n, $n \in \mathbb{N}_0$, are called the *Catalan numbers*.

Certainly, $C_0 = 1$ because we then have only one factor x_1. To illustrate this we give the list for $1 \leq n \leq 3$ explicitly:

$$n = 1 : x_1 x_2$$
$$n = 2 : (x_1 x_2)x_3, \ x_1(x_2 x_3)$$
$$n = 3 : ((x_1 x_2)x_3)x_4, \ (x_1 x_2)(x_3 x_4), \ (x_1(x_2 x_3))x_4, \ x_1((x_2 x_3)x_4), x_1(x_2(x_3 x_4)).$$

Hence, $C_1 = 1$, $C_2 = 2$, $C_3 = 5$.

Theorem 9.64 (Recurrence relation).

$$C_0 = 1 \quad and \quad C_{n+1} = \sum_{k=0}^{n} C_k C_{n-k} \quad for \ n \geq 0.$$

Proof. This is clear for $n = 0$. Now let $n \geq 1$. An $(n + 2)$-fold product, provided accordingly with brackets, combines a $(k+1)$-fold product with an $(n+2)-(k+1)$-fold product, where $k \in \{0, 1, 2, \ldots, n\}$. For these partial products there are C_k and C_{n-k} possibilities, respectively. Hence, for the complete $(n + 2)$-fold product there are

$$C_{n+1} = \sum_{k=0}^{n} C_k C_{n-k} \quad \text{possibilities.} \qquad \square$$

There are many counting problems in combinatorics, solutions of which are given by the Catalan numbers. We give here three more examples:

(1) A Dyck word of length $2n$ is a string consisting of n X's and n Y's such that no initial (of the first half) segment of the string has more Y's than X's. C_n denotes the number of Dyck words of length $2n$. These are named after W. F. A. von Dyck (1856–1934).

For example, the following are the Dyck words of length 6:

$$XXXYYY, \quad XYXXYY, \quad XYXYXY, \quad XXYYXY, \quad XXYXYY.$$

(2) C_n is the number of rooted binary trees with n internal nodes. Here, we consider as binary trees those in which each node has zero or two children, and the internal nodes are those that have children.
The binary trees corresponding to $n = 3$ are the following, see Figure 9.8.

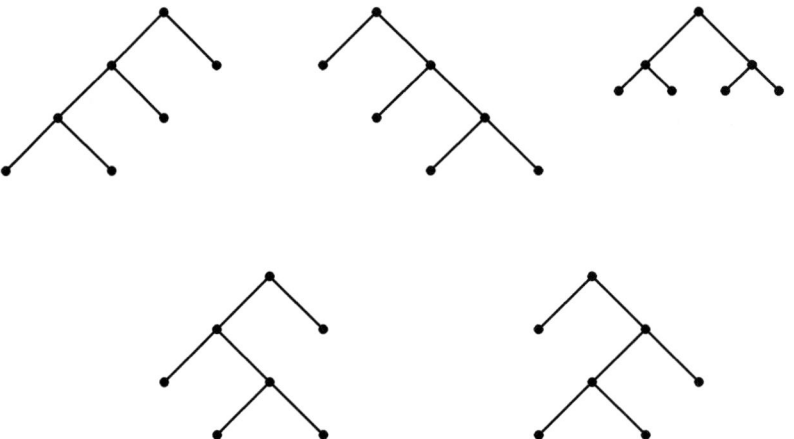

Figure 9.8: Binary trees corresponding to $n = 3$.

(3) C_n is the number of different ways a convex polygon with $n + 2$ sides can be cut into triangles by connecting vertices with straight lines. Figure 9.9 illustrates the case $n = 3$.

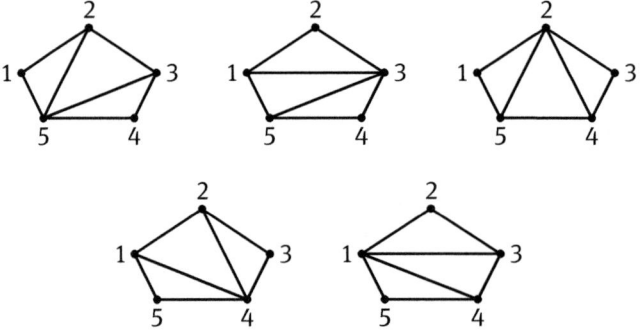

Figure 9.9: Cutting of convex polygons into triangles for $n = 3$.

For more details and information on the Catalan numbers we refer, for instance, to [18] or [10].

9.9 Generating Functions

9.9.1 Ordinary Generating Functions

The term *ordinary generating function* is used to describe an infinite sequence of (real or complex) numbers $(a_n)_{n \in \mathbb{N}_0}$ by treating them as the coefficients of a power series expression. This (formal) infinite series is the generating function:

$$a(t) := \sum_{n=0}^{\infty} a_n t^n.$$

When the term generating function is used without qualification, it is taken to mean an ordinary generating function. Unlike an ordinary series, this formal series is allowed to diverge, meaning that the generating function is not always a true function and the variable t is actually an indeterminate with

$$\sum_{n=0}^{\infty} a_n t^n + \sum_{n=0}^{\infty} b_n t^n = \sum_{n=0}^{\infty} (a_n + b_n) t^n$$

and

$$\left(\sum_{n=0}^{\infty} a_n t^n \right) \cdot \left(\sum_{n=0}^{\infty} b_n t^n \right) = \sum_{n=0}^{\infty} c_n t^n \quad \text{where } c_n = \sum_{\substack{i+j=n \\ i,j \geq 0}} a_i b_j,$$

the (formal) generating functions over $\mathbb{R}$ (or $\mathbb{C}$) form a commutative ring $\mathbb{R}[[t]]$ (or $\mathbb{C}[[t]]$) with unity 1. With the addition and the scalar multiplication

$$r \sum_{n=0}^{\infty} a_n t^n = \sum_{n=0}^{\infty} r a_n t^n$$

they also form a vector space over $\mathbb{R}$ (or $\mathbb{C}$).

In what follows we assume that there exists a real number $a > 0$ with $|a_n| < a^n$ for all $n \in \mathbb{N}_0$. In this case the generating function is a true function which converges absolutely for $|x| < \frac{1}{a}$, where $x \in \mathbb{R}$ (or $\mathbb{C}$). More precisely, if there exists

$$r = \left(\limsup_{n \to \infty} \sqrt[n]{|a_n|} \right)^{-1}$$

such that $r > 0$ then the generating function converges absolutely for all $x \in \mathbb{R}$ (or $\mathbb{C}$) with $|x| < r$. In addition, if $a_n \neq 0$ for almost all n, that is, for all n up to at most finitely many, and if $\lim_{n \to \infty} |\frac{a_n}{a_{n+1}}|$ exists, then also $r = \lim_{n \to \infty} |\frac{a_n}{a_{n+1}}|$. This r is called the radius of convergence. Hence, more precisely, we assume that the generating function has a positive radius of convergence. This includes the case that the radius is ∞. In that case we may apply the known results from real (or complex) analysis.

In particular, if $a(t)$, $b(t)$ are the generating functions of $(a_n)_{n \in \mathbb{N}_0}$, $(b_n)_{n \in \mathbb{N}_0}$, respectively, and if $r_a, r_b > 0$ for their radii of convergence, then the following hold:
(1) $a(x) = b(x)$ for all x with

$$|x| < \min(r_a, r_b) \Leftrightarrow a_n = b_n \quad \text{for all } n \in \mathbb{N}_0.$$

(2) $a_n = \frac{a^{(n)}(0)}{n!}$.

The latter is a consequence of the fact that the nth derivative of the generating function (around 0) is

$$a^{(n)}(x) = n! \sum_{k \geq n} \binom{k}{n} a_k x^{k-n}.$$

Example 9.65.
(1) Let a be a constant with $a \neq 0$, and $a_n = a^n$ for all $n \in \mathbb{N}_0$. Then $a(t) = \sum_{n=0}^{\infty} a^n t^n = \frac{1}{1-at}$ has radius of convergence $r = \frac{1}{|a|}$.
In particular,

$$a(t) = \sum_{n=0}^{\infty} t^n = \frac{1}{1-t} \quad \text{for } a = 1$$

and

$$a(t) = \sum_{n=0}^{\infty} (-1)^n t^n = \frac{1}{1+t} \quad \text{for } a = -1.$$

(2) If $a_n = n$ for all $n \in \mathbb{N}_0$, then

$$a(t) = \sum_{n=0}^{\infty} nt^n = t \cdot \sum_{n=1}^{\infty} nt^{n-1}$$

$$= t \left(\sum_{n=0}^{\infty} t^n \right)' = t \left(\frac{1}{1-t} \right)'$$

$$= \frac{1}{(1-t)^2}$$

with radius of convergence $r = 1$.
(3) $a_n = \binom{m}{n}$, $m \in \mathbb{N}$ fixed, $n \in \mathbb{N}_0$. Then $a_n = 0$ for $n > m$, and hence

$$a(t) = \sum_{n=0}^{\infty} \binom{m}{n} t^n = \sum_{n=0}^{m} \binom{m}{n} t^n = (1+t)^m$$

with radius of convergence $r = \infty$.

We now calculate the generating functions for some concrete examples.

Theorem 9.66. *Let N be a given (finite or infinite) subset of* $\mathbb{N}$*. The generating function for the number of all partitions of n in summands of N (without consideration of the order) is given by*

$$\prod_{m \in N} \frac{1}{1 - t^m}.$$

Proof. Let N consist of the numbers $m_1 < m_2 < m_3 < \cdots$.
We know that

$$\frac{1}{1 - t^m} = \sum_{n=0}^{\infty} (t^m)^n$$

has radius of convergence $r = 1$.
Hence

$$\prod_{m \in N} \frac{1}{1 - t^m} = (1 + t^{m_1} + t^{2m_1} + \cdots)(1 + t^{m_2} + t^{2m_2} + \cdots) \cdots.$$

Expansion of the product shows that t^n occurs as often as there exist sequences $(r_1, r_2, \ldots)$ of numbers from $\mathbb{N}_0$ with $n = r_1 m_1 + r_2 m_2 + \cdots$, and that is the number of all partitions of n in summands of N (without consideration of the order). □

Example 9.67 (Example for nostalgists). In how many ways may we change 1 DM into smaller coins?
Let c_n be the number for an amount of n pennies.
At the time of the German mark there existed coins with values of 1, 2, 5, 10 and 50 pennies.
Hence the generating function for the number of all partitions of n in summands of $\{1, 2, 5, 10, 50\}$ is given by

$$\sum_{n=0}^{\infty} c_n t^n = \frac{1}{(1 - t)(1 - t^2)(1 - t^5)(1 - t^{10})(1 - t^{50})}$$

$$= (1 + t + t^2 + \cdots)(1 + t^2 + t^4 + \cdots) \cdots (1 + t^{50} + t^{100} + \cdots).$$

We ask for c_{100} and get $c_{100} = 2498$.

Remark 9.68. To calculate c_{100} we may argue also successively.
We start with $\frac{1}{1-t} = 1 + t + t^2 + \cdots$ and divide successively the other polynomials subject to the general method:

$$(a_0 + a_1 t + a_2 t^2 + \cdots) : (1 - t^k) = b_0 + b_1 t + b_2 t^2 + \cdots$$

with

$$b_n = \begin{cases} a_n & \text{for } n = 0, 1, \ldots, k-1, \\ a_n + b_{n-k} & \text{for } n \geq k. \end{cases}$$

Corollary 9.69.
(a) *The generating function for the summary partition function $p(n)$ is $\sum_{m=1}^{\infty} \frac{1}{1-t^m}$.*
(b) *The generating function of the partitions of n in natural summands, which are not bigger than $k \in \mathbb{N}$, is*

$$\prod_{m=1}^{k} \frac{1}{1-t^m}.$$

(c) *The generating function for the partition numbers $p(n, k)$ is*

$$\prod_{m=1}^{k} \frac{t}{1-t^m}.$$

Proof. (a) and (b) are special cases of Theorem 9.66. For (a) we may take $N = \mathbb{N}$, and for (b) we may take $N = \{1, 2, \ldots, k\}$.

(c) From Ferrers diagram we know that $p(n, k)$ is the number of partitions of $n - k$ in natural summands which are not bigger than k. Hence (c) follows from (b) through

$$t^k \prod_{m=1}^{k} \frac{1}{1-t^m} = \prod_{m=1}^{k} \frac{t}{1-t^m}. \qquad \square$$

Remark 9.70. $\binom{n}{k}$ is the number of k-combinations of elements of an n-set.

Let $u_1, u_2, \ldots, u_n$ be n pairwise distinct real numbers and t be an indeterminate. Then

$$(1 + u_1 t)(1 + u_2 t) \cdots (1 + u_n t)$$

is a polynomial of $\mathbb{R}[t]$. Expansion of the product leads to

$$1 +$$
$$(u_1 + u_2 + \cdots u_n)t +$$
$$(u_1 u_2 + u_1 u_3 + \cdots + u_2 u_3 + \cdots + u_{n-1} u_n)t^2 +$$
$$\vdots$$
$$u_1 u_2 \cdots u_n t^n.$$

Hence, the coefficient of t^k is the sum of all k-combinations of elements from $\{u_1, u_2, \ldots, u_n\}$.

If we put

$$u_1 := u_2 := \cdots := u_n := 1,$$

then the coefficient of t^k is the number of these k-combinations, that is, $\binom{n}{k}$. This shows that sometimes we can give the generating function directly and afterwards we can calculate the coefficients.

We generalize this as follows.

Theorem 9.71. *The generating function for the number of those k-repetitions from an n-set ($1 \le k \le n$), for which the multiplicity of the jth element ($j = 1, 2, \ldots, n$) is from a given set $N_j \subset \mathbb{N}_0$, is*

$$\prod_{j=1}^{n}\left(\sum_{v \in N_j} t^v\right).$$

Proof. Let again $u_1, u_2, \ldots, u_n$ be pairwise distinct real numbers. Expansion of the product

$$\left(\sum_{v \in N_1} u_1^v t^v\right)\left(\sum_{v \in N_2} u_2^v t^v\right)\cdots\left(\sum_{v \in N_n} u_n^v t^v\right)$$

leads to a sum or (if at least one N_j is infinite) a series where the coefficient of t^k is the sum of all those repetitions of $\{u_1, u_2, \ldots, u_n\}$ for which the multiplicity of u_j ($j = 1, 2, \ldots, n$) is given by a number from the set N_j. If we put $u_1 = u_2 = \cdots = u_n = 1$, then the resulting coefficient is the number of those k-repetitions. $\qquad\square$

Example 9.72. Calculate the number of all 4-repetitions of elements from $\{a_1, a_2, a_3, a_4\}$ in which a_1 occurs 0, 1 or 2 times, a_2 occurs exactly twice, a_3 occurs arbitrarily often and a_4 does not occur.

Hence, we have $N_1 = \{0, 1, 2\}$, $N_2 = \{2\}$, $N_3 = \mathbb{N}_0$ and $N_4 = \{0\}$.

We get the generating function

$$(1 + t + t^2) \cdot t^2 \cdot \left(\sum_{v=0}^{\infty} t^v\right) = t^2 + 2t^3 + 3t^4 + 3t^5 + \cdots.$$

The number of acceptable 4-repetitions is 3, which we get by looking at $3t^4$, and we have

$$a_1 a_2 a_2 a_3, \quad a_2 a_2 a_3 a_3, \quad a_1 a_1 a_2 a_2.$$

Corollary 9.73. *The generating function for the number of all k-repetitions of elements from an n-set is*

$$\prod_{j=1}^{n}\left(\sum_{v=0}^{\infty} t^{v}\right) = \frac{1}{(1-t)^{n}} = \sum_{k=0}^{\infty}\binom{n+k-1}{k}t^{k}.$$

Theorem 9.74 (Fibonacci numbers). *The Fibonacci numbers f_n, $n \geq 0$, are given by $f_0 = 0$, $f_1 = 1$ and $f_n = f_{n-1} + f_{n-2}$ for $n \geq 2$ (see [12] for properties of the f_n). Let $f(t) = \sum_{n=0}^{\infty} f_n t^n$ be the generating function of $(f_n)_{n \in \mathbb{N}_0}$. Then*

(1) $f(t) = \frac{t}{1-t-t^2}$;

(2) $f_n = \frac{1}{\sqrt{5}}\left(\left(\frac{1+\sqrt{5}}{2}\right)^n - \left(\frac{1-\sqrt{5}}{2}\right)^n\right).$

Proof. Let $f(t) = \sum_{n=0}^{\infty} f_n t^n$. This series converges for $|x| < \frac{1}{2}$. To calculate $f(t)$ we use the recursion formula for f_n. Then

$$f(t) = t + \sum_{n=2}^{\infty} f_n t^n = t + \sum_{n=2}^{\infty} f_{n-1} t^n + \sum_{n=2}^{\infty} f_{n-2} t^n$$

$$= t + tf(t) + t^2 f(t),$$

and hence

$$f(t) = \frac{t}{1-t-t^2} = -\frac{t}{t^2+t-1}.$$

The zeros of the denominator of $f(t)$ are $-\frac{1}{2} + \frac{\sqrt{5}}{2}$ and $-(\frac{1}{2} + \frac{\sqrt{5}}{2})$.
Hence partial fraction decomposition leads to

$$f(t) = \frac{1}{\sqrt{5}}\left(\frac{1}{1-\frac{1+\sqrt{5}}{2}t} - \frac{1}{1-\frac{1-\sqrt{5}}{2}t}\right).$$

We apply

$$\frac{1}{1-at} = \sum_{n=0}^{\infty} a^n t^n = \sum_{n=0}^{\infty}\left(\frac{a}{2}\right)^n (2t)^n$$

to both terms in the bracket and get

$$f_n = \frac{1}{\sqrt{5}}\left(\left(\frac{1+\sqrt{5}}{2}\right)^n - \left(\frac{1-\sqrt{5}}{2}\right)^n\right). \qquad \square$$

Theorem 9.75. *Let the sequence $(a_n)_{n \in \mathbb{N}_0}$ be recursively defined by*

$$a_0 = 0, \quad a_n = 2a_{n-1} + 1, \quad n \geq 1.$$

Let $a(t) = \sum_{n=0}^{\infty} a_n t^n$ be its generating function. Then

(a) $a(t) = \frac{t}{(1-t)(1-2t)}$,

(b) $a_n = 2^n - 1$ for $n \geq 0$.

Proof. Let $a(t) = \sum_{n=0}^{\infty} a_n t^n$. This series converges for $|x| < \frac{1}{2}$. Then

$$a(t) = \sum_{n=0}^{\infty} a_n t^n = \sum_{n=0}^{\infty} (2a_n + 1)t^{n+1}$$

$$= 2ta(t) + \frac{t}{1-t}.$$

Therefore,

$$a(t) = \frac{t}{(1-t)(1-2t)} = -\frac{1}{1-t} + \frac{1}{1-2t}$$

$$= -\sum_{n=0}^{\infty} t^n + \sum_{n=0}^{\infty} 2^n t^n$$

$$= \sum_{n=0}^{\infty} (2^n - 1)t^n,$$

and so $a_n = 2^n - 1$. $\square$

Theorem 9.76. *Let the sequence $(a_n)_{n\in\mathbb{N}_0}$ be recursively defined by $a_0 = 2$, $a_1 = 5$ and $a_{n+2} = 5a_{n+1} - 6a_n$ for $n \geq 0$. Let $a(t) = \sum_{n=0}^{\infty} a_n t^n$ be the generating function. Then*

(a) $a(t) = \frac{2-5t}{1-5t+6t^2}$.

(b) $a_n = 2^n + 3^n$.

Proof. Let $a(t) = \sum_{n=0}^{\infty} a_n t^n$. The series converges for $|x| < \frac{1}{3}$. Now

$$a(t) = a_0 + a_1 t + \sum_{n=2}^{\infty} (5a_{n-1} - 6a_{n-2})t^n$$

$$= 2 - 5t + 5ta(t) - 6t^2 a(t).$$

It follows that

$$a(t) = \frac{2-5t}{1-5t+6t^2} = \frac{1}{1-2t} + \frac{1}{1-3t}.$$

Therefore,

$$a_n = 2^n + 3^n. \quad \square$$

Theorem 9.77 (Stirling numbers of the second kind). *Let $k \in \mathbb{N}_0$ be fixed. For this fixed k we define $S_k(t) = \sum_{n=0}^{\infty} S(n,k)t^n$, where $S(n,k)$ are the Stirling numbers of second kind, and call $S_k(t)$ the generating function for $S(n,k)$ (with fixed k). Then $S_0(t) = 1$ and $S_k(t) = \frac{t^k}{(1-t)(1-2t)\cdots(1-kt)}$ for $k \geq 1$.*

Proof. Certainly $S_0(t) = S(0,0) = 1$ because $S(n,0) = 0$ for $n \geq 1$. Now, let $k \geq 1$ and

$$S_k(t) = \sum_{n=0}^{\infty} S(n,k)t^n.$$

The series converges for $|x| < \frac{1}{k}$, since at least $S(n,k) \leq \frac{k^n}{k!}$ (see Theorem 9.42). Now

$$S_k(t) = \sum_{n=0}^{\infty} S(n,k)t^n = \sum_{n=k}^{\infty} S(n,k)t^n$$

$$= t^k + \sum_{n=k+1}^{\infty} S(n-1,k-1)t^n + k \sum_{n=k+1}^{\infty} S(n-1,k)t^n$$

$$= t^k + t \sum_{n=k}^{\infty} S(n,k-1)t^n + kt \sum_{n=k}^{\infty} S(n,k)t^n$$

$$= kt S_k(t) + t S_{k-1}(t),$$

that is,

$$S_{k(t)} = \frac{t S_{k-1}(t)}{1-kt}.$$

From this we get recursively

$$S_k(t) = \frac{t^k}{(1-t)(1-2t)\cdots(1-kt)}. \qquad \square$$

The Stirling numbers of the first kind may be considered analogously.

Remark 9.78 (The Chebyshev polynomials (of the second kind)). The Chebyshev polynomials $T_n(x)$, $n \in \mathbb{N}_0$, are defined recursively by $T_0(x) = 0$, $T_1(x) = 1$ and $T_n(x) = xT_{n-1}(x) - T_{n-2}(x)$ for $n \geq 2$. These polynomials satisfy some interesting identities:
(1) $T_{n+m}(x) = T_n(x)T_{m+1}(x) - T_m(x)T_{n-1}(x)$,
(2) $T_n^2(x) - T_{n+1}(x)T_{n-1}(x) = 1$ and
(3) $T_{mn}(x) = T_m(T_{n+1}(x) - T_{n-1}(x))T_n(x)$.

If now $x \in \mathbb{R}$, $x \geq 0$, then we have the following:
(a) If $0 \leq x < 2$ then there is a $\Theta \in \mathbb{R}$, $0 < \Theta \leq \frac{\pi}{2}$ with $x = 2\cos(\Theta)$, and we have

$$T_n(x) = \frac{\sin(n\Theta)}{\sin(\Theta)}, \quad n \in \mathbb{N}_0.$$

(b) If $x = 2$ then $T_n(x) = n$ for $n \in \mathbb{N}_0$.
(c) If $x > 2$ then there is a $\Theta \in \mathbb{R}$ with $x = 2\cosh(\Theta)$, and we have

$$T_n(x) = \frac{\sinh(n\Theta)}{\sinh(\Theta)}.$$

Let $U(x)(t) = \sum_{n=0}^{\infty} T_n(x)t^n$, $x \in \mathbb{R}$, be the generating function of the sequence $(T_n(x))_{n \in \mathbb{N}_0}$, $x \in \mathbb{R}$.

Theorem 9.79. *Let $U(x)(t) = \sum_{n=0}^{\infty} T_n(x)t^n$, $x \in \mathbb{R}$, be the generating function of the sequence $(T_n(x))_{n \in \mathbb{N}_0}$, $x \in \mathbb{R}$. Then*

$$U(x)(t) = \frac{t}{t^2 - xt + 1}, \quad x \in \mathbb{R}.$$

Proof. Let $U(x)(t) = \sum_{n=0}^{\infty} T_n(x)t^n$, $x \in \mathbb{R}$. From the definition (and properties) of $T_n(x)$, we get that for a fixed $x \in \mathbb{R}$ the series $U(x)(t)$ has a positive radius of convergence. Then

$$U(x)(t) = t + \sum_{n=2}^{\infty} T_n(x)t^n$$

$$= t + \sum_{n=2}^{\infty} (xT_{n-1}(x) - T_{n-2}(x))t^n$$

$$= t + xtU(x)(t) - t^2 U(x)(t),$$

and hence

$$U(x)(t) = \frac{t}{t^2 - tx + 1}. \qquad \square$$

Remark 9.80 (The Catalan numbers C_n). Recall that, by Theorem 9.64, $C_0 = 1$ and $C_{n+1} = \sum_{k=0}^{n} C_k C_{n-k}$, $n \geq 0$. Let $C(t) = \sum_{n=0}^{\infty} C_n t^n$ be the generating function for the series $(C_n)_{n \in \mathbb{N}_0}$. From the recurrence relation we get inductively that $C_n \leq 4^n$. Hence the series converges for $x < \frac{1}{4}$.

Now we form

$$C^2(t) = \sum_{n=0}^{\infty} \left(\sum_{k=0}^{n} C_k C_{n-k} \right) t^n.$$

Since $C_0 = 1$ and $C_{n+1} = \sum_{k=0}^{n} C_k C_{n-k}$ for $n \geq 0$, we get

$$tC^2(t) = \sum_{n=0}^{\infty} C_{n+1} t^{n+1} = -1 + \sum_{n=0}^{\infty} C_n t^n = C(t) - 1,$$

hence $C(t)$ satisfies the functional equation $tC^2(t) - C(t) + 1 = 0$.
Therefore either

$$C(t) = \frac{1 + \sqrt{1 - 4t}}{2t} \quad \text{or} \quad C(t) = \frac{1 - \sqrt{1 - 4t}}{2t}.$$

Only the function

$$C(t) = \frac{1 - \sqrt{1 - 4t}}{2t} = \frac{2}{1 + \sqrt{1 - 4t}}$$

has a power series at 0 and its coefficients must therefore be the Catalan numbers. Therefore

$$C(t) = \frac{1 - \sqrt{1 - 4t}}{2t}.$$

The chosen solution satisfies

$$\lim_{\substack{x \to 0 \\ x > 0}} C(x) = C_0 = 1.$$

The square-root term can be expanded as a Taylor power series using the identity

$$\sqrt{1 + y} = \sum_{n=0}^{\infty} \frac{(-1)^{n+1}}{4^n (2n-1)} \binom{2n}{n} y^n = 1 + \frac{1}{2} y - \frac{1}{8} y^2 + \cdots,$$

see [17].

Setting $y = -4t$ and substituting this power series into the expression for $C(t)$ and shifting the summation index n by 1, the expansion simplifies to

$$C(t) = \sum_{n=0}^{\infty} \binom{2n}{n} \frac{t^n}{n+1}.$$

The coefficients are the C_n, hence $C_n = \frac{1}{n+1} \binom{2n}{n}$, $n \in \mathbb{N}_0$. Hence we get the following.

Theorem 9.81. *Let $C(t) = \sum_{n=0}^{\infty} C_n t^n$ be the generating function of the sequence $(C_n)_{n \in \mathbb{N}_0}$ of the Catalan numbers. Then*
(a) $C(t) = \frac{1}{2t}(1 - \sqrt{1 - 4t})$.
(b) $C_n = \frac{1}{n+1} \binom{2n}{n}$, $n \in \mathbb{N}_0$.

We remark that also

$$C_n = \binom{2n}{n} - \binom{2n}{n+1}, \quad n \in \mathbb{N}_0$$

because

$$\binom{2n}{n+1} = \frac{n}{n+1} \binom{2n}{n}.$$

This last equation gives also

$$(n+2) C_{n+1} = \binom{2n+2}{2+1} = \frac{n+2}{n+1} \binom{2n+2}{n+2} = \frac{2(2n+1)}{n+1} \binom{2n}{n},$$

hence we have the following.

Corollary 9.82.

$$C_0 = 1 \quad and \quad C_{n+1} = \frac{2(2n+1)}{n+2} C_n \quad for \ n \geq 0.$$

9.9.2 Exponential Generating Functions

If we have a combinatorial sequence $(a_n)_{n \in \mathbb{N}_0}$, then the values a_n grow often very fast. This happens, for instance, if the values are related to permutations. In such a situation the ordinary generating function for $(a_n)_{n \in \mathbb{N}_0}$ diverges for each $x \in \mathbb{R}, x > 0$.

Hence in such a case we do not get new information for the a_n. One obvious case for this is $a_n = n!$ for $n \in \mathbb{N}_0$. This leads to the concept of exponential generating functions.

Let $(a_n)_{n \in \mathbb{N}_0}$ be a sequence of real (or complex) numbers. Then the (formal) series

$$A(t) = \sum_{n=0}^{\infty} \frac{a_n}{n!} t^n$$

is the *exponential generating function* for the series $(a_n)_{n \in \mathbb{N}_0}$. The rules for addition and multiplication are given (formally) by

$$\sum_{n=0}^{\infty} \frac{a_n}{n!} t^n + \sum_{n=0}^{\infty} \frac{b_n}{n!} t^n = \sum_{n=0}^{\infty} \frac{(a_n + b_n)}{n!} t^n$$

and

$$\left(\sum_{n=0}^{\infty} \frac{a_n}{n!} t^n \right) \left(\sum_{n=0}^{\infty} \frac{b_n}{n!} t^n \right) = \sum_{n=0}^{\infty} \frac{1}{n!} \left(\sum_{k=0}^{n} \binom{n}{k} a_k b_{n-k} \right) t^n.$$

In what follows we assume that there exists a real number $a > 0$ with $|a_n| \leq (an)^n$ for all $n \in \mathbb{N}_0$. In this case the exponential generating function is a true function which converges absolutely for $|x| < \frac{1}{ae}$ (because $\frac{n^n}{n!} \leq ne^n$). In particular, the series has a positive radius of convergence (this includes the case that the radius is ∞). Then the derivative of the exponential generating function (around 0) is

$$A'(x) = \sum_{n=0}^{\infty} \frac{a_{n+1}}{n!} x^n.$$

Examples 9.83.

(1) Let $a_n = n!$ for all $n \in \mathbb{N}_0$. Then $A(t) = \sum_{n=0}^{\infty} t^n = \frac{1}{1-t}$.

(2) Let $I(n, m)$, $m, n \in \mathbb{N}_0$, be the number of injective maps from an n-set into an m-set. Then, by Theorem 9.15 and the interpretation, $I(n, m) = (m)_n = n!\binom{m}{n}$. We choose

m to be fixed, and then $(a_n)_{n\in\mathbb{N}_0}$, $a_n = I(n, m)$, is a sequence in n, and we get

$$A(t) = \sum_{n=0}^{\infty} \frac{I(n, m)}{n!} t^n = \sum_{n=0}^{\infty} \binom{m}{n} t^n = (1 + t)^m.$$

(3) **The Bell numbers B_n.** We remind that $B_n = \sum_{k=0}^{n} S(n, k)$ where $S(n, k)$ are the Stirling numbers of the second type.

From $S(n, k) \le \frac{k^n}{k!}$ (see Theorem 9.42) we see that the series

$$B(t) = \sum_{n=0}^{\infty} \frac{B_n}{n!} t^n$$

has a positive radius of convergence. We already know that

$$B_{n+1} = \sum_{k=0}^{n} \binom{n}{k} B_k$$

for all $n \in \mathbb{N}_0$. Using the above rules we get

$$B'(t) = \sum_{n=0}^{\infty} \frac{B_{n+1}}{n!} t^n$$

$$= \sum_{n=0}^{\infty} \left(\frac{1}{n!} \sum_{k=0}^{n} \binom{n}{k} B_k \right) t^n$$

$$= \left(\sum_{n=0}^{\infty} \frac{t^n}{n!} \right) \left(\sum_{n=0}^{\infty} \frac{B_n}{n!} t^n \right)$$

$$= e^t B(t).$$

Hence $B'(t) = e^t B(t)$. We now want to solve this type of differential equation. The function e^{e^t} also satisfies this differential equation. Now let $B_1(t)$ and $B_2(t)$ be two solutions of this differential equation with $B_1(x) > 0$ and $B_2(x) > 0$ for all $x \in \mathbb{R}$, $x > 0$. Then

$$(\ln \circ B_1)'(t) = \frac{B_1'(t)}{B_1(t)} = \frac{B_2'(t)}{B_2(t)} = e^t.$$

Hence, $B_1(t)$ and $B_2(t)$ differ only by a constant factor because the derivative of $\frac{B_1(t)}{B_2(t)}$ is zero.

Hence $B(t) = c e^{e^t}$ for a constant $c \in \mathbb{R}$, $c > 0$. Because of $B(0) = 1$ we get $c = \frac{1}{e}$. Therefore $B(t) = e^{e^t - 1}$.

In particular, the series

$$B(t) = \sum_{n=0}^{\infty} \frac{B_n}{n!} t^n$$

converges for all $x \in \mathbb{R}$.

Now

$$B(t) = e^{e^t-1} = \frac{1}{e} \sum_{k=0}^{\infty} \frac{1}{k!} \left(\sum_{n=0}^{\infty} \frac{t^n k^n}{n!} \right)$$
$$= \sum_{n=0}^{\infty} \left(\frac{1}{e} \sum_{k=0}^{\infty} \frac{k^n}{n!} \right) \frac{t^n}{n!}.$$

If we compare the coefficients, we get again the Dobinski formula (Theorem 9.46):

$$B_n = \frac{1}{e} \sum_{k=0}^{\infty} \frac{k^n}{k!}.$$

Altogether we have the following beautiful result.

Theorem 9.84. *Let $(B_n)_{n\in\mathbb{N}}$ be the sequence of the Bell numbers. Then*
(a) *The exponential generating function of $(B_n)_{n\in\mathbb{N}_0}$ is $B(t) = e^{e^t-1}$.*
(b) $B_n = \frac{1}{e} \sum_{k=0}^{\infty} \frac{k^n}{k!}.$

Exercises

1. (a) Consider a triangle in the plane $\mathbb{R}^2$ with vertices A, B, C. A line g intersects the triangle, but not at one of the vertices A, B, C.
   Show that two of the vertices are in one of the half-planes determined by g.
   (b) 9 students are sitting in a row with 12 chairs.
   Show that there are at least three consecutive chairs which are occupied.
   (c) Consider 12 pairwise distinct binary natural numbers.
   Show that there are two among them whose difference is a binary number whose both digits are equal.
2. (a) There are 3 routes from Passau to Dortmund and 4 routes from Dortmund to Hamburg. How many routes do there exist altogether to drive from Passau to Hamburg via Dortmund?
   Realize the situation in a tree diagram.
   (b) How many 0–1-sequences of length 8 do there exist?
3. (a) Let p be a prime number. Show that $p \mid \binom{p}{k}$ for all $1 \le k < p$. Conclude from this that

$$(x + y)^p \equiv x^p + y^p \mod p$$

   for all $x, y \in \mathbb{Z}$.
   (b) Let $0 \le k \le n$. Show that

$$\binom{n}{k} = \frac{n-k+1}{k} \binom{n}{k-1}.$$

4. Let M be a nonempty finite set. Let E be the set of subsets of M with an even number of elements and O be that with an odd number of elements.
   Show that $|E| = |O|$.

5. (a) How many natural numbers with 9 digits do there exist, in which each digit between 0 and 9 occurs at most once and the 0 occurs at least once?
   (b) How many possibilities do there exist to have at least four correct numbers in the lottery 6 from 49?

6. In a group of 20 persons there are 7 chosen which should form a working team.
   (a) How many different teams with 7 persons do there exist?
   (b) Two persons A and B refuse to cooperate in one team. How many different teams do there exist which do not contain both A and B?
   (c) Suppose that the group has 12 men and 8 women.
       - How many different teams do there exist with 5 men and 2 women?
       - How many different teams do there exist with at least 1 man?
       - How many different teams do there exist with at most 1 man?

7. (a) We consider the natural numbers between 1001 and 2000.
       How many of these numbers are divisible by 3 or 5 or 8?
   (b) Let S be the set of all students which registered for all exams.
       Let D be the set of students from S registered for Discrete Mathematics, A the set of students from S registered for Algebra and G the set of students from S registered for Geometry.
       Further, let

$$|D| = 60, \quad |A| = 50, \quad |G| = 40,$$
$$|D \cap A| = 40, \quad |D \cap G| = 30, \quad |A \cap G| = 20 \quad \text{and}$$
$$|D \cap A \cap G| = 10.$$

   - How many students are registered for at least one exam in Discrete Mathematics, Algebra or Geometry?
   - How many students are registered for exactly two of the exams in Discrete Mathematics, Algebra or Geometry?
   - How many students are registered for exactly one of the exams in Discrete Mathematics, Algebra or Geometry?

8. In a class, a group of 28 students celebrate the pre-Christmas Secret Santa. Each student puts one gift into a big box. If all gifts are in the box then each student randomly takes one gift from the box.
   How many possibilities do there exist that at least one student gets that gift he put himself/herself into the box?

9. Let $n \in \mathbb{N}$. Show that
   - $S(n, 3) = \frac{1}{2}(3^{n-1} - 2^n + 1)$ for $n \geq 3$,
   - $S(2n, n) \geq n^n$

(*Hint*: Compare the surjective maps $\{1, 2, \ldots, 2n\} \rightarrow \{1, 2, \ldots, n\}$ with the combinations of permutations of $\{1, 2, \ldots, n\}$ and arbitrary maps $\{n+1, n+2, \ldots, 2n\} \rightarrow \{1, 2, \ldots, n\}$),

- $s(n, n-1) = \frac{1}{2}n(n-1)$,
- $s(n, 2) = (n-1)!(1 + \frac{1}{2} + \cdots + \frac{1}{n-1})$ for $n \geq 2$,
- $S(n, k) \leq s(n, k)$ for $0 \leq k \leq n$.

10. Let $y_1, y_2, \ldots, y_n$, $n \in \mathbb{N}$, be real numbers and

$$f(x) = (x - y_1)(x - y_2) \cdots (x - y_n)$$

$$= \sum_{i=0}^{n} (-1)^i a_i x^{n-i}, \quad a_0 = 1.$$

Show that

(a)

$$a_1 = y_1 + y_2 + \cdots + y_n,$$

$$a_2 = y_1 y_2 + y_1 y_3 + \cdots + y_{n-1} y_n,$$

$$a_3 = y_1 y_2 y_3 + \cdots + y_{n-2} y_{n-1} y_n,$$

$$\vdots$$

$$a_n = y_1 y_2 \cdots y_n.$$

(b)

$$a_{n-k} = s(n, k), \quad 0 \leq k \leq n,$$

if we take

$$y_1 = 0, \quad y_2 = 1, \quad y_3 = 2, \quad \ldots, \quad y_n = n - 1.$$

11. Let $n \in \mathbb{N}$. Show that
- $p(n, n) = p(n, 1) = 1$,
- $p(n, 1) = 1$ for $n \geq 2$,
- $p(n, 2) = \lfloor \frac{n}{2} \rfloor$ where $\lfloor \frac{n}{2} \rfloor$ is the biggest natural number less than or equal to $\frac{n}{2}$,
- $p(n, n-2) = 2$ for $n \geq 4$,
- $p(n + k, k) = \sum_{j=1}^{k} p(n, j)$ for $k \geq 1$.

12. (a) Let $q \in \mathbb{R}$, $q \neq 1$, and $n \in \mathbb{N}$. Let $S(n) = 1 + q + q^2 + \cdots + q^n$. Show that
- $S(n) = \frac{1-q^{n+1}}{1-q}$,
- $\lim_{n \to \infty} S(n) = \frac{1}{1-q}$ if $|q| < 1$.

(b) Let $f(x) = x^2 + px + q \in \mathbb{R}[x]$. Suppose that $f(x)$ has two distinct real zeros x_1 and x_2. Show that there exist real numbers a and b with

$$\frac{1}{f(x)} = \frac{a}{x - x_1} + \frac{b}{x - x_2}.$$

13. (a) Let a_n be the number of words of length n from an alphabet with m letters. Show that

$$\sum_{n=0}^{\infty} a_n t^n = \frac{1}{1-mt}.$$

(b) Assume that the sequence $(a_n)_{n \in \mathbb{N}_0}$ has the generating function

$$A(t) = \sum_{n=0}^{\infty} a_n t^n = \frac{1}{t^2 - 5t + 6}.$$

Show that

$$a_n = \frac{1}{2^{n+1}} - \frac{1}{3^{n+1}},$$

and conclude from that

$$a_0 = \frac{1}{6}, \quad a_1 = \frac{5}{36} \quad \text{and} \quad a_n = \frac{5}{6}a_{n-1} - \frac{1}{6}a_{n-2} \quad \text{for } n \geq 2.$$

14. (a) The Lucas numbers ℓ_n, $n \in \mathbb{N}_0$, are defined by

$$\ell_0 = 2, \quad \ell_1 = 1 \quad \text{and} \quad \ell_n = \ell_{n-1} + \ell_{n-2} \quad \text{for } n \geq 2.$$

Show that:
(i) $\ell_n = f_{n-1} + f_{n+1}$ for $n \geq 1$, where the f_i are the Fibonacci numbers.
(ii) The generating function for the sequence $(\ell_n)_{n \in \mathbb{N}_0}$ is

$$L(t) = \sum_{n=0}^{\infty} \ell_n t^n = \frac{2-t}{1-t-t^2}.$$

Conclude from this that

$$\ell_n = \left(\frac{1+\sqrt{5}}{2}\right)^n + \left(\frac{1-\sqrt{5}}{2}\right)^n.$$

(b) The number W_n of possibilities to construct a wall of length n and height 2 with dominoes is given by the recursive formula

$$w_0 = 1, \quad w_1 = 1, \quad w_n = w_{n-1} + w_{n-2} \quad \text{for } n \geq 2.$$

Show that the generating function for the sequence $(w_n)_{n \in \mathbb{N}_0}$ is

$$W(t) = \sum_{n=0}^{\infty} w_n t^n = \frac{1}{1-t-t^2} = \frac{1}{2\sqrt{5}}\left(\frac{1+\sqrt{5}}{1-\frac{1+\sqrt{5}}{2}t} + \frac{\sqrt{5}-1}{1-\frac{1-\sqrt{5}}{2}t}\right).$$

Conclude that w_n is approximately $\frac{1+\sqrt{5}}{2}f_n$ for large n.

15. (a) Given the sequence $(v_n)_{n\in\mathbb{N}_0}$ with

$$v_0 = 2, \quad v_1 = 3 \quad \text{and} \quad v_n = 3v_{n-1} - 2v_{n-2} \quad \text{for } n \geq 2.$$

Show that $v_n = 2^n + 1$ for $n \in \mathbb{N}_0$.
Give two proofs, one by induction and the other with help of the generating function $V(t)$.
Prove that

$$V(t) = \sum_{n=0}^{\infty} v_n t^n = \frac{2 - 3t}{2t^2 - 3t + 1}.$$

(b) Given the sequence $(a_n)_{n\in\mathbb{N}_0}$ with

$$a_0 = 0, \quad a_1 = 1, \quad \text{and} \quad a_n = 7a_{n-1} - 12a_{n-2} \quad \text{for } n \geq 2.$$

Show that the generating function $A(t)$ for the sequence $(a_n)_{n\in\mathbb{N}_0}$ is

$$A(t) = \frac{1}{1 - 7t + 12t^2}.$$

Conclude from this that

$$a_n = 4^n - 3^n \quad \text{for } n \in \mathbb{N}_0.$$

(c) Consider the sequence $(b_n)_{n\in\mathbb{N}_0}$ with

$$b_0 = 0, \quad b_1 = 1 \quad \text{and} \quad b_n = 5a_{n-1} - 6a_{n-2} \quad \text{for } n \geq 2.$$

Determine the generating function $B(t)$ of the sequence $(b_n)_{n\in\mathbb{N}_0}$ and show that

$$b_n = 3^n - 2^n \quad \text{for } n \in \mathbb{N}_0.$$

16. (a) Determine the number of possibilities to write 32 as a sum of natural numbers (without the consideration of the natural order), first with help of the recursive formula and second with help of the generating function.

(b) Determine the number of possibilities to exchange a 20-cent coin into smaller coins.

10 Finite Probability Theory and Bayesian Analysis

10.1 Probabilities and Probability Spaces

Probability theory deals with the modeling and analysis of random events. It has a long history: Gambling is one of the oldest human endeavors, and anyone who gambles must consider probabilities. Insurance also depends upon probabilities, and insurance was already being offered during the Renaissance. The beginnings of the mathematical theory of probability were in a series of letters called the Pascal–Fermat letters between P. Fermat (1607–1665) and B. Pascal (1623–1662). These letters explored the heavy reliance of finite probabilities on combinatorics. During the eighteenth and nineteenth centuries these ideas were extended to uses of probability in science and continuous spaces by T. Bayes (1702–1761), P. S. Laplace (1749–1827), C. F. Gauss (1777–1855), P. L. Chebyshev (1821–1899) and others. A rigorous mathematical classification of the concept of probabilities was first worked out by A. N. Kolmogorov (1903–1987).

In this chapter we introduce basic probability theory and Bayesian analysis. For the most part we will be dealing with finite probability spaces.

The concept of probability begins with *probability spaces* and *probability functions*.

Definition 10.1. A *probability space* consists of a triple $(S, \mathcal{E}, P)$ where
(1) S is a set called the *sample space*. The individual elements of S are called *outcomes*.
(2) $\mathcal{E}$ is a distinguished collection of subsets of S called the *class of events* (which contain S and are closed under complements and unions). $\mathcal{E}$ is often called the *event space*.
(3) P is a function $P : \mathcal{E} \rightarrow \mathbb{R}$ called a *probability measure* on $\mathcal{E}$ satisfying:
    (a) $P(E) \geq 0$ for all events $E \in \mathcal{E}$;
    (b) $P(S) = 1$;
    (c) $P(E_1 \cup E_2 \cup \cdots \cup E_n \cup \cdots) = \sum_i P(E_i)$ for any finite or countable collection of mutually disjoint events.

Remark 10.2. Here, if the sample space S is finite then each individual element $s \in S$ is an event.

In probability theory the empty set $\emptyset$ is called the *empty event*. The following theorem summarizes the basic properties of probability functions and probability spaces. Here E^c indicates the complement of the event E.

Theorem 10.3. *Let $(S, \mathcal{E}, P)$ be a probability space then*
(1) $0 \leq P(E) \leq 1$ *for any every event E.*
(2) $P(\emptyset) = 0$.
(3) $P(E) + P(E^c) = 1$ *for any event E.*

https://doi.org/10.1515/9783110740783-010

(4) *If $E \subset F$ then $P(E) \le P(F)$.*
(5) $P(E \cup F) = P(E) + P(F) - P(E \cap F)$ *for any two events E, F (general rule of addition).*

Proof. Since $E \cup \emptyset = E$ for any event E, we have

$$P(E \cup \emptyset) = P(E) = P(E) + P(\emptyset) \implies P(\emptyset) = 0,$$

proving (2).

Further $E \cup E^c = S$ and $E \cap E^c = \emptyset$ so $P(E) + P(E^c) = P(S) = 1$.
If $E \subset F$ then $F = E \cup (E^c \cap F)$. Because $E \cap (E^c \cap F) = \emptyset$, we have

$$P(F) = P(E) + P(E^c \cap F) \ge P(E) \quad \text{since } P(E^c \cap F) \ge 0.$$

We leave the remainder of the proof as an exercise. □

A *discrete sample space* is one which is countable, otherwise it is a *continuous sample space*. A *finite sample space* has a finite number of outcomes.

If S is a finite sample space with $|S| = n$ where as before $|S|$ stands for the number of elements in S; S then has *equiprobable outcomes* if $P(\{s\}) = \frac{1}{n}$ for all $s \in S$. A finite sample space with equiprobable outcomes is often called a *Laplace space*. The following is straightforward.

Theorem 10.4. *If S is a finite sample space with equiprobable outcomes and E is an event then*

$$P(E) = \frac{|E|}{|S|}.$$

In this chapter we will concentrate for the most part on finite sample spaces with equiprobable outcomes. Thus computing probabilities involves finding the sizes of sample spaces and the relevant subsets. Hence the use of combinatorics.

Before moving on, we mention that probability theory is actually a part of real analysis where it is a special case of *measure theory*. Let S be a set. A collection of subsets $\mathcal{E}$ of S is a *σ-algebra* if it contains S and is closed under complements and countable unions (it follows also that it is closed under countable intersections).

Definition 10.5. A *measure space* consists of a set S, a σ-algebra $\mathcal{E}$ of subsets of S and a function $m : \mathcal{E} \to \mathbb{R}$ called a *measure* satisfying
(1) $m(E) \ge 0$ for all $E \in \mathcal{E}$;
(2) $m(\emptyset) = 0$;
(3) $m(E_1 \cup E_2 \cup \cdots \cup E_n \cup \cdots) = \sum_i m(E_i)$ for any finite or countable collection of pairwise disjoint sets in $\mathcal{E}$.

There are many examples of measure spaces, for example, consider the real line with the length measure. In this more general context a probability space is just a general measure space where the measure of the whole space is 1.

10.2 Some Examples of Finite Probabilities

In this section we present some examples of finite probabilities.

(1) *The dice problem*

Suppose we roll a pair of fair dice as in Monopoly or backgammon. Then the possible numbers are 2 through 12. We find the probabilities of each. A dice has six faces numbered 1 through 6. We generally throw two dice and consider the sum. This becomes crucial in many gambling games such as craps and backgammon. Since each dice has six faces, the size of the sample space is the size of pairs of numbers (n_1, n_2) with each $n_i = 1, \ldots, 6$. The size of the sample space is then $6 \cdot 6 = 36$. To get a 2, for example, the only possibility is $(1, 1)$ and therefore the probability of a 2 is the $\frac{1}{36}$. We summarize all the probabilities:

$$P(2) = \frac{1}{36}, \quad P(12) = \frac{1}{36},$$
$$P(3) = \frac{2}{36}, \quad P(11) = \frac{2}{36},$$
$$P(4) = \frac{3}{36}, \quad P(10) = \frac{3}{36},$$
$$P(5) = \frac{4}{36}, \quad P(9) = \frac{4}{36},$$
$$P(6) = \frac{5}{36}, \quad P(8) = \frac{5}{36},$$
$$P(7) = \frac{6}{36}.$$

(2) *Matches in lotto*

We ask about the possibilities to have exactly 4 correct matches in a lotto. This means that there are 4 of the 6 winning numbers marked with a cross, and 2 of the marked numbers are from the remaining $49 - 6 = 43$ non-winning numbers. By the multiplication rule, the number of possibilities to have marked 4 of the 6 winning numbers is

$$\binom{6}{4}\binom{43}{2} = \frac{6!}{4! \cdot 2!} \cdot \frac{43!}{2! \cdot 41!} = 15 \cdot 903 = 13545.$$

Hence the probability to have exactly 4 of the 6 winning numbers is $\frac{13545}{13983816} \approx 0.00097$.

If we have exactly 5 of the 6 winning numbers and the bonus number, then the probability is

$$\frac{\binom{6}{5}\binom{1}{1}\binom{42}{0}}{13983816} = \frac{6}{13983816} \approx 0.000000429.$$

(3) *The birthday problem*

In a group of n people we ask for the probability that at least two of them have the same birthday. It is higher than most people expect. In a group of 40 people there is almost a 90 % probability that at least two have the same birthday. The complete solution is given in the following theorem.

Theorem 10.6 (The birthday problem). *The probability $p(n)$ that out of n persons at least 2 have birthday at the same day is*

$$p(n) = 1 - \prod_{k=1}^{n-1}\left(1 - \frac{k}{365}\right),$$

where we assume that there is an equipartition of the birthdays of all persons over a year with 365 days.

As a first approximation we get

$$p(n) \approx 1 - e^{\frac{-n(n-1)}{730}}.$$

Proof. Altogether there are 365^n possibilities for the birthdays of n persons, and we have

$$365 \cdot 364 \cdots (365 - n + 1)$$

possibilities that no person out of the n has the same birthday as one of the others. Hence the probability that all n persons have different birthdays is

$$P(B) = \frac{365!}{(365 - n)!365^n}$$

where B is the subset of the set M of all n persons which have different birthdays (here we assumed that we have a sample space with equiprobable outcomes). Let $B^c = M \setminus B$ be the complement of B in M. Then we get

$$p(n) = P(M \setminus B) = 1 - P(B) = 1 - \frac{365!}{(365 - n)!365^n} = 1 - \prod_{k=1}^{n-1}\left(1 - \frac{k}{365}\right).$$

If we take the linear approximation

$$1 - \frac{k}{365} \approx e^{-\frac{k}{365}},$$

then we get

$$p(n) \approx 1 - e^{-\frac{(1+2+\cdots+n-1)}{365}} = 1 - e^{-\frac{n(n-1)}{730}}.$$
□

Numerically we get:

$$p(23) \approx 0.507,$$
$$p(30) \approx 0.706,$$
$$p(50) \approx 0.970.$$

10.3 Random Variables, Distribution Functions and Expectation

A crucial concept in dealing with probabilities is that of a *random variable*. If S is a probability space then a *random variable* on S is a function $X : S \to \mathbb{R}$ which is *measurable*, that is, for all real numbers t the set $\{s \in S | X(s) \leq t\}$ is an event in S.

Intuitively, a random variable assigns real numbers to the outcomes of a chance phenomenon.

Example 10.7.
(1) A *Bernoulli process*, named after J. Bernoulli (1655–1705), is any random event with only two outcomes that we call success and failure. Consider a Bernoulli process with n independent trials and probability p of success on each trial. Let X be the number of successes obtained. Then X is called a *binomial random variable* with parameters n, p.
(2) Suppose we have a container with M balls of which R are red and $M - R$ are non-red. Suppose n are sampled and let X be the number of red balls in the sample. Then X is called a *hypergeometric random variable* with parameters M, R, n.
(3) Suppose we have a process which generates discrete occurrences over a continuous interval with a fixed average λ per unit interval. This is called a *Poisson process*, named after S. D. Poisson (1781–1840), with parameter λ. Let X be the number of occurrences in the unit interval. Then X is a *Poisson random variable* with parameter λ.
(4) Let $\mathcal{P}$ be a population of some measured variable such as height, weight or time. Let X be the particular measurements. Then X is the *population random variable*.

If X is a random variable then X is a *discrete random variable* if the range of X is a discrete set in $\mathbb{R}$ (contains no interval); X is a *continuous random variable* if the range of X is either an interval or a union of intervals. If the range of X contains both discrete and continuous sections, X is called a *mixed random variable*.

Actually we are not interested in the random variable itself but rather the probabilities that it takes on its values. To this end we study several related functions: *distribution functions*, *mass functions* for discrete random variables and *density functions* for continuous random variables.

The *distribution function* of the random variable X is the function $F : \mathbb{R} \to \mathbb{R}$ defined by

$$F(x) = P(X \leq x) = \text{Probability that } X \text{ takes on a value } \leq x.$$

That is, $F(x)$ measures how much of the random variable accumulates less than or equal to x. A distribution function is also called a *cumulative distribution function* abbreviated *c.d.f.*

The following theorem gives a complete characterization of a distribution function.

Theorem 10.8. *A function $F : \mathbb{R} \to \mathbb{R}$ is the distribution function of some random variable if and only if it satisfies the following five properties:*
(1) $0 \leq F(x) \leq 1$.
(2) $F(x)$ *is monotonically non-decreasing.*
(3) $\lim_{x \to \infty} F(x) = 1$.
(4) $\lim_{x \to -\infty} F(x) = 0$.
(5) $F(x)$ *is right continuous, that is,*

$$\lim_{h \to 0^+} F(x + h) = F(x).$$

Example 10.9.
(1) If $\lambda > 0$ show that $F(x) = 1 - e^{-\lambda x}$ if $x > 0$ and 0 elsewhere is the distribution function for some random variable X. We must show that $F(x)$ satisfies each of the 5 properties in the theorem. Showing this is direct.
(2) If $\lambda = 2$, what is $P(X \leq 4)$ and what is the *median* of this random variable.

$$P(X \leq 4) = F(4) = 1 - e^{-2 \cdot 4} = 0.997.$$

The median is the value m such that $F(m) = 0.5$. Then

$$F(m) = 0.5 = 1 - e^{-2m} \implies m = \frac{\ln(0.5)}{-2} = 0.3466.$$

Two random variables X, Y are *identically distributed* if they have the same distribution function. Often it is easier to work with other functions rather than directly with the distribution function.

If X is a discrete random variable assuming the values $x_1, x_2, \ldots, x_n, \ldots$, then its *mass function*, also called its *probability mass function*, abbreviated *p.m.f.*, is the function

$$p(x) = \begin{cases} P(X = x_i), & \text{if } x = x_i, \\ 0, & \text{otherwise.} \end{cases}$$

As with distribution functions there is a complete characterization of mass functions.

Theorem 10.10. *A function* $p : \mathbb{R} \to \mathbb{R}$ *is the mass function of some discrete random variable if and only if it satisfies the following three properties:*
(1) $p(x) = 0$ *except at discrete points* $x_1, x_2, \ldots, x_n, \ldots$.
(2) $p(x) \geq 0$.
(3) $\sum_{x_i} p(x_i) = 1$.

If X is a continuous random variable with distribution function $F(x)$ then $f(x)$ is its *density function*, also called its *probability density function*, abbreviated *pdf*, if for all $x \in \mathbb{R}$

$$F(x) = \int_{-\infty}^{x} f(t)dt.$$

Roughly, the density function is the function whose curve is the *normalized frequency curve*.

Theorem 10.11. *A function* $f : \mathbb{R} \to \mathbb{R}$ *is the density function of some random variable if and only if it satisfies the following properties:*
(1) $f(x) \geq 0$,
(2) $\int_{-\infty}^{\infty} f(x)dx = 1$.

Notice that if X is a continuous random variable with distribution function $F(x)$ then its density function $f(x)$ is given by the derivative of the distribution function

$$f(x) = F'(x).$$

Example 10.12. Let X be the random variable which picks a point at random from the interval $[a, b]$ and all points are equally likely. X is called a *uniform random variable* with parameters a, b.
(1) The distribution function for X, called the *uniform distribution*, is given by

$$F(x) = \begin{cases} 0, & \text{if } x \leq a, \\ \frac{x-a}{b-a}, & \text{if } a < x \leq b, \\ 1, & \text{if } x > b. \end{cases}$$

(2) The density function for X, called the *uniform density*, is given by the derivative of the distribution function. Therefore:

$$f(x) = \begin{cases} \frac{1}{b-a}, & \text{if } a \leq x \leq b, \\ 0, & \text{otherwise.} \end{cases}$$

Finally, random variables are described by their *expectation* or *expected value*.

Definition 10.13. If X is a random variable then its *expectation* or *expected value*, denoted $E(X)$, is

$$E(X) = \sum_x xp(x)$$

if X is discrete and $p(x)$ is its mass function, or

$$E(X) = \int_{-\infty}^{\infty} xf(x)dx$$

if X is continuous and $f(x)$ is its density function.

If $g(x) : \mathbb{R} \to \mathbb{R}$ then $g(X)$ is a random variable and

$$E(g(X)) = \sum_x g(x)p(x)$$

if X is discrete and $p(x)$ is its mass function, or

$$E(g(X)) = \int_{-\infty}^{\infty} g(x)f(x)dx$$

if X is continuous and $f(x)$ is its density function.

Expectation is a *linear operator* on random variables. That is,

$$E(\alpha g(X) + \beta h(X)) = \alpha E(g(X)) + \beta E(h(X))$$

for real numbers α, β and functions $g(X), h(X)$.

Further the expectation of a constant random variable is that constant, that is, $E(c) = c$.

The *kth population moment* of a random variable denoted $m_k(X)$ is $E(X^k)$. The first population moment $E(X)$ is called the *population mean* or *population average* denoted μ.

$E((X-\mu)^k)$ is called the *kth central moment*. The second central moment $E((X-\mu)^2)$ is the *variance* denoted $\mathrm{var}(X)$ or $\sigma^2(X)$. The square root of the variance is the *standard deviation* denoted $\sigma(X)$.

10.4 The Law of Large Numbers

In ordinary language we often say that *things tend to average out*. By this we mean mathematically that in a limiting sense random events converge to their mean. Formally, this is called the law of large numbers which we will describe in this section.

First, we show the importance of the mean as described in the next theorem, called Chebyshev's theorem. It says in essence that for any population with finite mean and variance the population will cluster about the mean.

Theorem 10.14 (Chebyshev's Theorem). *Let X be a random variable with finite expectation $E(X) = \mu$ and finite variance $V(X) = \sigma^2$. Let $k > 0$ be any positive real number. Then*

$$P(|X - \mu| \geq k\sigma) \leq \frac{1}{k^2}.$$

Proof. We first prove the discrete case and then do the continuous case. Let X be a discrete random variable with discrete values $\{x_i\}$, mass function $p(x) = P(X = x)$ and finite mean μ and finite standard deviation σ. Let $k > 0$ be any positive real number. Then

$$\sigma^2 = E((X - \mu)^2) = \sum_{x_i} (x_i - \mu)^2 p(x_i).$$

Now the set $\{x_i \mid |x_i - \mu| > k\sigma\}$ is contained in the whole range $\{x_i\}$, so adding over this set the result is smaller than when adding over the whole range. Hence

$$\sigma^2 \geq \sum_{\{x_i \mid |x_i - \mu| > k\sigma\}} (x_i - \mu)^2 p(x_i).$$

However, on this set $(x_i - \mu)^2 > k^2\sigma^2$ and therefore replacing this in the sum we get

$$\sigma^2 \geq \sum_{\{x_i \mid |x_i - \mu| > k\sigma\}} k^2\sigma^2 p(x_i) = k^2\sigma^2 \sum_{\{x_i \mid |x_i - \mu| > k\sigma\}} p(x_i).$$

However, $p(x)$ is the mass function so

$$\sum_{\{x_i \mid |x_i - \mu| > k\sigma\}} p(x_i) = P(|X - \mu| > k\sigma).$$

Combining these gives

$$\sigma^2 \geq k^2\sigma^2 P(|X - \mu| > k\sigma)$$

and hence

$$\frac{1}{k^2} \geq P(|X - \mu| > k\sigma),$$

proving the theorem in the discrete case.

The continuous case is analogous with the mass function replaced by the density function and summation replaced by integration. Let X be a continuous random variable with density function $f(x)$ and finite mean μ and finite standard deviation σ. Let $k > 0$ be any positive real number. Then

$$\sigma^2 = E((X - \mu)^2) = \int_{\mathbb{R}} (x - \mu)^2 f(x)dx.$$

Now the set $\{x \mid |x - \mu| > k\sigma\}$ is contained in the whole range $(-\infty, \infty)$, so integrating over this set the result is smaller than when integrating over the whole real line. Hence

$$\sigma^2 \geq \int_{\{x||x-\mu|>k\sigma\}} (x - \mu)^2 f(x)dx.$$

However, on this set $(x - \mu)^2 > k^2\sigma^2$ and therefore replacing this in the integral we get

$$\sigma^2 \geq \int_{\{x||x-\mu|>k\sigma\}} k^2\sigma^2 f(x)dx = k^2\sigma^2 \int_{\{x||x-\mu|>k\sigma\}} f(x)dx.$$

However, $f(x)$ is the density function, so integrating over a set gives the probability of that set. Hence

$$\int_{\{x||x-\mu|>k\sigma\}} f(x)dx = P(|X - \mu| > k\sigma).$$

Combining these yields

$$\sigma^2 \geq k^2\sigma^2 P(|X - \mu| > k\sigma)$$

and hence

$$\frac{1}{k^2} \geq P(|X - \mu| > k\sigma),$$

proving the theorem in the continuous case. □

From Chebyshev's theorem we get the law of large numbers.

Theorem 10.15 (The law of large numbers). *Let $(X_i)_{i\in\mathbb{N}}$ be a sequence of independent and identical distributed random variables X_i with common expectation $E(X_i) = \mu < \infty$ and common variance $V(X_i) = \sigma^2 < \infty$. If $S_n = X_1 + X_2 + \cdots + X_n$, then*

$$\lim_{n\to\infty} P\left(\left|\frac{S_n}{n} - \mu\right| \geq \epsilon\right) = 0$$

for all real $\epsilon > 0$.

Proof. From the linearity of the expectation we get

$$E\left(\frac{S_n}{n}\right) = \frac{E(S_n)}{n} = \frac{n\mu}{n} = \mu.$$

From the properties of the variance and independence of the X_i, we get

$$V\left(\frac{S_n}{n}\right) = \frac{V(S_n)}{n^2} = \frac{n\sigma^2}{n^2} = \frac{\sigma^2}{n}.$$

Let $\epsilon > 0$ be given. Then for some $k > 0$ we have $\epsilon = k\sqrt{V(\frac{S_n}{n})} = \frac{k\sigma}{\sqrt{n}}$. From Chebyshev's Theorem we obtain

$$P\left(\left|\frac{S_n}{n} - \mu\right| \geq \epsilon\right) = P\left(\left|\frac{S_n}{n} - \mu\right| \geq k\frac{\sigma}{\sqrt{n}}\right) \leq \frac{\sigma^2}{n\epsilon^2}$$

Taking the limit as $n \to \infty$ finishes the proof. $\square$

Remark 10.16.

(1)   Theorem 10.15 goes back to J. Bernoulli (1654–1705).

(2)   Theorem 10.15 is sometimes called the *weak* law of large numbers. A stronger result can be found in [3].

Example 10.17. We roll with fair dice. Let the (discrete) random variable X be 1 if we roll a six, and 0 otherwise. We get an expectation

$$E(X) = \frac{1}{6}$$

and the variance

$$V(X) = \frac{5}{6}\left(0 - \frac{1}{6}\right)^2 + \frac{1}{6}\left(1 - \frac{1}{6}\right)^2 = \frac{5}{36}.$$

We throw the dice n times. Then $B_n = X_1 + X_2 + \cdots + X_n$ is the number of sixes that appear. The expectation is $E(B_n) = \frac{n}{6}$, and the variance is $V(B_n) = \frac{5n}{36}$.

We consider the relative frequency $\frac{B_n}{n}$ of the sixes, and we get

$$E\left(\frac{B_n}{n}\right) = \frac{1}{6} \quad \text{and} \quad V\left(\frac{B_n}{n}\right) = \frac{5}{36n}.$$

From Theorem 10.14 and the proof of Theorem 10.15, we obtain

$$P\left(\left|\frac{B_n}{n} - \frac{1}{6}\right| \geq \epsilon\right) \leq \frac{5}{36n\epsilon^2}.$$

Hence the probability that the relative frequency of the sixes is by at least ϵ away from $\frac{1}{6}$ goes to 0 with an increasing quantity of the casts of the dice. In this sense, the relative frequency converges to the expectation.

10.5 Conditional Probabilities

In the following we use the notation $A \cap B$ for the event that both events A and B occur. Analogously, $A_1 \cap A_2 \cap \cdots \cap A_n$ denotes the event that all events $A_1, A_2, \ldots, A_n$ occur. Again, if $A \subset M$ then we write A^c for the complement $M \setminus A$ of A.

Definition 10.18. Let P be a probability on a set M. If A and B are events then the *conditional probability* of A given B with $P(B) > 0$, which is denoted by $P(A \mid B)$, is defined as

$$P(A \mid B) = \frac{P(A \cap B)}{P(B)}.$$

The conditional probability $P(A \mid B)$ can be interpreted as the probability that A will occur given that B has occurred. To see how this becomes an appropriate definition, recall that $A \cap B$ is the event that both A and B will occur. If B has occurred and A is to occur then $A \cap B$ will occur. This describes the proposition of times that $A \cap B$ occurs among all possible times that B has occurred. This would then be the ratio in the definition. Conditional probability can also be thought of as the effect the event B (which has occurred) has on the event A.

Example 10.19. Table 10.1 below represents a cross-classification of male and female smokers.

Table 10.1: Male and female smokers.

	Smokers	Non-Smokers	Total
Male	34	66	100
Female	23	87	110
Total	57	153	210

Then we get

$$P(Male) = \frac{100}{210},$$

$$P(Male \mid Smoker) = \frac{34}{57},$$

$$P(Male \cap Smoker) = \frac{34}{210},$$

which agrees with the definition.

The events A and B are independent if and only if

$$P(A) = P(A \mid B).$$

This follows from the definition and the fact that A and B are independent if and only if

$$P(A \cap B) = P(A) \cdot P(B)$$

using the multiplication rule. In this case, the event B has no influence for the event A.

Using these definitions we get the Bayes Expansion and then Bayes Theorem. These were originally written by T. Bayes who was a British clergyman and amateur mathematician. His original motivation was to use conditional probability to prove the existence of God. The original motivation was the *clockmaker argument*. Suppose someone is stranded on a desert island and finds a clock. There are two possible explanations. There was a clockmaker who was on the island first or random particles came together to form a clock. The probability is much higher that there was a clockmaker. Putting this in probability terms and applying it to the universe, much more complicated than a clock, leads to an extremely high probability, in Bayes view, of the existence of God.

Bayes expansion given below expresses the probability of an event A in terms of another event B and its complement B^c.

Theorem 10.20 (Bayes Expansion). *If A, B are any two events with $P(B) > 0$, we have*

$$P(A) = P(A \mid B)P(B) + P(A \mid B^c)P(B^c).$$

Proof. We have $A = (A \cap B) \cup (A \cap B^c)$. Since $A \cap B$ and $A \cap B^c$ are disjoint, we then get

$$P(A) = P(A \cap B) + P(A \cap B^c).$$

From the definition we have

$$P(A \cap B) = P(A \mid B)P(B) \quad \text{and} \quad P(A \cap B^c) = P(A \mid B^c)P(B^c). \qquad \square$$

This can be extended as follows. Let $\mathcal{E}$ be the event space and let the collection of events B_i partition the sample space so that $S = \bigcup_i B_i$ with $B_i \cap B_j = \emptyset$ if $i \neq j$. Then for any event A we have

$$P(A) = \sum_i P(A \mid B_i)P(B_i).$$

Theorem 10.21 (Bayes Formula). *Let A, B be two events with $P(B) > 0$. Then*

$$P(A \mid B) = \frac{P(B \mid A) \cdot P(A)}{P(B)}.$$

Proof. If $P(A) = 0$ then also $P(A \cap B) = 0$ and $P(A \mid B) = 0$, and the equation holds. Now, let $P(A) > 0$. Then

$$P(A \mid B) = \frac{P(A \cap B)}{P(B)} = \frac{\frac{P(A \cap B)}{P(A)} \cdot P(A)}{P(B)}$$

$$= \frac{P(B \mid A) \cdot P(A)}{P(B)},$$

and the equation holds. □

For illustration see Figure 10.1, where $A^c = M \setminus A$.

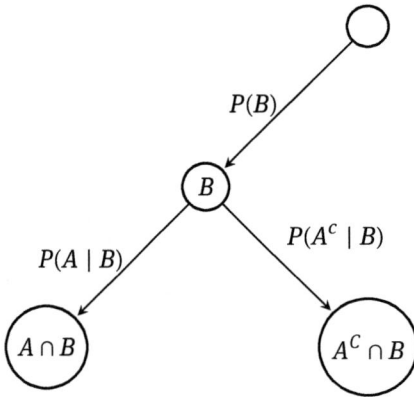

Figure 10.1: Illustration of Bayes Formula.

Corollary 10.22. *Let $\{B_i \mid i = 1, 2, \ldots, n\}$ be a partition of the sample space S, that is,*

$$S = \bigcup_{i=1}^{n} B_i \quad and \quad B_i \cap B_j = \emptyset \quad for \ i \neq j$$

where B_i are events. Then

$$P(B_j \mid A) = \frac{P(A \mid B_i)P(B_i)}{\sum_{j=1}^{n} P(A \mid B_j)P(B_j)} \quad if \ P(A) > 0.$$

In particular, if A is an event with $P(A) > 0$, then

$$P(B \mid A) = \frac{P(A \mid B)P(B)}{P(A \mid B)P(B) + P(A \mid B^c)P(B^c)}.$$

Proof. We have

$$P(A) = P\left(\bigcup_{i=1}^{n}(B_i \cap B)\right) = \sum_{i=1}^{n} P(B_i \cap A)$$

$$= \sum_{i=1}^{n} P(A \mid B_i)P(B_i).$$

Further,

$$P(B_i \mid A) = \frac{P(A \cap B_i)}{P(A)} = \frac{P(A \mid B_i)P(B_i)}{P(A)}$$

by Theorem 10.21.

Now, we plug in this value of $P(A)$ into this last equation.

The additional statement follows because $S = B \cup B^c$ and $B \cap B^c = \emptyset$. □

Example 10.23.

(1) Consider two urns A and B, each containing 10 balls. In A there are 7 red and 3 white balls, in B there is one red and 9 white balls. We move one ball from a randomly chosen urn.

Here the probability for picking urn A and urn B is the same. The result of the drawing is that the ball is red.

We want to find the probability that this red ball is from urn A.

We define:

– A is the event that the ball is from the urn A.
– B is the event that the ball is from urn B.
– R is the event that the ball is red.

Then we have:

$$P(A) = P(B) = \frac{1}{2}, \quad P(R \mid A) = \frac{7}{10}, \quad P(R \mid B) = \frac{1}{10}.$$

Hence,

$$P(R) = P(R \mid A)P(A) + P(R \mid B)P(B) = \frac{7}{10} \cdot \frac{1}{2} + \frac{1}{10} \cdot \frac{1}{2} = \frac{2}{5}$$

is the total probability to move a red ball.

Then

$$P(A \mid R) = \frac{P(R \mid A) \cdot P(A)}{P(R)} = \frac{\frac{7}{10} \cdot \frac{1}{2}}{\frac{2}{5}} = \frac{7}{8}.$$

Hence the conditional probability that the red ball is from urn A comes to $\frac{7}{8}$.

(2) Assume that there are two coins in a pot. One coin (A_1) is fair with heads and tails, the other coin (A_2) has two heads. Now we choose randomly one coin and throw it. Suppose we get heads (K).

The probability that the coin is fair comes to

$$P(A_1 \mid K) = \frac{P(A_1)P(K \mid A_1)}{P(A_1)P(K \mid A_1) + P(A_2)P(K \mid A_2)} = \frac{\frac{1}{2} \cdot \frac{1}{2}}{\frac{1}{2} \cdot \frac{1}{2} + \frac{1}{2} \cdot 1} = \frac{1}{3}.$$

Hence the probability that we get the coin with two heads comes to $P(A_2 \mid K) = \frac{2}{3}$.

Remark 10.24. The formula of Bayes allows us in a certain sense to reverse the conclusion.

We start with a known value $P(B \mid A)$, but actually we are interested in the value $P(A \mid B)$. For instance, it is of interest to know how big the probability is that a person carries a certain illness if a special quick test for this illness has a positive result.

From empirical studies one normally knows the probability for a positive test if the person carries the illness. The desired conversion is possible if one knows the (absolute) probability for the presence of the illness in the whole population.

Example 10.25. A certain illness is carried with a frequency 20 of 100000 persons. The event K, a person carries the illness, has therefore the probability

$$P(K) = 0.0002.$$

Then

$$P(K^c) = 1 - P(K) = 0.9998$$

where K^c is the complement of K, the event that a person does not carry the illness.

Let T be the event that the test is positive.

It is known that $P(T \mid K) = 0.95$, which means that if a person carries the illness then the test is positive with a probability of 0.95.

Sometimes the test is positive for a person who does not carry the illness. In fact,

$$P(T \mid K^c) = 0.01$$

for the respective conditional probability.

Then

$$1 - P(T \mid K^c) = 1 - 0.01 = 0.99.$$

We are interested in the conditional probability $P(K \mid T)$ that a person carries the illness if the test is positive.

By the Bayes formula

$$
\begin{aligned}
P(K \mid T) &= \frac{P(T \mid K)P(K)}{P(T \mid K)P(K) + P(T \mid K^c)P(K^c)} \\
&= \frac{0.95 \cdot 0.0002}{0.95 \cdot 0.0002 + 0.01 \cdot 0.9998} \\
&\approx 0.0186.
\end{aligned}
$$

Example 10.26 (Hemophilia problem). A woman with no known history of hemophilia in her family has a son with hemophilia. She reasons that a gene must have mutated, which has a tiny probability m. What is the probability that her second son has the disease given the first son has it.

The second son will have hemophilia if one of her X chromosomes mutated to the complement X^c. Let A_1 be the event that the first son has it, A_2 the event that the second son has it and B the event that she is a carrier, that is, one of her chromosomes has mutated.

Let $P(B)$ be the probability that one of her chromosomes has mutated. Each woman has two X chromosomes, and they would mutate independently. The probability that one chromosome does not mutate is $1 - m$. Hence the probability that both do not mutate is $(1 - m)^2$. Therefore, the probability that at least one chromosome mutates is $1 - (1 - m)^2$. However,

$$1 - (1 - m)^2 = -m^2 + 2m \approx 2m,$$

since m is tiny. Then

$$P(B) \approx 2m \quad \text{and} \quad P(B^c) \approx 1 - 2m.$$

Now

$$P(A_1 \mid B) = P(A_2 \mid B) = \frac{1}{2} \quad \text{and} \quad P(A_1 \mid B^c) = P(A_2 \mid B^c) = m.$$

This can be seen as follows. We have

$$P(A_1 \mid B) = P(A_2 \mid B) = \frac{1}{2}$$

because the mother has two X chromosomes, and one is mutated. We have

$$P(A_1 \mid B^c) = P(A_2 \mid B^c) = m$$

because, if none of her chromosomes is mutated, the only way a son is hemophilic is that a gene is mutated in the child which has probability m. Therefore,

$$P(A_1 \cap A_2 \mid B) = \frac{1}{4} \quad \text{and} \quad P(A_1 \cap A_2 \mid B^c) = m^2.$$

We want

$$P(A_2 \mid A_1) = \frac{P(A_1 \cap A_2)}{P(A_1)}.$$

Then, using Bayes expansion,

$$P(A_1) = P(A_1 \mid B)P(B) + P(A_1 \mid B^c)P(B^c) \approx \frac{1}{2}2m + m(1 - 2m) \approx 2m$$

and

$$P(A_1 \cap A_2) = P(A_1 \cap A_2 \mid B)P(B) + P(A_1 \cap A_2 \mid B^c)P(B^c) \approx \frac{1}{4}(2m) + (1 - 2m)m^2 \approx \frac{m}{2}.$$

Then our final computation is

$$P(A_2 \mid A_1) \approx \frac{\frac{m}{2}}{2m} = \frac{1}{4},$$

which is surprising. For medical reasons the $\frac{1}{4}$ is easy to understand.

Example 10.27 (Pareto analysis). Pareto analysis is a method used in quality control to reason backwards to find the cause of an industrial problem.

Suppose we are constructing an item and D is the event that there is a defect. Suppose that the defect can be caused by
- bad raw material, denoted by B,
- bad settings, denoted by S,
- worker error, denoted by W, or
- everything else, denoted by E.

Suppose that by experiment (Pareto analysis) you find the following conditional probabilities:

$$P(D \mid B) = 0.15,$$
$$P(D \mid S) = 0.10,$$
$$P(D \mid W) = 0.05 \quad \text{and}$$
$$P(D \mid E) = 0.005.$$

Suppose that we have for the total probabilities:

$$P(B) = 0.01,$$
$$P(S) = 0.05,$$
$$P(W) = 0.05 \quad \text{and}$$
$$P(E) = 0.89.$$

Then the probability for a defect is

$$P(D) = P(D \mid B)P(B) + P(D \mid S)P(S) + P(D \mid W)P(W) + P(D \mid W)P(E)$$
$$= 0.1345,$$

using Bayes expansion.

Hence, given a defect D, using Bayes formula gives

$$P(B \mid D) = \frac{P(D \mid B)P(B)}{P(D)} = \frac{0.15 \cdot 0.1}{0.01345} = 0.112,$$

$$P(S \mid D) = \frac{P(D \mid S)P(S)}{P(D)} = \frac{0.10 \cdot 0.05}{0.01345} = 0.372,$$

$$P(W \mid D) = \frac{P(D \mid W)P(W)}{P(D)} = \frac{0.05 \cdot 0.05}{0.01345} = 0.186,$$

$$P(E \mid D) = \frac{P(D \mid E)P(E)}{P(D)} = \frac{0.005 \cdot 0.89}{0.01345} = 0.331.$$

Therefore, one checks bad settings first, then everything else, then worker error, and finally bad material.

10.6 The Goat or Monty Hall Problem

The Monty Hall or goat problem is a brain teaser, loosely based on the game show "Let's Make a Deal" and named after its host, Monty Hall.

The problem was originally posed in a letter by S. Selvin to the American Statistician in 1975. It became famous as a question from a letter quoted in M. vos Savant's column in Parade magazine in 1990:

> Suppose you are on a game show, and you have the choice of three doors: Behind one door is a car; behind the others are goats. You pick a door, say No. 1, and the host, who knows what is behind the doors, opens another door, say No. 3, which has a goat. He then says to you "Do you want to pick now door No. 2?" Is it to your advantage to switch your choice?

The answer is that the contestant should switch to the other door. Under the standard assumptions, contestants who switch have a $\frac{2}{3}$ chance of winning the car, while contestants who stick to their initial choice have only a $\frac{1}{3}$ chance.

We now give the solution using the formula of Bayes:

We have the followings events:

- G_i, the win is behind door i, $i = 1, 2, 3$.
- M_j, the host opens door M_j, $j = 1, 2, 3$.

The contestant opens one door, and for his first choice we have

$$P(G_1) = P(G_2) = P(G_3) = \frac{1}{3}.$$

Say, the contestant chooses door No. 1.

Now, the host chooses a door such that a goat is behind it. If the win is behind door No. 1, then the host has two possibilities, namely door No. 2 or door No. 3. Hence, for the choice of door No. 3 we have

$$P(M_3 \mid G_1) = \frac{1}{2}.$$

If the win is behind door No. 2, then the host has to choose door No. 3, that is,

$$P(M_3 \mid G_2) = 1.$$

If the win is behind door No. 3 then the host cannot choose door No. 3. Hence,

$$P(M_3 \mid G_3) = 0.$$

Now, the contestant switches to door No. 2.

We ask for the conditional probability $P(G_2 \mid M_3)$ if the host chooses door No. 3. By the Bayes formula, we get

$$
\begin{aligned}
P(G_2 \mid M_3) &= \frac{P(M_3 \mid G_2)P(G_2)}{P(M_3 \mid G_1)P(G_1) + P(M_3 \mid G_2)P(G_2) + P(M_3 \mid G_3)P(G_3)} \\
&= \frac{1 \cdot \frac{1}{3}}{\frac{1}{2} \cdot \frac{1}{3} + 1 \cdot \frac{1}{3} + 0 \cdot \frac{1}{3}} \\
&= \frac{2}{3}.
\end{aligned}
$$

If the contestant continues with his initial choice, then he has only a $\frac{1}{3}$ chance to win.

We now give a different proof for the goat problem using probability trees together with the multiplication and addition rules, see Figure 10.2.

10.7 Bayes Nets

Bayes formula can be used to utilize amounts of data (big data) when we describe the data in a Bayes net.

At this stage, to introduce Bayes nets we need some more graph theory than we had in Chapter 5 to classify Platonic solids.

Definition 10.28. A *directed graph* or *digraph* is a pair $G = (V,E)$ consisting of a nonempty set V of vertices and a set $E \subset V \times V$ of *oriented edges* or *arcs*.

We write $k = (x,y)$ for $k \in E$.

Remark 10.29.

(1) In other words, a digraph is a nonempty set V equipped with a relation in V.

Constant chooses door	Car is behind door	Host opens door	Success with switching	Success without switching
		No. 2 $\frac{1}{6}$	no	yes
No. 1 $\frac{1}{2}$ C $\frac{1}{2}$		No. 3 $\frac{1}{6}$	no	yes
No. 1 $\frac{1}{3}$	No. 2 D $\frac{1}{3}$ 1	No. 3 $\frac{1}{3}$	yes	no
$\frac{1}{3}$	No. 3 1	No. 2 $\frac{1}{3}$	yes	no
			$\frac{2}{3}$	$\frac{1}{3}$

Figure 10.2: Proof for the goat problem using probability trees together with the multiplication and addition rules.

Representation.

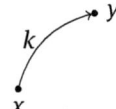

for $k = (x, y) \in E$, $x \neq y$.

if $k = (x, x) \in E$ (here $k = (x, x)$ is called a *loop*).

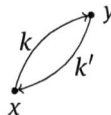

if $k = (x, y) \in E$ and $k' = (y, x) \in E$, $x \neq y$.

(2) We do not allow multiple oriented edges between two vertices $x, y \in V$, that is, we do not allow situations given in Figure 10.3.
Then we call the digraph *simple*.

(3) If V is finite then we say that the digraph is *finite*.

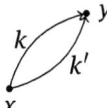

Figure 10.3: Oriented edges.

Agreement. In what follows we only consider finite, simple digraphs.

Definition 10.30. Let $G = (V, E)$ be a digraph and $x \in V$. The *output degree* $d_+(x)$ (*input degree* $d_-(x)$) of x is defined as the number of directed edges with x as beginning vertex (ending vertex, respectively). We define $d(x) = d_+(x) + d_-(x)$. If we have a loop at x then this loop contributes two times for $d(x)$.

If $G = (V, E)$ is a digraph, then

$$|E| = \sum_{x \in V} d_+(x) = \sum_{x \in V} d_-(x) = \frac{1}{2} \sum_{x \in V} d(x).$$

Remark 10.31. In case of a digraph we use the notations edge sequence, edge line, edge path, edge circle of arcs disregarding the direction of the arcs.

If the beginning vertex of each arc is equal to the ending vertex of the previous arc, that is, all arcs are directed in the running through direction, then we talk about directed edge sequences, directed edge line, directed edge path, directed edge circle.

Example 10.32. In Figure 10.4 we have
- $u_1 u_4 u_5 u_3 u_6$ is an (undirected) line,
- $u_1 u_3 u_7 u_1 u_6$ is a directed edge sequence,
- $u_1 u_3 u_2 u_6$ is a directed edge line and
- $u_1 u_6$ is a directed edge path from a to b.

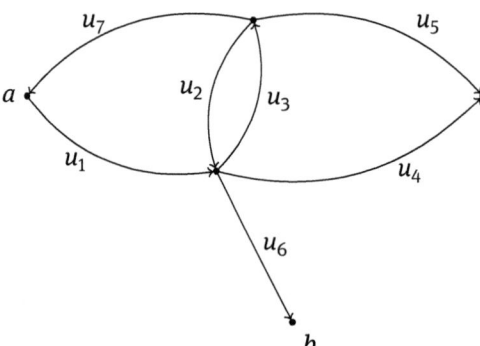

Figure 10.4: Example of a digraph.

Analogously, connected components and connection of a digraph are defined as in an undirected graph.

Definition 10.33. A digraph is called *strongly connected* if for each pair a, b of vertices there exists a directed path from a to b.

In general, an Euler line and Euler cycle are defined for a digraph if we do not consider the direction of the graph.

Theorem 5.40 of Chapter 5 holds analogously for digraphs if we consider them as undirected graphs with, eventually, two edges between two vertices.

Definition 10.34. A closed directed edge line C in a digraph is called a *directed Euler cycle* if C contains each arc of G exactly one time.

Analogously as in Theorem 5.40 of Chapter 5 we get the following.

Theorem 10.35. *A digraph $G = (V,E)$ has a directed Euler cycle if and only if G is strongly connected and $d_+(x) = d_-(x)$ for all $x \in V$.*

We now give an application.

Theorem 10.36. *For each $n \in \mathbb{N}$ there exists a cyclic ordering of 2^n numbers 0 and 1 such that the 2^n n-tuples of consecutive elements represent each integer k with $0 \le k < 2^n$ exactly once as binary numeral.*

Example 10.37. Let $n = 3$, thus $2^3 = 8$. Therefore we get in this example the circle given in Figure 10.5.

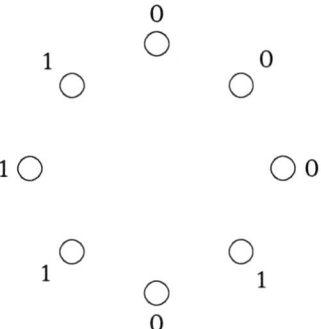

Figure 10.5: Cyclic ordering of 2^3 with numbers 0 and 1.

The triples of consecutive elements are

$$(0,0,0), \quad (0,0,1), \quad (0,1,0), \quad (1,0,1),$$
$$(0,1,1), \quad (1,1,1), \quad (1,1,0), \quad (1,0,0)$$

where (x,y,z) means $x \cdot 2^2 + y \cdot 2 + z$.

Proof. We construct a digraph $G = (V,E)$ with V the set of the 2^{n-1} $(n-1)$-tuples $a = (a_1, a_2, \ldots, a_{n-1})$, $a_i \in \{0,1\}$ for $i = 1, 2, \ldots, n-1$, and E the set of the pairs (a,b) with $a_2 = b_1, a_3 = b_2, \ldots, a_{n-1} = b_{n-2}$.

Then $d_+(a) = d_-(a)$ in G for all vertices a; indeed, the neighbors of a are exactly as given in Figure 10.6.

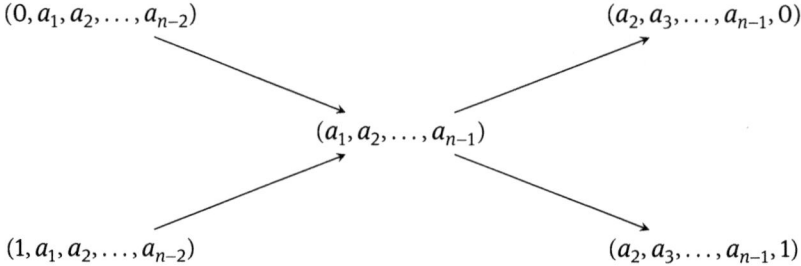

Figure 10.6: Neighbors of a.

Then we have $|E| = 2^n$.

Graph G is strongly connected because for arbitrary $a, b \in V$ we have the directed connecting line

$$(a_1, a_2, \ldots, a_{n-1}) \to (a_2, \ldots, a_{n-1}, b_1) \to (a_3, \ldots, a_{n-1}, b_1, b_2) \to \cdots \to (b_1, b_2, \ldots, b_{n-1}).$$

Hence, by Theorem 10.35, there exists a directed Euler cycle. The allocation

$$(a, b) \mapsto (a_1, a_2, \ldots, a_{n-1}, b_{n-1})$$

defines a mapping $f : E \to \{0, 1\}^n$. We have that f is injective because $f(a, b) = f(c, d)$ means $a = c$ and $b_{n-1} = d_{n-1}$, which gives $b = d$ by definition, and therefore $(a, b) = (c, d)$. Hence, f is bijective because $|E| = 2^n$. By writing one behind the other coordinates, we get the desired ordering, corresponding to the iteration of the directed Euler cycle. $\square$

Example 10.38. Let $n = 3$ and see Figure 10.7 with the directed Euler cycle abhgcdef.

$$(00, 00) \mapsto (0, 0, 0),$$
$$(00, 01) \mapsto (0, 0, 1),$$
$$(01, 10) \mapsto (0, 1, 0),$$
$$(01, 11) \mapsto (0, 1, 1),$$
$$(11, 11) \mapsto (1, 1, 1),$$
$$(11, 10) \mapsto (1, 1, 0),$$
$$(10, 00) \mapsto (1, 0, 0).$$

We now come to the final definition we need for a digraph.

Definition 10.39. Let $G = (V, E)$ be a digraph.

(1) G is called a *directed cyclic graph*, or *cyclic digraph*, if G contains a closed directed edge line, or, in other words, if G contains a directed edge circle.

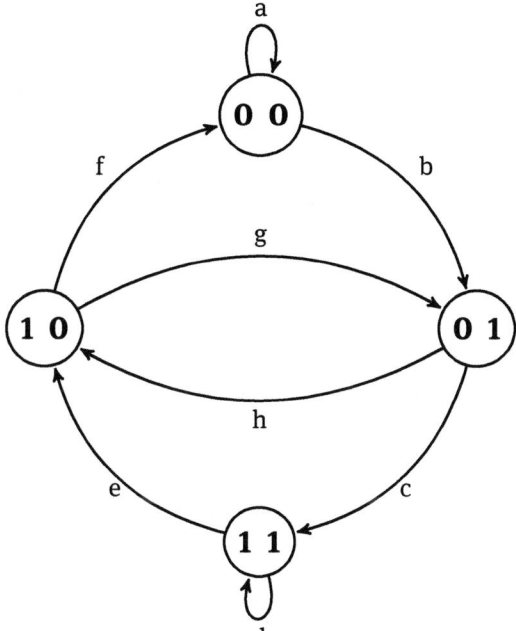

Figure 10.7: Directed Euler cycle is the path abhgcdef.

(2) *G* is called a *directed acyclic graph,* or *acyclic digraph*, if *G* does not contain a closed directed edge line.

Example 10.40. We give three examples, see Figures 10.8, 10.9 and 10.10.

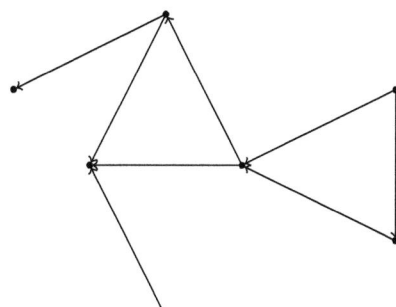

Figure 10.8: Directed acyclic graph.

 Figure 10.9: A directed loop is a directed circle.

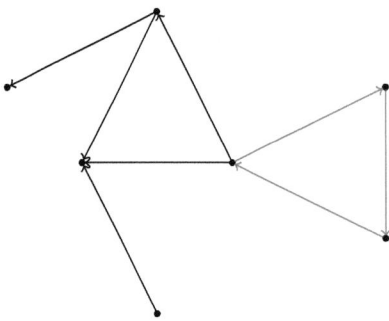

Figure 10.10: Digraph with a closed directed line in gray.

Remark 10.41. An acyclic digraph does not contain a directed loop and, more generally, does not contain a directed circle.

Now we are prepared to introduce Bayes nets.

A *Bayes net,* or *Bayes network,* is a probabilistic graphical model that represents a set of random events and their conditional dependence via a directed acyclic graph (DAG) or acyclic digraphs. For instance, it can represent the probabilistic relationship between diseases and symptoms. Given symptoms, the net can be used to compute the probabilities of the presence of various diseases (see also the examples for the use of Bayes formula for the probability to carry a certain illness). More generally, Bayes nets are a probabilistic method to conclude results under uncertainties.

Practically, in all real-life problems certain unknown and random influences play an essential role. These influences obliterate technically the connection to models. The resulting uncertainties can be regarded by modeling with Bayes probabilities. The Bayes probability method makes available, on the one hand, a consistent mathematical basis and, on the other hand, a comprehensible possibility to describe stochastic coherences.

We now give a more formal definition.

Definition 10.42. A *Bayes net* is made of an acyclic digraph G and conditional probabilities such that:
– The vertices describe events and the edges conditional probabilities;
– The edges are directed and establish respectively the causal effect of the event at the beginning vertex of the edge on the event at the ending vertex of the edge.

Remark 10.43.
(1) It is possible that there exist several undirected paths from one event to another event.
(2) Since unconnected subgraphs are completely independent from each other, we can always assume that the graph is connected as an undirected graph.

From the graph structure, an event (vertex) A has a set of *parent events* (*parent vertices*) pa(A). This set involves all those events from which an edge starts with ending vertex A. It is possible that pa(A) = ∅, which has to be for at least one vertex.

Analogously, there exists for each event A a set of *child events* ch(A) which involves all events which are ending events of an edge which starts at A.

For each event A of G, the conditional probability distribution $P(A \mid$ pa(A)) has to be given. With $P(A \mid$ pa(A)) we mean the following.

If pa(A) = $\{X_1, X_2, \ldots, X_n\}$ then

$$P(A \mid \text{pa}(A)) = P(A \mid X_1 \cap X_2 \cap \cdots \cap X_n) =: P(A \mid X_1, X_2, \ldots, X_n), \tag{10.1}$$

which is the probability of A under the condition that $X_1, X_2, \ldots, X_n$ arose. If pa(A) = ∅ then the total probability $P(A)$ must be given.

If $X_1, X_2, \ldots, X_n$ are events in G (which are closed among adding of parent events), then we calculate the common occurrence as follows:

$$P(X_1 \cap X_2 \cap \cdots \cap X_n) =: P(X_1, X_2, \ldots, X_n) = \prod_{i=1}^{n} P(X_i \mid \text{pa}(X_i)). \tag{10.2}$$

If some X_i has no parent events then we take the total probability for X_i.

This formula is a consequence of the multiplication rule.

We now give an example which explains in detail the above description.

Example 10.44. Suppose that there are two events which could cause grass to be wet: either the sprinkler is on or it is raining. Also, suppose that the rain has direct effect on the use of the sprinkler. If it rains, the sprinkler is usually not turned on.

The situation can be modeled with a Bayes net.

All three events have two possible values, T (for true) and F (for false).

We denote the event grass wet by W, the event sprinkler by S and the event rain by R.

The Bayes net looks like that in Figure 10.11.

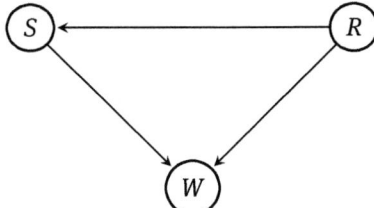

Figure 10.11: Bayes net.

We have the following conditional probability tables see Tables 10.2, 10.3 and 10.4.

	S	
R	T	F
F	0.4	0.6
T	0.01	0.99

Table 10.2: Conditional probabilities for the sprinkler.

R	
T	F
0.2	0.8

Table 10.3: Probabilities of rain.

		W	
S	R	T	F
F	F	0.0	1.0
F	T	0.8	0.2
T	F	0.9	0.1
T	T	0.99	0.01

Table 10.4: Conditional probabilities of wet grass.

From (10.2), the joint probability function is

$$P(W, S, R) = P(W \mid S, R)P(S \mid R)P(R).$$

Now the model can answer the question "what is the probability that it is raining, given the grass is wet?"

By Bayes formula, we get

$$P(R = T \mid W = T) = \frac{P(W = T \mid R = T) \cdot P(R = W)}{P(W = T)} = \frac{P(W = T, R = T)}{P(W = T)}.$$

By the above rule (10.1) we have to sum over the nuisance variables and get

$$P(W = T, R = T) = \sum_{S \in \{T,F\}} P(W = T, S, R = T)$$

and

$$P(W = T) = \sum_{S,R \in \{T,F\}} P(W = T, S, R).$$

Hence,

$$P(R = T \mid W = T) = \frac{\sum_{S \in \{T,F\}} P(W = T, S, R = T)}{\sum_{S,R \in \{T,F\}} P(W = T, S, R)}.$$

Then the numerical results (subscripted by the empirical values from the tables above) are:

$$P(W = T, S = T, R = T) = P(W = T \mid S = T, R = T)P(S = T \mid R = T)P(R = T)$$
$$= 0.99 \cdot 0.01 \cdot 0.2 = 0.00198,$$
$$P(W = T, S = F, R = T) = 0.1584,$$
$$P(W = T, S = W, R = F) = 0.288 \quad \text{and}$$
$$P(W = T, S = F, R = F) = 0.$$

Using these values we obtain

$$P(W = T \mid G = T) = \frac{0.00198 + 0.1584}{0.00198 + 0.288 + 0.1584 + 0} = \frac{891}{2491}$$
$$\approx 0.3577.$$

Hence, the probability that it would rain, given the grass is wet, is 0.3577.

Exercises

1. (a) We throw four identical fair six-side dice and take care of the order of the realized numbers. What is the probability that all four numbers are different? What is the probability that all numbers are equal?
   (b) What is the probability
       (i) to have at least four correct figures in the lottery 6 of 49?
       (ii) to have exactly four correct figures in the lottery 6 of 49?
2. (a) A group of four women and one man are arranged randomly for a group photo. What is the probability that the only man stands in the center?
   (b) A fair coin will be tossed three times. What is the probability to get
       (i) three times heads?
       (ii) at least two times heads?
       (iii) at least one time head?
3. Complete the proof of Theorem 10.3.
4. (a) A hunter has the marksmanship $\frac{1}{2}$. What is the probability that he has at least 3 hits after 10 shots?
   (b) A family has four children. The probability to have a girl is $\frac{1}{2}$. Compute the probability that
       (i) the family has exactly four girls?
       (ii) the first and the second child of the family is a boy?
       (iii) the family has at least two boys?
5. (a) There are 5 red, 3 white and 2 green balls in a box. What is the probability to have

(i) 3 white balls after three drawings with replacement?

(ii) 1 red, 1 white and 1 green ball after three drawings with replacement and considering the order?

(iii) 1 red, 1 white and 1 green ball after three drawings with replacement and not considering the order?

(b) A person writes 12 letters and the related envelops. He puts the letters randomly into the envelopes. What is the probability that no recipient gets the intended letter for him/her?

6. Let $m, n \in \mathbb{N}$ with $n < m$. Alice and Bob independently each make up a number from the set $M = \{1, 2, \ldots, m\}$.

What is the probability that the two numbers differ in at most n?

(*Hint*: Calculate the cardinality of the set

$$\{(a, b) \mid a, b \in M \text{ and } |a - b| \le n\}.$$

Make for this purpose the case analysis

– $a = b$,

– $a < b \le n$,

– $a < b$ and $n + 1 \le b \le m$.)

7. (a) A single roll of a fair six-sided dice produces one of the numbers 1, 2, 3, 4, 5 or 6. Show, using the law of large numbers, that the relative frequency (sometimes called the sample mean) is likely to be close to 3.5 if a large number of such dice are rolled.

(b) We consider the case of tossing a coin n times with S_n the number of heads that turn up. Show that for large n the relative frequency is close to $\frac{1}{2}$.

8. Prove Theorem 10.35.

9. (a) A schoolgirl takes the bus on 70 % of the school days. If she takes the bus then she is on time in school on 80 % of days. On average, she arrives on time at the school only on 60 % of the school days. Today she is on time. What is the probability that she took the bus?

(b) The red–green blindness is an inborn defect of sight which appears in 9 % of the boys but only in 0.6 % of the girls.

We assume that a newborn is a boy for 51 % and a girl for 49 % of instances. A mother is having her baby which has the red–green blindness.

What is the probability that her baby is a boy?

10. A machine can recognize a counterfeit banknote. We define event A as "The machine gives an alarm" and event F as "The banknote is counterfeit". The machine was tested by means of many real banknotes and many fake banknotes. Thereby one got $P(A \mid F) = 0.96$ and $P(A \mid F^c) = 0.01$ for the relative probabilities.

Additionally, it is known that $P(F) = 0.001$, that is, 0.1 % of the circulating banknotes are counterfeit.

Determine $P(F \mid A)$, that is, the probability that a banknote is a counterfeit if the machine gives an alarm.

Give an explanation for the terrifying low value for $P(F \mid A)$.

11. Three death row prisoners – Anton, Bernd and Clemens – are located in individual cells when the governor decides to pardon one of them. He writes their names on three slips of paper, puts them into a hat and pulls one randomly out. He makes known the name of the lucky prisoner to the jail guard. Rumors about that reach Anton.

    He asks the jail guard to tell him who will be given a pardon. The jail guard refused. Now Anton makes the following suggestions to the jail guard:

    Give me the name of one of the other prisoners who will be executed. If Bernd will be pardoned then name Clemens. If Clemens will be pardoned then name Bernd. If I will be pardoned then toss a fair coin to decide between naming Bernd or Clemens. After a while the jail guard agrees with the procedure and tells Anton that Bernd will be executed. What is the probability that Anton survives? What is the probability that Clemens survives?

12. Students' humor in the afternoon often depends on the weather and the quality of the lunch in the commons. We have three events: weather (W), lunch in the commons (M) and humor (H).

    We say that W is true if the sun is shining and false if it is raining. We say that M is true if lunch is eatable and false if lunch is not eatable.

    Finally, we say that H is true if humor is good and false if humor is bad. As usual, we denote true by T and false by F. We have the following graphical situation with the respective conditional dependence, see Figure 10.12.

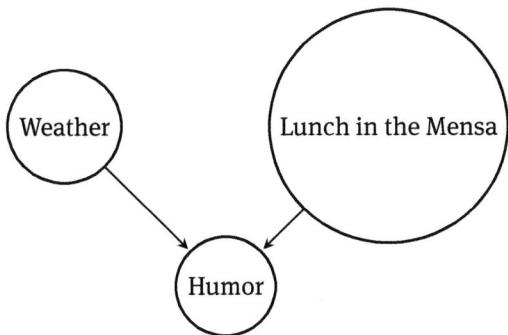

Figure 10.12: Graphical situation for Exercise 12.

Consider Tables 10.5, 10.6 and 10.7.

$P(W = T)$	$P(W = F)$
0.4	0.6

Table 10.5: Probabilities for weather.

P(M = T)	P(M = F)
0.9	0.1

Table 10.6: Probabilities for lunch at the commons.

Table 10.7: Conditional probabilities for humor depending on weather and lunch at the commons.

W	M	P(H = T | W, M)	P(H = F | W, M)
T	T	0.95	0.05
T	F	0.70	0.30
F	T	0.75	0.25
F	F	0.1	0.9

What is the probability that the sun is shining if humor is good? What is the probability that lunch in the commons is eatable if humor is good?

13. We consider a person who might suffer from a back injury, an event represented by the back (B). Such an injury can cause a backache, an event represented by ache (A). The back injury might result from a wrong sport activity, an event represented by sport (S) or from new uncomfortable chairs installed at the person's office, represented by chair (C). In the latter case, it is reasonable to assume that a coworker will suffer and report a similar backache syndrome, an event represented by worker (W).

All events are either true, denoted by T, or false, denoted by F. We have the following graphical situation together with respective conditional dependence, see Figure 10.13:

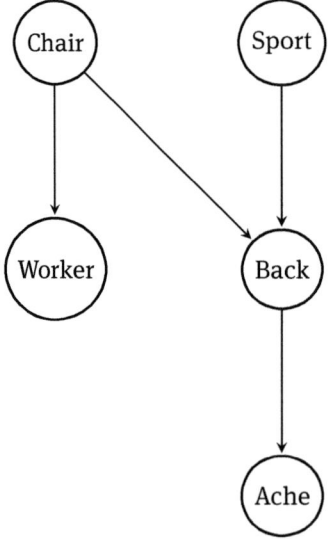

Figure 10.13: Graphical situation for Exercise 13.

Consider Tables 10.8, 10.9, 10.10, 10.11 and 10.12.

$P(C = T)$	$P(C = F)$
0.8	0.2

Table 10.8: Probabilities for chair.

$P(S = T)$	$P(S = F)$
0.02	0.98

Table 10.9: Probabilities for sport.

C	$P(W = T \mid C)$	$P(W = F \mid C)$
T	0.9	0.1
F	0.01	0.99

Table 10.10: Conditional probabilities for chair and worker.

B	$P(A = T \mid B)$	$P(A = F \mid B)$
T	0.7	0.3
F	0.1	0.9

Table 10.11: Conditional probabilities for back and ache.

C	S	$P(B = T \mid C, S)$	$P(B = F \mid C, S)$
T	T	0.9	0.1
T	F	0.2	0.8
F	T	0.9	0.1
F	F	0.01	0.99

Table 10.12: Conditional probabilities for chair, sport and back.

Calculate the common occurrences

$$P(A, B, W, S, C) = P(A \mid B) \cdot P(B \mid S, C) \cdot P(W \mid C) \cdot P(S) \cdot P(C).$$

What is the probability that the chair is uncomfortable in case of a backache? (*Hint*: Make the calculation in steps.)

11 Boolean Lattices, Boolean Algebras and Stone's Theorem

11.1 Boolean Algebras and the Algebra of Sets

In mathematics and mathematical logic, *Boolean algebra* is a branch of algebra in which the values of elements and variables are the truth values *true* and *false*, usually denoted 1 and 0, respectively. The main operations of Boolean algebra are the conjunction *and*, the disjunction *or*, and the negation *not*. Boolean algebra is a formalism for describing logical relations in the same way that ordinary algebra describes numeric relations.

Boolean algebra was introduced by G. Boole (1815–1864) in his book *The Mathematical Analysis of Logic* (1847), and set forth more fully in his next book *An Investigation of the Laws of Thought* (1854). The name *Boolean algebra* was first suggested by H. M. Sheffer (1882–1964) in 1913.

The operation of almost all modern digital computers is based on two-valued or binary systems. Binary systems were known in the ancient Chinese civilization and by the classical Greek philosophers who created a well structured binary system, called *propositional logic*. Propositions may be TRUE or FALSE, and are stated as functions of other propositions which are connected by the three basic logical connectives: AND, OR, and NOT.

In the 1930s, while studying switching circuits, C. Shannon (1916–2001) observed that one could apply the rules of Boolean algebra in this setting, and he introduced switching algebra as a way to analyze and design circuits by algebraic means in terms of logic gates. Shannon already had at his disposal the abstract mathematical apparatus, thus he cast his switching algebra as the two-element Boolean algebra. Because of Shannon's work, the operation of all digital computers depends on Boolean algebras. Boolean algebras are provided in all modern programming languages. They are also used in set theory and statistics.

Boolean algebras are related to the algebra of sets with disjunction analogous to union, conjunction analogous to intersection and negation analogous to complement. A highlight of this chapter will be to prove a celebrated theorem of M. Stone (1903–1989) which is that any Boolean algebra is equivalent to an algebra of sets.

11.2 The Algebra of Sets and Partial Orders

In this section we start with a set M with a partial order and then develop Boolean algebras step by step. As a first highlight we will show in Section 11.5 the theorem of Stone that each finite Boolean algebra can be realized as a power set algebra. In particular each finite Boolean algebra with at least two elements has exactly 2^n elements for some natural number n.

https://doi.org/10.1515/9783110740783-011

We start with a set M and let $\mathcal{P}(M)$ be its power set, that is, the collection of all subsets of M. The set operations intersection $\cap$ and union $\cup$ provide binary operations on $\mathcal{P}(M)$ that satisfy a whole collection of nice properties which we describe in detail in this section. We start first by discussing *partial orders*. We need the basic properties of relations.

Recall that a *relation* $\mathcal{R}$ on a set M is a subset of the Cartesian product $M \times M$. If $(a, b) \in \mathcal{R}$ then we say that a is related to b and write $a \sim b$.

Let $\mathcal{R}$ be a relation on a set M. Then:

(a) $\mathcal{R}$ is *reflexive* if and only if $(x, x) \in \mathcal{R}$ for all $x \in M$. That is, an element x is related to itself.

(b) $\mathcal{R}$ is *symmetric* if and only if whenever $(x, y) \in \mathcal{R}$ then $(y, x) \in \mathcal{R}$ for all $x, y \in M$. That is, if an element x is related to y then y is related to x.

(c) $\mathcal{R}$ is *transitive* if and only if whenever $(x, y) \in \mathcal{R}$ and $(y, z) \in \mathcal{R}$ then $(x, z) \in \mathcal{R}$ for all $x, y, z \in M$. That is, if an element x is related to y and y is related to z then x is related to z.

(d) $\mathcal{R}$ is an *equivalence relation* if and only if it is reflexive, symmetric and transitive. Equivalence relations mimic the properties of equality and define a partition of the set M.

(e) $\mathcal{R}$ is *antisymmetric* if and only if whenever $(x, y) \in \mathcal{R}$ and $(y, x) \in \mathcal{R}$ then $x = y$.

(f) If $\mathcal{R}$ is a relation on M then the *inverse relation* to $\mathcal{R}$, denoted $\mathcal{R}^{-1}$, is defined by

$$\mathcal{R}^{-1} = \{(y, x) \mid (x, y) \in \mathcal{R}, \ x, y \in M\}.$$

For Boolean algebras we need partial order and order relations.

Definition 11.1. A *partial order* on M is a relation $\mathcal{R}$, which is reflexive, antisymmetric and transitive. A set M with a partial order $\mathcal{R}$ is called a *partially ordered set* or *poset*.

Two elements x, y in a partially ordered set M with partial order $\mathcal{R}$ are *comparable* if either $(x, y) \in \mathcal{R}$ or $(y, x) \in \mathcal{R}$.

A partial order is called an *order* if any two elements are comparable.

A set with an order $\mathcal{R}$ on it is called an *ordered set*.

Remarks 11.2.

(1) If not stated otherwise, we denote a partial order by $\leq$, that is, $x \leq y$ if $(x, y) \in \mathcal{R}$. If $x \leq y$ and $x \neq y$ then we write also $x < y$.

(2) If $\mathcal{R}$ is a partial order then the inverse relation $\mathcal{R}^{-1}$ is also a partial order called the dual partial order of $\mathcal{R}$. We write $\geq$ for the inverse relation of $\leq$. In particular we use $x \geq y$ as the equivalent formulation for $y \leq x$.

We have the *duality principle* for order theory. If we interchange in a valid theorem for each partial order consistently $\leq$ and $\geq$ then we again get a valid theorem for each partial order.

Example 11.3.

(1) The division relation on the natural numbers $\mathbb{N}$. Here $x|y$ if and only if $y = kx$ for some $k \in \mathbb{N}$. This is a partial order relation on $\mathbb{N}$.

(2) For each set M, inclusion $\subset$ is a partial order on the power set $\mathcal{P}(M)$.

(3) In the set of real numbers $\mathbb{R}$ the usual size $\leq$ is an order relation.

(4) Let $M = \mathbb{R}^n$. Let $x = (x_1, x_2, \ldots, x_n), y = (y_1, y_2, \ldots, y_n)$. Define $x \leq y$ if and only if $x_k \leq y_k$ for $k = 1, 2, \ldots, n$. This defines a partial order on M.

(5) Let $(M, \leq)$ be an ordered set and let $F(M)$ be the set of all finite sequences of elements from M. The *lexicographic order* on M is a partial order on $F(M)$ defined by

$$(x_1, x_2, \ldots, x_m) \leq (y_1, y_2, \ldots, y_n)$$

$$= \begin{cases} m \leq n & \text{and} \\ x_k = y_k, & \text{for } k = 1, 2, \ldots, m \text{ or} \\ x_k < y_k & \text{for the smallest } k \text{ with } x_k \neq y_k, 1 \leq k \leq m. \end{cases}$$

This is the alphabetic order of words in a dictionary.

The element a is called a *lower-neighbor* of b, written as $a <_N b$, if and only if $a < b$ and $\{x \mid x \in M \text{ and } a < x < b\} = \emptyset$; b is then called an *upper-neighbor* of a denoted by $b >_N a$.

Remark 11.4.

(1) It is possible that there do not exist neighboring elements. For example, in $(\mathbb{R}, \leq)$ for the usual size $\leq$.

(2) If M is finite then there do not exist neighboring elements if and only if $\leq$ is a subset of the relation I_M defined by $(x, y) \in I_M$ if and only if $x = y$. Here certainly I_M is a partial order.

(3) We may represent a finite partially ordered set by a mathematical diagram, the Hasse diagram, named after H. Hasse (1898–1979). For a finite partially ordered set $(M, \leq)$, one represents each element of M as a vertex in the plane $\mathbb{R}^2$ and draws a line segment upward from a to b whenever $a <_N b$. These line segments may cross each other but must not touch any vertices other than their endpoints. Such diagrams with labeled vertices uniquely determine its partial order $a < b$ if and only if, starting from the bottom up there is a sequence of line segments from a to b.

Example 11.5.

(1) Let $M = \{1, 2, 3\}$. Then we have the following possible Hasse diagrams in Figure 11.1.

(2) Let $M = \{1, 2, \ldots, 10\}$ be equipped with the division relation as the partial order. Then we have the following possible Hasse diagram, see Figure 11.2.

We now discuss maximal and minimal elements together with upper and lower bounds.

Definition 11.6. Let $(M, \le)$ be a partially ordered set and $a \in M$, $D \subset M$. Then:

(a) a is called a *maximal element* if $\{x \mid x \in M, x > a\} = \emptyset$.

(b) a is called a *greatest element* if $\{x \mid x \in M, x \le a\} = M$.

(c) a is called a *minimal element* if $\{x \mid x \in M, x < a\} = \emptyset$ and a is a *smallest element* if $\{x \mid x \in M, x \ge a\} = M$.

(d) If a is a smallest element then the elements $b \in M$ with $a <_N b$ are called the *atoms* of M.

(e) An element $b \in M$ is an *upper bound* for D if $x \le b$ for all $x \in D$. An element $b \in M$ is a *least upper bound*, written as lub(D), or sup(D), if b is an upper bound for D and $b \le c$ for any other upper bound c for D.

(f) An element $b \in M$ is a *lower bound* for D if $x \ge b$ for all $x \in D$. An element $b \in M$ is a *greatest lower bound*, written as glb(D), or inf(D), if b is a lower bound for D and $b \ge c$ for any other lower bound c for D.

(g) The subset D is called *directed* if for any $x, y \in D$ there exists a $z \in D$ with $x \le z$ and $y \le z$.

(h) The partially ordered set $(M, \le)$ is *complete* if each directed subset D of M has a least upper bound in M.

Remarks 11.7.

(1) If $D \subset M$ then it is possible that sup(D) or inf(D) does not exist. For instance, sup($\emptyset$) exists if and only if M has a smallest element and inf($\emptyset$) exists if and only if M has a greatest element.

(2) It is possible that there do not exist a smallest or greatest element and also not a minimal or maximal element. For instance this is the case for $(\mathbb{R}, \le)$ with the usual size $\le$.

(3) In each partially ordered set there exists at most one smallest and at most one greatest element.

(4) A complete partially ordered set $(M, \le)$ has a uniquely determined smallest element sup($\emptyset$) = 0.

(5) Each finite partially ordered set contains minimal elements. If it contains exactly one minimal element then this is the smallest element. The analogous statement holds for maximal elements.

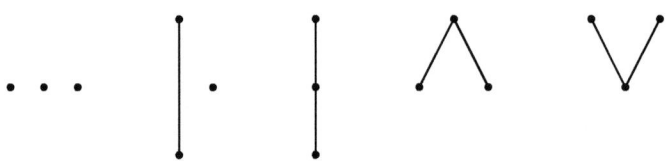

Figure 11.1: All Hasse diagrams with three elements.

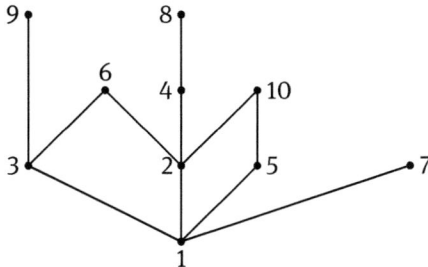

Figure 11.2: Hasse diagram with divisor relation as partial order on $\{1, 2, \ldots, 10\}$.

Theorem 11.8 (Adjunction of a smallest element). *Let $(M, \leq)$ be a partially ordered set. Let N be a set with $N \neq M$ and let $a \in N \setminus M$. We extend the partial order $\leq$ to the set $M \cup \{a\}$ by setting $a \leq a$ and $a \leq x$ for all $x \in M$. Then $M \cup \{a\}$ is a partially ordered set which has a as the smallest element. Analogously, we may adjoin a greatest element to M.*

Example 11.9. $\mathbb{N} \cup \{0\} \cup \{\infty\}$ is a partially ordered set with smallest element 0 and greatest element ∞.

If we adjoin the smallest element a to a finite partially ordered set $(M, \leq)$ then we draw a under the Hasse diagram for $(M, \leq)$ and connect a with each minimal element of $(M, \leq)$ by a line segment.

In the following we call a partially ordered set $(M, \leq)$ just a partial order.

Definition 11.10. Let $(M, \leq)$ be a partial order.
(1) We call a partial order $(N, \leq)$ a *partial suborder* of $(M, \leq)$ if $N \neq \emptyset$, $N \subset M$ and if and only if $a \leq b$ in N then $a \leq b$ in M for $a, b \in N$.
(2) For $a, b \in M$ we call the subset

$$[a, b] = \{x \mid x \in M \text{ and } a \leq x \leq b\}$$

an *interval* in M.
(3) A *chain* in M is a subset $C \subset M$ in which any two elements are comparable.
(4) A chain C in M is called *maximal* if and only if whenever a, b are neighbors in C then a, b are neighbors in M.
(5) The length of a chain C denoted $\ell(C)$ is the number of elements in C minus 1.
(6) If M has a smallest element 0 then

$$d(x) = \sup\{\ell(C) \mid C \text{ a chain from } 0 \text{ to } x\} \in \mathbb{N}_0 \cup \{\infty\}$$

is called the *dimension of $x \in M$*.

Intervals and chains are special partial suborders of $(M, \leq)$.

Example 11.11.

(1) Suppose that we have the Hasse diagram given in Figure 11.3.

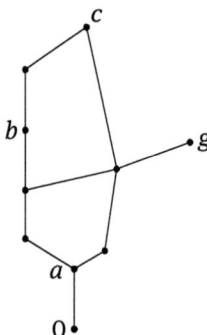

(smallest element)　　　**Figure 11.3:** A Hasse diagram with maximal dimension 6.

Then

$$d(c) = 6, \quad d(g) = 5, \quad d(b) = 4, \quad d(a) = 1, \quad d(0) = 0.$$

The induced partial suborder on $\{a, b, c, g\}$ is given by the Hasse diagram in Figure 11.4.

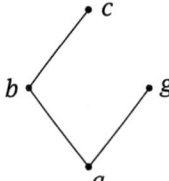

Figure 11.4: Induced partial suborder on $\{a, b, c, g\}$.

The interval $[a, c]$ is given by Figure 11.5.

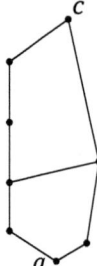

Figure 11.5: Interval $[a, c]$.

There are three maximal chains from a to c with lengths 3, 4, 5, respectively.

(2) We consider $\mathbb{N}$ with the divisor relation. Here 1 is the smallest element.
If $x = p_1^{n_1} p_2^{n_2} \cdots p_k^{n_k}$ is the prime factorization of $x \in \mathbb{N}$ then

$$d(x) = n_1 + n_2 + \cdots + n_k$$

is the dimension of x.

Definition 11.12. Let $(M, \le)$ be a finite partial order. A map $g : M \to \mathbb{Z}$ is called a *gradation* of $(M, \le)$ if and only if $x <_N y$ implies that $g(x) + 1 = g(y)$ for all $x, y \in M$.

Theorem 11.13. *Let $(M, \le)$ be a finite partial order with smallest element $0 \in M$. Then the following are equivalent:*

(1) *Any two maximal chains with the same endpoints have the same length.*
    This is called the Jordan–Hölder chain condition named after C. Jordan (1838–1922) and O. Hölder (1853–1937).
(2) *The dimension is a gradation.*

Proof. Suppose that $a <_N b$ and that C is a maximal chain from 0 to a. Then $C \cup \{b\}$ is a maximal chain from 0 to b. By definition then $\ell(C) = d(a)$ and $\ell(C \cup \{b\}) = d(b)$. This gives $d(b) = d(a) + 1$, showing that (1) implies (2).

Conversely, we first show that if $a, b \in M$ and C is a maximal chain from a to b, then $\ell(C) = d(b) - d(a)$. This means that $\ell(C)$ is independent of the particular chain C. For this we let $a \in M$ be arbitrary but fixed. Let $b \in M$ and let C be a maximal chain from a to b. If $d(a) - d(b) = 0$ then $a = b$ and certainly $d(b) - d(a)$ is the length of C.

Now if $d(b) - d(a) \ge 1$, and let the statement hold for all $b_1 \in M$ and all maximal chains C from a to b_1 with $d(b_1) \le d(b)$. Let $b_2 \in C$ with $b_2 <_N b$. Then $C_2 = C \setminus \{b\}$ is a maximal chain from a to b_2. By assumption $d(b_2) - d(a)$ is the length of C_2. From the second induction principle we get that $d(b_2) + 1 - d(a)$ is the length of C. Now d is a gradation, that is $d(b_2) + 1 = d(b)$ and therefore $d(b) - d(a)$ is the length of C. $\square$

We now define various types of maps between partial orders.

Definition 11.14. Let $(M, \le)$ and $(N, \le)$ be partial orders. A map $f : M \to N$ is called

(1) an *order homomorphism*, or *isotone map*, if $a \le b$ implies $f(a) \le f(b)$ for all $a, b \in M$.
(2) an *order antihomomorphism*, or *antitone map*, if $a \le b$ implies $f(a) \ge f(b)$ for all $a, b \in M$.
(3) If f is a bijective order homomorphism and if also f^{-1} is an order homomorphism then F is called an *order isomorphism*. Analogously, we define an *order antihomomorphism*.
(4) If $(M, \le)$ and $(N, \le)$ are both complete and $f : M \to N$ is an order homomorphism then f is called *continuous* if $\sup(f(D)) = f(\sup(D))$ for each nonempty directed subset $D \subset M$.

Example 11.15.
(1) The identity map id : $\mathbb{N} \to \mathbb{N}$ gives an order homomorphism from $(\mathbb{N}, |)$ to $(\mathbb{N}, \leq)$ where $|$ is the divisor relation and $\leq$ is the usual size. This is not an order isomorphism because the inverse function is not an order homomorphism.
(2) Let M be a set and $(\mathcal{P}(M), \subset)$ be the partial order on the power set of M with containment. Then $\phi : (\mathcal{P}, \subset) \to (\mathbb{N}_0, \leq)$ given by $\phi(A) = |A|$ for each $A \subset M$ is an order homomorphism.
(3) Let $(M, \leq)$ be a partial order. The identity map gives an order isomorphism of the partial order.

Theorem 11.16. *Let $(M, \leq)$ be a finite partial order. Then there is an isotone and injective map $(M, \leq) \to (\mathbb{N}, \leq)$ where $\leq$ is the usual size on $\mathbb{N}$.*

Proof. Without loss of generality, we may assume from Theorem 11.8 that M has a smallest element a_0. Now let $d : M \to \mathbb{N}_0$ be the dimension map. The relation $x \sim y$ if and only if $d(x) = d(y)$ defines an equivalence relation on M. We have $x < y$ if and only if $d(x) < d(y)$ and any two disjoint elements from the same equivalence class are not comparable. Let n be the maximal dimension in M and let A_v denote the equivalence class of those elements with dimension v.

We now define

$$k_v = |A_0| + |A_1| + \cdots + |A_v| \quad \text{for } v = 0, 1, \ldots, n.$$

We have $|A_v| = k_v - k_{v-1}$ for $v = 1, 2, \ldots, n$. The map $a_0 \mapsto 1$ and A_v with $v = 1, 2, \ldots, n$ bijective onto the number set $\{k_{v-1} + 1, k_{v-1} + 2, \ldots, k_v\}$. This defines an isotone and bijective map $M \to \{1, 2, \ldots, k_n\}$. $\square$

Note that in the theorem we may replace finite by countable. Further we may map each finite partial order isotonely and injectively into an ordered set. With this procedure we map elements with the same dimension onto a block of consecutive integers. We may also construct an isotone and injective map successively by the following rule:
(a) Map a minimal element of $(M, \leq)$ onto 1.
(b) If the images of a nonempty subset $N \subset M$ are already defined, then map a minimal element of the partial suborder $(M \setminus N, \leq)$ onto $|N| + 1$.

11.3 Lattices

Let $(M, \leq)$ be a partial order. We now call an element b an *upper element* of a and a a *lower element* of b if $a \leq b$.

Definition 11.17. A partial order $(M, \leq)$ is called a *lattice* if for any two elements a, b of M there exists in M a greatest common lower element called the *conjunction* of a and

b and denoted by $a \wedge b$ and a smallest common upper element called the *disjunction* of a and b and denoted by $a \vee b$.

Example 11.18.

(1) Let $n \in \mathbb{N}$ and $M = \{m \mid m \in \mathbb{N}, m|n\}$. In $\mathbb{N}$ we have a partial order with the divisor relation $m_1 \leq m_2$ if and only if $m_1|m_2$. The given two natural numbers the greatest common divisor $\gcd(m_1, m_2)$ is the greatest common lower element and the least common multiple $\mathrm{lcm}(m_1, m_2)$ is the smallest common upper element for $m_1, m_2 \in M$. Recall that $\gcd(m_1, m_2)$ and $\mathrm{lcm}(m_1, m_2)$ are elements of M.

(2) Suppose $M \subset \mathbb{R}$ with the usual size is a lattice with $a \wedge b = \min(a, b)$ and $a \vee b = \max(a, b)$ for $a, b \in M$. Analogously each order $(M, \leq)$ is a lattice with $a \wedge b$ being the smaller element of the two and $a \vee b$ the larger element of the two for $a, b \in M$.

(3) Let M be a set and $(\mathcal{P}(M), \subset)$ the partial order on the power set with the order relation containment. This is then a lattice since any two subsets A, B of M one can define $A \wedge B = A \cap B$ and $A \vee B = A \cup B$.

We note that $(\mathcal{M}, \subset)$ with $\mathcal{M} \subset \mathcal{P}(M)$ is not a lattice, in general.

(4) The Hasse diagrams for all lattices with at most five elements are given in Figure 11.6.

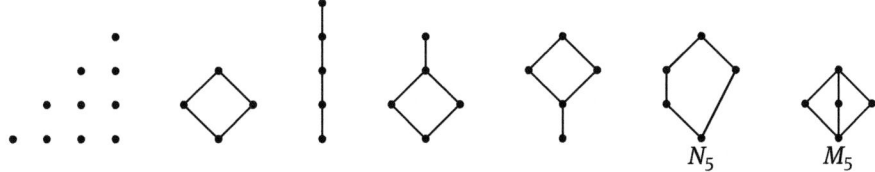

Figure 11.6: Hasse diagrams for all lattices with at most five elements.

(5) If $(M_1, \leq)$ and $(M_2, \leq)$ are partial orders then the Cartesian product $M_1 \times M_2$ together with the componentwise comparison

$$(x_1, x_2) \leq (y_1, y_2) \quad :\Leftrightarrow \quad x_1 \leq y_1 \text{ in } M_1 \text{ and } x_2 \leq y_2 \text{ in } M_2$$

is also a partial order. If both $(M_1, \leq)$ and $(M_2, \leq)$ are lattices then $(M_1 \times M_2, \leq)$ is also a lattice with

$$(x_1, x_2) \wedge (y_1, y_2) = (x_1 \wedge x_2, y_1 \wedge y_2)$$

and

$$(x_1, x_2) \vee (y_1, y_2) = (x_1 \vee x_2, y_1 \vee y_2).$$

An analogous result follows for more than two factors $M_1, M_2, \ldots, M_n$.

Theorem 11.19. *Let $(M, \leq)$ be a lattice. Then the following hold:*

(1) *If M is finite then there exists a smallest element 0 and a greatest element 1 in M.*

(2) *Let $x, y, z \in M$. Then we have the following rules:*

   (L1) $x \wedge x = x$ and $x \vee x = x$. *This is called idempotence.*

   (L2) $x \wedge y = y \wedge x$ and $x \vee y = y \vee x$. *This is commutativity.*

   (L3) $(x \wedge y) \wedge z = x \wedge (y \wedge z)$ and $(x \vee y) \vee z = x \vee (y \vee z)$. *This is associativity.*

   (L4) $x \wedge (y \vee z) = x$ and $x \vee (y \wedge z) = x$. *This is absorbtion.*

(3) *For $x, y \in M$ then $x \leq y$ if and only if $x \wedge y = x$ if and only if $x \vee y = y$.*

Proof. Lattices are sets with special partial orders. Hence by the definition of $a \wedge b$ and $a \vee b$ we use the duality principle for $\wedge$ and $\vee$. For this we have to interchange $\leq$ and $\geq$, $\wedge$ and $\vee$ and also 0 and 1 if they exist. Therefore we must only prove the existence of 0 if M is finite, and for (L1) through (L4) only one equation and for (3) only the first equivalence.

(1) Since M is finite, there exists at least one minimal element and at most one minimal element because we have $a \wedge b$ for all $a, b \in M$. Altogether there must exist a minimal element 0 in M.

(2) The rules (L1) and (L2) are obvious. We prove (L3). Here $y \wedge z$ is a common lower element of y and z and hence $x \wedge (y \wedge z)$ is a common lower element of x, y and z. By the definition of the greatest common lower element, we must have $x \wedge (y \wedge z) \leq (x \wedge y) \wedge z$. Analogously, we get the same result for $\geq$.

   Rule (L4) follows directly from (3) since $x \leq x \vee y$ and $x \wedge y \leq x$.

(3) We now prove the first part of (3) and show that $x \leq y$ if and only if $x = x \wedge y$. Suppose that $x \leq y$. Then we get that $x \leq x \wedge y$. Further, $x \wedge y \leq x$ from the definition. Hence $x = x \wedge y$.

   Suppose that $x \wedge y = x$. Then $x \leq y$ follows from $x \wedge y \leq y$. $\square$

Lattices are not only sets with special partial orders, they are also algebraic structures with two binary operations $\wedge$ and $\vee$. We may take this idea as a characterization of lattices.

Theorem 11.20. *Let M be a nonempty set equipped with two binary operations $\wedge : M \times M \to M$ and $\vee : M \times M \to M$ which satisfy rules (L1) through (L4). If we define $x \leq y$ if and only if $x \wedge y = x$ for all $x, y \in M$ then $(M, \leq)$ is a lattice such that $\wedge$ and $\vee$ have the meaning of the greatest common lower element and smallest common upper element, respectively.*

Proof. We first show that $x \wedge y = x$ if and only if $x \vee y = y$ for all $x, y \in M$ $(*)$. We then have $y = y \vee (y \wedge x) = y \vee (x \wedge y)$ by (L2) and (L4). But then $y = y \vee x = x \vee y$ from $x \wedge y = x$, which proves the result in one direction.

In the other direction, from $x \vee y = y$ we get that $x = x \wedge (x \vee y) = x \wedge y$. This shows that $(*)$ holds.

We now show the properties of a partial order for $\leq$.

From $(L1)$ we get that $\leq$ is reflexive.

If $x \leq y$ and $y \leq x$, that is, $x = x \wedge y$ and $y = y \wedge x = x \wedge y$, we get $x = y$, and hence $\leq$ is antisymmetric.

If $x \leq y$ and $y \leq z$ then $x \leq z$ because

$$x = x \wedge y = x \wedge (y \wedge z) = (x \wedge y) \wedge z = x \wedge z,$$

and hence $\leq$ is transitive.

We finally have to show that $\wedge$ and $\vee$ have the meaning of greatest common lower element and smallest common upper element, respectively. By the duality principle, it is enough to consider the statement for $\wedge$.

Let $x, y \in M$. Then:

(1) $(x \wedge y) \wedge x = x \wedge (x \wedge y) = (x \wedge x) \wedge y = x \wedge y$, that is, $x \wedge y \leq x$.
(2) $(x \wedge y) \wedge y = x \wedge (y \wedge y) = x \wedge y$, that is, $x \wedge y \leq y$.

Therefore $x \wedge y$ is a common lower element of x and y.

Now let $z \in M$ be any common lower element of x and y. Then $z \leq x$ and $z \leq y$. Therefore $y \wedge z = z$ and $x \wedge z = z$. Then we get

$$z \wedge (x \wedge y) = (x \wedge y) \wedge z = x \wedge (y \wedge z) = x \wedge z = z,$$

and hence $z \leq x \wedge y$. □

Theorem 11.21 (Calculations in lattices). *Let $(M, \leq)$ be a lattice. Then for all $a, b, c, d \in M$ the following hold:*

(1) *$a \leq b$ implies that $a \wedge c \leq b \wedge d$ and $a \vee c \leq b \vee d$.*
(2) *$a \leq b$ and $c \leq d$ imply that $a \wedge c \leq b \wedge d$ and $a \vee c \leq b \vee d$.*
(3) *The distributivity inequalities*

$$a \wedge (b \vee c) \geq (a \wedge b) \vee (a \wedge c),$$
$$a \vee (b \wedge c) \leq (a \vee b) \wedge (a \vee c).$$

(4) *The modular inequality*

$$a \leq c \quad implies \quad a \vee (b \wedge c) \leq (a \vee b) \wedge c.$$

Proof. By the duality principle, it is enough to show only one inequality in (3).

(1) Since always $x \leq x \vee y$, we get that

$$a \leq b \leq b \vee c \quad \text{and} \quad c \leq c \vee b = b \vee c \quad \text{if } a \leq b.$$

Therefore $a \vee c \leq b \vee c$ since $a \vee c$ is the smallest common upper element of a and c.

Analogously we get

$$a \wedge c \le a \le b \quad \text{and} \quad a \wedge c \le c \quad \text{if } a \le b$$

and

$$b \wedge c \le b \quad \text{and} \quad b \wedge c \le c.$$

Therefore $a \wedge c \le b \wedge c$ by the maximality of $b \wedge c$.

(2) This follows from (1) by doing it twice.

(3) By (1) we have

$$(a \wedge b) \vee (a \wedge c) \le a \vee (a \wedge c) = a.$$

Then by (2) we have

$$(a \wedge b) \vee (a \wedge c) \le b \vee c.$$

Combining these we get the statement of (3) because $a \wedge (b \vee c)$ is the greatest common lower element of a and $b \vee c$.

(4) We have $a \le c$ and also $a \le a \vee b$. Therefore $a \le (a \vee b) \wedge c$. Further from (1) we have $b \wedge c \le (a \vee b) \wedge c$. Together this means that

$$a \vee (b \wedge c) \le (a \vee b) \wedge c$$

by the minimality of $a \vee (b \wedge c)$. $\qquad\qquad\qquad\qquad\qquad\qquad\qquad\qquad$ □

We note that there are already lattices with five elements, for which equality in (3) or (4) of Theorem 11.21 does not hold, see Figure 11.7.

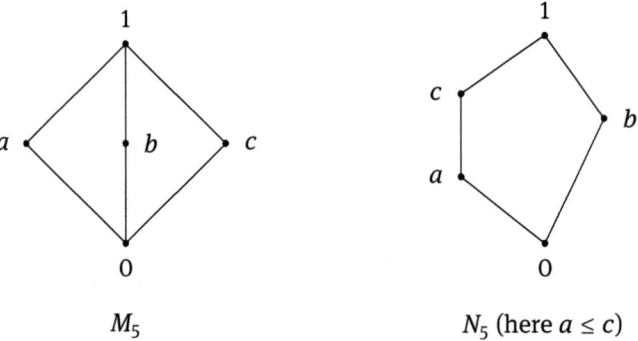

$$M_5 \qquad\qquad\qquad\qquad N_5 \text{ (here } a \le c\text{)}$$

Figure 11.7: Lattices with five elements for which equality in (3) or (4) of Theorem 11.21 does not hold.

In M_5 we have

$$0 = (a \wedge c) \vee (a \wedge b) < a \wedge (b \vee c) = a.$$

In N_5 we have

$$a = a \vee (b \wedge c) < (a \vee b) \wedge c = c.$$

Definition 11.22. Let $(V, \leq)$ and $(V', \leq)$ be lattices. Then a map $f : V \to V'$ is a *lattice homomorphism* if and only if

$$f(x \wedge y) = f(x) \wedge f(y) \quad \text{and} \quad f(x \vee y) = f(x) \vee f(y).$$

Lemma 11.23. *A lattice homomorphism is also an order homomorphism.*

Proof. From $x \leq y$ if and only if $x = x \wedge y$ if and only if $y = x \vee y$ we have

$$f(x) = f(x \wedge y) = f(x) \wedge f(y),$$
$$f(y) = f(x \vee y) = f(x) \vee f(y),$$

and hence $f(x) \leq f(y)$. $\qquad\square$

A *lattice isomorphism* is a bijective lattice homomorphism.

Lemma 11.24. *A lattice isomorphism is also an order isomorphism.*

Proof. Since a lattice isomorphism is also an order homomorphism, we have to show that f^{-1} is also an order homomorphism.

Let $f : V \to V'$ be a bijective lattice homomorphism and let $a, b \in V'$. Suppose that $f(x) = a$ and $f(y) = b$ with $x, y \in V$ and $f(x) \leq f(y)$. Then

$$f^{-1}(a \wedge b) = f^{-1}(f(x) \wedge f(y)) = f^{-1}(f(x \wedge y)) = x \wedge y = f^{-1}(a) \wedge f^{-1}(b).$$

Analogously $f^{-1}(a \vee b) = f^{-1}(a) \vee f^{-1}(b)$. $\qquad\square$

Definition 11.25. Let $(V, \leq)$ be a lattice and $V' \subset V$. Then V' is a *sublattice* of V if V' is closed under the binary operations $\wedge$ and $\vee$.

Example 11.26.
(1) Let $(V, \leq)$ be a lattice. Then:
    (a) Each interval $[a, b]$ in V is a sublattice. From $a \leq x \leq b$ and $a \leq y \leq b$ we get that $a \leq x \wedge y \leq x \vee y \leq b$.
    (b) Each chain in V is a sublattice since $x \leq y$ implies $x = x \wedge y$ and $y = x \vee y$.
(2) Let M be a nonempty set and $M \subset \mathcal{P}(M)$. Then M is a sublattice of $\mathcal{P}(M)$ if it is closed under intersection and union.

11.4 Distributive and Modular Lattices

In this section we discuss certain special types of lattices.

Theorem 11.27. *Let $(V, \leq)$ be a lattice.*
(1) *The following two conditions are equivalent:*
   (D) $x \vee (y \wedge z) = (x \vee y) \wedge (x \vee z)$ *for all* $x, y, z \in V$.
   (D') $x \wedge (y \vee z) = (x \wedge y) \vee (x \wedge z)$ *for all* $x, y, z \in V$.
(2) *If (D) holds, and hence also (D'), then we have*
   (M) $x \leq z \implies x \vee (y \wedge z) = (x \vee y) \wedge z$ *for all* $x, y, z \in V$.

Proof.
(1) It is enough to show that $(D) \implies (D')$. The implication $(D') \implies (D)$ follows by the duality principle.
   Suppose (D) is true and let $a, b, c \in V$ with $x = a \wedge b$, $y = a$ and $z = c$. Then

$$(a \wedge b) \vee (a \wedge c) \underset{(D)}{=} ((a \wedge b) \vee a) \wedge ((a \wedge b) \vee c)$$

$$\underset{(L2),(L4)}{=} a \wedge ((a \wedge b) \vee c)$$

$$\underset{(L2)}{=} a \wedge (c \vee (a \wedge b)).$$

We apply (D) once more, this time with $x = c$, $y = a$ and $z = b$, to obtain

$$a \wedge (c \vee (a \wedge b)) \underset{(D),(L2)}{=} a \wedge ((a \vee c) \wedge (b \vee c))$$

$$\underset{(L3)}{=} (a \wedge (a \vee c)) \wedge (b \vee c)$$

$$\underset{(L4)}{=} a \wedge (b \vee c).$$

This gives (D') with a, b, c for x, y, z.
(2) If $x \leq z$ then (M) follows from (D):

$$x \leq z \Leftrightarrow z = x \vee z \Rightarrow x \vee (y \wedge z)$$

$$\underset{(D)}{=} (x \vee y) \wedge (x \vee z) = (x \vee y) \wedge z. \qquad \square$$

Definition 11.28. A lattice $(V, \leq)$ is a *distributive lattice* if (D), and therefore (D'), holds. It is a *modular lattice* if (M) holds.

In general, a modular lattice is not distributive. As an example, we have already seen that M_5 in Figure 11.7 is not distributive but modular. M_5 is also shown in Figure 11.8.

M_5 is not distributive since

$$0 = (a \wedge c) \vee (a \wedge b) < a \wedge (b \vee c) = a,$$

but it is modular.

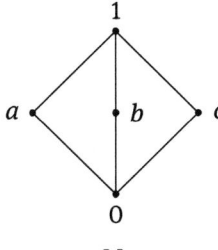

M_5 **Figure 11.8:** Lattice M_5.

Because of the equivalence of (D) and (D') and the self-duality of (M) (replacing $\leq$ with $\geq$ and $\wedge$ by $\vee$ gives the same equality with z and x interchanged) we also have the duality principle for both distributive lattices and modular lattices.

Theorem 11.29 (J. W. R. Dedekind (1831–1916)). *Let $(V, \leq)$ be a lattice. Then the following are equivalent:*
(1) *V is modular.*
(2) *The modular cancellation rule holds:*

$$\textit{If } x \leq y \textit{ and } a \wedge x = a \wedge y \textit{ and } a \vee x = a \vee y \quad \textit{then } x = y.$$

(3) *V does not contain a sublattice that is lattice-isomorphic to N_5.*

Proof. (1) $\implies$ (2) Let $x \leq y$. Then

$$x = x \vee (x \wedge a) = x \vee (y \wedge a)$$
$$\underset{(M)}{=} (x \vee a) \wedge y = (y \vee a) \wedge y = y.$$

(2) $\implies$ (3) We have to show that the modular cancellation rule does not hold in N_5, see Figure 11.9.

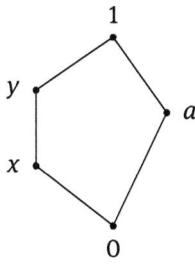

N_5 **Figure 11.9:** Lattice N_5.

We have $x \leq y$. Now $a \wedge x = a \wedge y$ and $a \vee x = a \vee y$ but $x \neq y$.

(3) $\implies$ (1) Assume that $(V, \leq)$ is not modular. Then there exist $a, b, c \in V$ with

$$a \leq c \quad \text{and} \quad a \vee (b \wedge c) < (a \vee b) \wedge c.$$

Let $x = a \vee (b \wedge c)$, $y = b$, $z = (a \vee b) \wedge c$, $n = b \wedge c$ and $e = a \vee b$. We then have $n \leq x < z \leq e$ and $n \leq y \leq e$. We cannot have $x \leq y$ because if $x \leq y$ then $a \leq b$ and, together with $a \leq c$, we get

$$b \wedge c = a \vee (b \wedge c) < (a \vee b) \wedge c = b \wedge c.$$

We cannot have $y \leq z$ because if $y \leq z$ then $b \leq c$ and, together with $a \leq c$, we get

$$a \vee b = a \vee (b \wedge c) < (a \vee b) \wedge c = a \vee b.$$

From these facts we get that $y \notin \{n, x, z, e\}$ and that y is comparable neither with x nor z. In particular, the five elements n, x, y, z, e are pairwise distinct.
Further

$$x \vee y = a \vee (b \wedge c) \vee b = a \vee b = e,$$
$$z \wedge y = (a \vee b) \wedge c \wedge b = b \wedge c = n.$$

This means, on the one hand, that y is comparable with neither x nor with z and, on the other hand, that $\{n, x, y, z, e\}$ is closed under $\wedge$ and $\vee$.
Therefore $\{n, x, y, z, e\}$ is a sublattice of V which is lattice-isomorphic to N_5. ☐

We saw that the lattice M_5 is modular but not distributive. If both M_5 and N_5 do not appear then the lattice is distributive. This characterizes distributive lattices.

Theorem 11.30 (G. Birkhoff (1911–1996)). *Let $(V, \leq)$ be a lattice. Then the following are equivalent:*
(1) *V is distributive.*
(2) *In V the cancellation rule holds:*

$$\text{if } a \wedge x = a \wedge y \text{ and } a \vee x = a \vee y \quad \text{then } x = y$$

for all $x, y, a \in V$.
(3) *V does not contain a sublattice which is lattice-isomorphic to either M_5 or N_5.*
Note that for the cancellation rule we need both equations $a \wedge x = a \wedge y$ and $a \vee x = a \vee y$.

Proof. (1) $\implies$ (2)

$$x \underset{(L4)}{=} x \vee (x \wedge a) = x \vee (a \wedge y)$$
$$\underset{(D)}{=} (x \vee a) \wedge (x \vee y) = (a \vee y) \wedge (x \vee y)$$

$$\underset{(D)}{=} (a \wedge x) \vee y$$

$$\underset{(L4)}{=} y.$$

(2) $\implies$ (3) We have to show that the cancellation rule does not hold in M_5 and N_5, which are given in Figure 11.10.

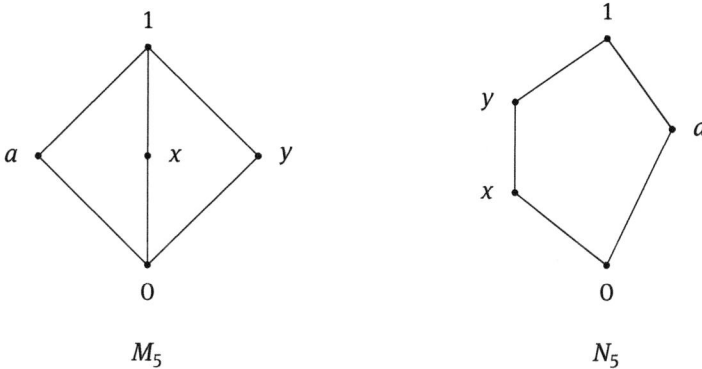

1

a $\quad\quad x \quad\quad y$

0

M_5

1

y

a

x

0

N_5

Figure 11.10: Lattices M_5 and N_5.

In both cases we have $a \wedge x = a \wedge y$ and $a \vee x = a \vee y$ but $x \neq y$.

(3) $\implies$ (1) Assume that $(V, \leq)$ is not distributive and does not contain a sublattice that is lattice isomorphic to N_5. Hence $(V, \leq)$ is modular by Theorem 11.29. Since $(V, \leq)$ is not distributive there exist $a, b, c \in V$ with

$$(a \wedge b) \vee (a \wedge c) < a \wedge (b \vee c).$$

We define

$$n = (a \wedge b) \vee (a \wedge c) \vee (b \wedge c),$$
$$e = (a \vee b) \wedge (a \vee c) \wedge (b \vee c),$$
$$x = (a \wedge e) \vee n,$$
$$y = (b \wedge e) \vee n,$$
$$z = (c \wedge e) \vee n.$$

Then we have $n \leq x, y, z$. From (M), which holds since $(V, \leq)$ is modular, we get

$$a \wedge n = a \wedge ((a \wedge b) \vee (a \wedge c) \vee (b \wedge c)) = ((a \wedge b) \vee (a \wedge c)) \vee (a \wedge (b \wedge c)) = (a \wedge b) \vee (a \wedge c).$$

This, together with $a \wedge e = a \wedge (b \vee c)$, leads to $n < e$. From this we get $x, y, z \leq e$. To see, for instance, $x \leq e$, we note that $a \wedge e \leq e$ and $n < e$.

To show that $\{n, e, x, y, z\}$ forms a sublattice of V which is lattice-isomorphic to M_5, it is enough to show that

$$x \wedge y = x \wedge z = y \wedge z = n$$

and

$$x \vee y = x \vee y = x \vee z = e.$$

We show that $x \wedge y = n$. The other cases are analogous.

We have

$$
\begin{aligned}
x \wedge y &= ((a \wedge e) \vee n) \wedge ((b \wedge e) \vee n) \\
&\underset{(M)}{=} ((a \wedge e) \wedge ((b \wedge e) \vee n)) \vee n \\
&\underset{(M)}{=} ((a \wedge e) \wedge ((b \vee n) \wedge e)) \vee n \\
&\underset{(L2)}{=} ((a \wedge e) \wedge e \wedge (b \vee n)) \vee n \\
&\underset{(L1)}{=} ((a \wedge e) \wedge (b \vee n)) \vee n \\
&\underset{(L4)}{=} ((a \wedge (b \vee c)) \wedge (b \vee (a \wedge c))) \vee n \\
&\underset{(M)}{=} (a \wedge (b \vee ((b \vee c) \wedge (a \wedge c)))) \vee n \\
&= (a \wedge (b \vee (a \wedge c))) \vee n && \text{(using } a \wedge c \leq c \leq b \vee c) \\
&\underset{(L1)}{=} n.
\end{aligned}
$$

This proves Theorem 11.30. □

11.5 Boolean Lattices and Stone's Theorem

We are now almost ready to describe the Boolean lattices. The standard example is the lattice $(\mathcal{P}(M), \subset)$ where M is a set. Here the empty set $\emptyset$ is the smallest and M the greatest element. It is possible that $M = \emptyset$. In what follows we always assume that a lattice $(V, \leq)$ has a smallest element 0 and a greatest element 1. If V is finite then these elements exist by Theorem 11.19.

Definition 11.31. Let $(V, \leq)$ be a lattice with smallest element 0 and greatest element 1. If V is finite these exist automatically.
(a) Elements $x, y \in V$ are called *complementary* to each other if and only if $x \wedge y = 0$ and $x \vee y = 1$. We call y a *complement* of x.
(b) $(V, \leq)$ is called a *complementary lattice* if each element in V has at least one complement.
(c) $(V, \leq)$ is a *Boolean lattice,* or just *Boolean,* if $(V, \leq)$ is distributive and complementary.

Remark 11.32.

(1) For distributive lattices we know that the duality principle holds. From the definition it also holds for complementary lattices if we interchange ∧ and ∨ and also 0 and 1. Hence the duality principle holds for Boolean lattices.

(2) A chain in a lattice is always a distributive sublattice since if, for instance, $y \leq z$ then $x \vee y \leq x \vee z$ and $x \vee (y \wedge z) = x \vee y = (x \vee y) \wedge (x \vee z)$.

(3) A complementary lattice is in general not distributive; M_5 provides an example, see Figure 11.11.

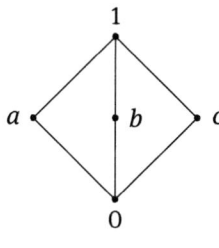

M_5 **Figure 11.11:** Lattice M_5.

We know that M_5 is not distributive but modular. M_5 is complementary. Certainly $0 \wedge 1 = 0$ and $0 \vee 1 = 1$ and, for example, $a \wedge b = 0$ and $a \vee b = 1$.

(4) A chain with more than two elements is distributive but not complementary.

Theorem 11.33. *Let $(V, \leq)$ be a distributive lattice with the smallest element 0 and the greatest element 1. Then each element of V has at most one complement.*

Proof. Let y_1 and y_2 be complementary to x. Then

$$x \wedge y_1 = x \wedge y_2 = 0 \quad \text{and} \quad x \vee y_1 = x \vee y_2 = 1.$$

Therefore $y_1 = y_2$ by the cancellation rule. ☐

It follows from the theorem that in a Boolean lattice for each element a there exists exactly one complement $\bar{a}$. Hence in a Boolean lattice we have besides $(L1)$–$(L4)$ and (D)–(D') also the rule: For each $a \in V$ a unique complement $\bar{a}$ exists with $a \wedge \bar{a} = 0$, $a \vee \bar{a} = 1$.

Further in a Boolean lattice we get *De Morgan's laws*, named after A. De Morgan (1806–1871), relating conjunction, disjunction and complement.

Theorem 11.34 (De Morgan's Laws). *In a Boolean lattice $(V, \leq)$ we have*

(1) $\overline{(a \wedge b)} = \bar{a} \vee \bar{b}$,

(2) $\overline{(a \vee b)} = \bar{a} \wedge \bar{b}$.

Proof. Here

$$(a \wedge b) \wedge (\overline{a} \vee \overline{b}) = a \wedge ((b \wedge \overline{a}) \vee (b \wedge \overline{b})) = a \wedge \overline{a} \wedge b = 0 \wedge b = 0$$

and

$$(a \wedge b) \vee (\overline{a} \vee \overline{b}) = ((a \vee \overline{a}) \wedge (\overline{a} \vee b)) \vee \overline{b} = (\overline{a} \vee b) \vee \overline{b} = \overline{a} \vee 1 = 1.$$

Hence $\overline{a} \vee \overline{b} = \overline{(a \wedge b)}$ by the uniqueness of the complement. The second of De Morgan's laws follows from the duality principle. □

We now define a *Boolean algebra*.

Definition 11.35. A *Boolean algebra* consists of a nonempty set B together with the following axioms:
(1) There exist in B two distinguished elements, the *zero element* 0 and the *unity element* 1.
(2) There exists in B a unary operation $^-$, which assigns to each element $a \in B$ a unique element $\overline{a}$, called the *complement of a*.
(3) There exists in B a binary operation $\wedge$.
(4) There exists in B a second binary operation $\vee$.
(5) The operations $\wedge$, $\vee$ and $^-$ satisfy the following axioms:
  (a) $a \wedge a = a$ and $a \vee a = a$ for all elements $a \in B$.
  (b) $a \wedge b = b \wedge a$ and $a \vee b = b \vee a$ for all $a, b \in B$.
  (c) $a \wedge (b \wedge c) = (a \wedge b) \wedge c$ and $a \vee (b \vee c) = (a \vee b) \vee c$ for all $a, b, c \in B$.
  (d) $a \wedge (b \vee c) = (a \wedge b) \vee (a \wedge c)$ and $a \vee (b \wedge c) = (a \vee b) \wedge (a \wedge c)$ for all $a, b, c \in B$.
  (e) $a \wedge (a \vee b) = a$ and $a \vee (a \wedge b) = a$ for all $a, b \in B$.
  (f) $a \wedge \overline{a} = 0$ and $a \vee \overline{a} = 1$ for all $a \in B$.
  (g) $a \wedge 0 = 0$, $a \vee 0 = a$, $a \vee 1 = 1$, $a \wedge 1 = a$ for all $a \in B$.

Remark 11.36.
(1) As an example, if M is a set, then the power set $\mathcal{P}(M)$ is a Boolean algebra with $0 = \emptyset$, $1 = M$ and $\overline{A} = A^c = M \setminus A$. Here $\wedge = \cap$ and $\vee = \cup$.
  By definition the duality principle holds for Boolean algebras. If we interchange consistently $\wedge$ and $\vee$ as well as 0 and 1 in a statement which is true for all Boolean algebras then we again get a true statement for all Boolean algebras.
(2) With the same proof as for Boolean lattices, for Boolean algebras we get De Morgan's laws:
  (a) $\overline{(a \wedge b)} = \overline{a} \vee \overline{b}$,
  (b) $\overline{(a \vee b)} = \overline{a} \wedge \overline{b}$.
(3) Boolean lattices and Boolean algebras denote the same (discrete) algebraic structure. We talk about lattices if we want to emphasize the order, and we talk about algebras if we want to emphasize the algebraic operations.

In a Boolean algebra with the operations $\wedge$, $\vee$ and $^-$, we get a partial order by defining

$$a \leq b \quad \text{if and only if} \quad a = a \wedge b.$$

In what follows we want to describe and classify the finite Boolean lattices (finite Boolean algebras). We show that they are lattice-isomorphic to a Boolean lattice $(\mathcal{P}(M), \subset)$ for a nonempty set M. This is the theorem of M. H. Stone (1903–1989).

Definition 11.37. Let $(V, \leq)$ be a finite lattice. An element $a \in V$ is called *irreducible (or more concrete V-irreducible)* if and only if a is not a minimal element and for all $b, c \in V$ we get from $a = b \vee c$ that $a = b$ or $a = c$.

Let $\mathcal{I}(V)$ denote the set of irreducible elements of V.

Remark 11.38. Each finite lattice $(V, \leq)$ with more than one element contains irreducible elements. Recall that a finite lattice V has exactly one minimal element 0 and the elements a with dimension $d(a) = 1$ are irreducible.

There do exist infinite lattices without irreducible elements. An example for this is the Cartesian product $\mathbb{Z} \times \mathbb{Z}$ with componentwise comparison. This follows from $(m, n) = (m - 1, n) \vee (m, n - 1)$ for $m, n \in \mathbb{Z}$.

Example 11.39.
(1) In a chain each $a \neq 0$ is irreducible.
(2) In the Hasse diagram in Figure 11.12 all elements except 0 are irreducible:

$$a = x \vee y \quad \Rightarrow \quad x = y = a \text{ or } x = 0 \text{ or } y = 0,$$

that is, $a = x$ or $a = y$.

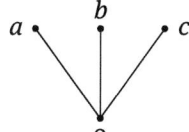

$$0 \qquad \text{\textbf{Figure 11.12:} Hasse diagram with three irreducible elements.}$$

(3) In the Hasse diagram in Figure 11.13 the irreducible elements are the four circled elements.

Definition 11.40.
(1) A lattice $(V, \leq)$ is called a *set lattice* if V is lattice-isomorphic to a sublattice of a power set lattice $\mathcal{P}(M)$ for some set M.

Each set lattice is certainly distributive since distributivity follows in sublattices.

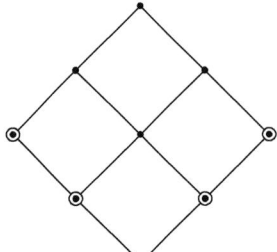

Figure 11.13: Hasse diagram in which irreducible elements are the four circled elements.

(2) A *Boolean set lattice* or *Boolean set algebra* is a set lattice V of a power set lattice $\mathcal{P}(M)$ for some set M with $\emptyset, M \in V$ and for $A, B \in V$ then $A^c, A \vee B$ and $A \wedge B$ are all in V.

Theorem 11.41. *Let* $(V, \leq)$ *be a finite distributive lattice. For* $a \in V$ *we define*

$$\rho(a) = \{x \mid x \in \mathcal{I}(V) \text{ and } x \leq a\}$$

if $a \neq 0$ *and* $\rho(a) = \emptyset$ *if* $a = 0$.
  Then the following hold:
(1) $\rho(V) = \{\rho(a) \mid a \in V\}$ *is a set lattice with respect to the inclusion* $\subset$ *and with the union* $\cup$ *and the intersection* $\cap$ *as the operations.*
(2) *The allocation* $a \mapsto \rho(a)$ *defines a lattice isomorphism* $(V, \leq) \to (\rho(V), \subset)$.

Proof.
(1) In any case $(\rho(V), \subset)$ is a partial suborder of the lattice $(P, \subset)$. It may be that $\rho(V) = \emptyset$, namely if $V = \emptyset$. However, for each $a \neq 0$ we get $\rho(a) \neq \emptyset$ because either $\dim(a) = 1$ or if $\dim(a) \geq 2$ then a maximal chain from 0 to a contains an element with dimension 1. (Such a maximal chain exists.) We have to show that $\rho(V)$ is closed under $\cap$ and $\cup$. Let $a, b \in V$ with $a \neq 0, b \neq 0$. Then

   (1) $\rho(a) \cap \rho(b) = \{x \mid x \in \mathcal{I}(V) \text{ and } x \leq a, x \leq b\} = \rho(a \wedge b) \in \rho(V).$
   (2) $\rho(a) \cup \rho(b) = \{x \mid x \in \mathcal{I}(V) \text{ and } x \leq a \text{ or } x \leq b\} \subset \{x \mid x \in \mathcal{I}(V) \text{ and } x \leq a \vee b\}$
        $= \rho(a \vee b).$

On the other hand, let $x \in \rho(a \vee b)$. Then by definition $x \leq a \vee b$. Now

$$x = x \wedge (a \vee b) \underset{(D)}{=} (x \wedge a) \vee (x \wedge b).$$

From $x \in \mathcal{I}(V)$ we get $x = x \wedge a$ or $x = x \wedge b$. Hence $x \leq a$ or $x \leq b$, and finally $x \in \rho(a) \cup \rho(b)$. Altogether we get $\rho(a) \cup \rho(b) = \rho(a \vee b) \in \rho(V)$.

(2) Since $\wedge$ and $\vee$ are associative, for finitely many $a_1, a_2, \ldots, a_n$ we may write

$$\bigvee_{i=1}^{n} a_i$$

instead of $a_1 \vee a_2 \vee \cdots \vee a_n$. Also if $A = \{a_1, a_2, \ldots, a_n\}$ we may just write

$$\bigvee_{a_i \in A} a_i.$$

The analogous representation we may use for $\wedge$.

With this setting we get

$$a = \bigvee_{x \in \rho(a)} x$$

for $a \in V \setminus \{0\}$.

This shows the injectivity of the map $\rho : V \rightarrow \rho(V)$ given by $a \mapsto \rho(a)$. The map ρ is surjective by definition.

The equations $\rho(a \wedge b) = \rho(a) \cap \rho(b)$ and $\rho(a \vee b) = \rho(a) \cup \rho(b)$ were shown already. $\qquad\qquad\square$

Definition 11.42. Let $(M, \leq)$ be a partial order with smallest element 0. The *atoms* of M are those elements b with $0 <_N b$. These are exactly those elements $b \in M$ with $d(b) = 1$.

Theorem 11.43. *Let $(V, \leq)$ be a finite Boolean lattice with $a \in V, a \neq 0$. Then a is irreducible if and only if a is an atom.*

Proof. It is clear that atoms are irreducible. Now let $a \in V, a \neq 0$, not be an atom. Then there exists an $a_1 \in V$ with $0 < a_1 < a$, and we get

$$a_1 \vee (a \wedge \overline{a_1}) \underset{(D)}{=} (a_1 \vee a) \wedge (a_1 \vee \overline{a_1}) = a \wedge 1 = a.$$

We claim that $a \wedge \overline{a_1} < a$.

For this claim first we have $a \wedge \overline{a_1} \leq a$. Assume that $a \wedge \overline{a_1} = a$. By De Morgan's law we get

$$\overline{a \wedge \overline{a_1}} = \overline{a} \vee \overline{\overline{a_1}} = \overline{a} \vee a_1 = \overline{a},$$

and therefore $a_1 \leq \overline{a}$. It follows that $a_1 \leq \overline{a} \wedge a = 0$, that is, $a_1 = 0$, which gives a contradiction. Hence $a \wedge \overline{a_1} < a$ and a is not irreducible. $\qquad\qquad\square$

We now come to the main result of this section Stone's theorem for finite lattices.

Theorem 11.44 (Stone's theorem). *Each finite Boolean lattice $(V, \leq)$ is lattice-isomorphic to a power set lattice $(\mathcal{P}(M), \subset)$.*

More concrete, let A be the set of atoms in V. Then $(\mathcal{P}(A), \subset)$ and $(V, \leq)$ are lattice-isomorphic.

Proof. V is distributive and hence by Theorem 11.41 we have that V is lattice-isomorphic to $(\rho(V), \subset)$ where

$$\rho(V) = \{\rho(a) \mid a \in V\}$$

with

$$\rho(a) = \{x \mid x \in \mathcal{I}(V), x \leq a\}.$$

By Theorem 11.43 we have $\mathcal{I}(V) = A$, and hence $\rho(V) \subset \rho(A)$. If $a \in A$ then $\rho(a) = \{a\}$, which means that all sets with one element from A belong to $\rho(V)$. Since V is closed under $\wedge$, we finally have $\rho(V) = \mathcal{P}(A)$. $\qquad\square$

Corollary 11.45. *Let V be a finite Boolean algebra. Then $|V| = 2^n$ where $n = |A|$ where A is the set of the atoms in V.*

Remark 11.46.
(1) In the opposite direction, if M is a finite set with $|M| = n$ then $(\mathcal{P}(M), \subset)$ is a Boolean lattice with 2^n elements. Hence for each $k = 2^n$, $n \in \mathbb{N}_0$, there exists a Boolean lattice with exactly k elements, and this is uniquely determined up to lattice isomorphism.
(2) We note further that Theorem 11.44 does not hold in general for infinite sets. We give an example. Let M be the set of all subsets of $\mathbb{N}$ which are either finite or cofinite, that is, have a finite complement.
We claim that M is a distributive lattice with respect to inclusion $\subset$ and with the operations $\cup$ and $\cap$. To see this we have:
(a) $A, B \in M \implies A \cup B \in M$. This is clear if both A, B are finite. If A is infinite then A^c is finite and hence

$$(A \cup B)^c = A^c \cap B^c \quad \text{is finite.}$$

Therefore $A \cup B \in M$.
The case where B is infinite is analogous.
(b) $A, B \in M \implies A \cap B \in M$. If either A or B is finite then $A \cap B$ is finite, and hence $A \cap B \in M$. If both A and B are infinite then both A^c and B^c are finite, and hence $(A \cap B)^c = A^c \cup B^c$ is finite. Therefore $A \cap B \in M$.
(c) Since $M \subset \mathcal{P}(\mathbb{N})$, all the rules for $\cup$ and $\cap$ hold in M, and hence M is a distributive lattice.

We next claim that M is a complementary lattice. For this claim we must show that if $A \in M$ then $A^c \in M$. If A is finite then A^c is cofinite, and hence $A^c \in M$. If A is infinite then A^c is finite, and hence $A^c \in M$.

Hence M is a distributive, complementary lattice, and therefore M is a Boolean lattice. Further M is an infinite Boolean lattice and M is countable.

Now let N be any set, then $\mathcal{P}(N)$ is either finite or uncountable. It follows that $(M, \subset)$ is not lattice-isomorphic to a power set lattice $(\mathcal{P}(N), \subset)$ for some set N.

We now state the general theorem of Stone which we will prove in Section 11.8.

Theorem 11.47 (General Stone's theorem). *Each Boolean lattice $(V, \leq)$ is a Boolean set lattice, that is, lattice-isomorphic to a Boolean sublattice of a power set lattice $(\mathcal{P}(N), \subset)$ for some set N.*

11.6 Construction of Boolean Lattices via 0–1 Sequences

Besides the trivial Boolean lattice where $\{0\} = \{1\}$, the Boolean lattice containing only the two elements 0 and 1 with $0 \neq 1$ is the simplest.

Let $B = \{0, 1\}$ with $0 \neq 1$. For the operations $\wedge$ and $\vee$ we give the following tables, see Table 11.1.

Table 11.1: Operations $\wedge$ and $\vee$.

$\wedge$	0	1		$\vee$	0	1
0	0	0		0	0	1
1	0	1		1	1	1

For the complement in B we have $\overline{1} = 0$ and $\overline{0} = 1$. We now consider $\{0, 1\}$ as a subset of $\mathbb{Z}$, and here $\{0, 1\}$ is closed under multiplication in $\mathbb{Z}$ and the S-sum defined by

$$a + b = \begin{cases} 0, & \text{if } a = b = 0 \text{ and} \\ 1, & \text{otherwise} \end{cases}$$

with $a, b \in \{0, 1\}$.

Now $\wedge$ coincides with the multiplication in $\{0, 1\}$ and $\vee$ with the S-sum or Boolean sum:

$$a \wedge b = ab \quad \text{and} \quad a \vee b = a + b \atop S$$

for $a, b \in \{0, 1\} \subset \mathbb{Z}$.

Starting with B we may construct each finite Boolean lattice.

Let B^n be the n-fold Cartesian product of B, that is, the set of n-digit 0–1 sequences $(a_1, a_2, \ldots, a_n)$ with $a_i \in \{0, 1\}, i = 1, 2, \ldots, n$, equipped with the componentwise multiplication and Boolean addition:

$$(a_1, a_2, \ldots, a_n) \wedge (b_1, b_2, \ldots, b_n) := (a_1, a_2, \ldots, a_n) \cdot (b_1, b_2, \ldots, b_n)$$

$$:= (a_1 b_1, a_2 b_2, \ldots, a_n b_n)$$

$$(a_1, a_2, \ldots, a_n) \vee (b_1, b_2, \ldots, b_n) := (a_1, a_2, \ldots, a_n) \underset{S}{+} (b_1, b_2, \ldots, b_n)$$

$$:= (a_1 \underset{S}{+} b_1, a_2 \underset{S}{+} b_2, \ldots, a_n \underset{S}{+} b_n)$$

with all $a_i, b_i \in \{0, 1\}$. We have $|B^n| = 2^n$ and define

$$(a_1, a_2, \ldots, a_n) \leq (b_1, b_2, \ldots, b_n)$$

if and only if

$$(a_1, a_2, \ldots, a_n) = (a_1, a_2, \ldots, a_n)(b_1, b_2, \ldots, b_n) = (a_1 b_1, a_2 b_2, \ldots, a_n b_n).$$

With $0 = (0, 0, \ldots, 0)$ and $1 = (1, 1, \ldots, 1)$ we get that $0 \leq (a_1, a_2, \ldots, a_n) \leq 1$ for all $(a_1, a_2, \ldots, a_n) \in B_m$ and certainly $(a_1, a_2, \ldots, a_n) \leq (b_1, b_2, \ldots, b_n)$ if and only if $a_i \leq b_i$ for $i = 1, 2, \ldots, n$.

Theorem 11.48. B^n with $n \geq 1$ is a Boolean algebra, that is, $(B^n, \leq)$ is a Boolean lattice with the zero element 0 and the unity 1. Further if $(a_1, a_2, \ldots, a_n) \in B^n$ then $\overline{(a_1, a_2, \ldots, a_n)} = (\overline{a_1}, \overline{a_2}, \ldots, \overline{a_n})$ is the complement of $(a_1, a_2, \ldots, a_n)$. Each Boolean algebra (lattice) with $2^n, n \geq 1$, elements is lattice-isomorphic to B^n.

Proof. Let $M = \{1, 2, \ldots, n\}$. For $A \subset M$ we define

$$\chi_A = (\chi_A(1), \chi_A(2), \ldots, \chi_A(n))$$

with

$$\chi_A(j) = \begin{cases} 0, & \text{if } j \notin A \text{ and} \\ 1, & \text{if } j \in A. \end{cases}$$

We then have the relations

$$\chi_{A \cap B} = \chi_A \cdot \chi_B$$

and

$$\chi_{A \cup B} = \chi_A \underset{S}{+} \chi_B.$$

The allocation $A \to \chi_A$ defines a bijective map $\phi : (\mathcal{P}(M), \leq) \to (B^n, \leq)$ with

$$\phi(A \cap B) = \phi(A) \cdot \phi(B)$$

and

$$\phi(A \cup B) = \phi(A) \underset{S}{+} \phi(B).$$

The rules for a Boolean lattice now come from $(P(M), \subset)$ to $(B^n, \leq)$ □

Definition 11.49. A *Boolean function* is a map $F : B^n \to B^m$ with $m, n \in \mathbb{N}$. A map $B^n \to B$ is called a *switching function*.

A Boolean function describes how to determine a Boolean-valued output based on some logical calculations from Boolean inputs. Such functions play a basic rule in questions of complexity theory as well as in the design of computer chips for digital computers.

The properties of Boolean functions also play a role in cryptology, particularly in the design of symmetric key protocols and their algorithms. Boolean arithmetic on $\{0, 1\}$ is called the *XOR operation*.

Each Boolean function $B^n \to B^m$ is uniquely determined by the m switching function which belong to the single coordinates in B^m.

The $2^{2^n} = 2^{|B^n|}$ switching functions form a Boolean lattice with respect to

$$(f \cdot g)(x) = f(x) \cdot g(x),$$
$$(f \underset{S}{+} g)(x) = f(x) \underset{S}{+} g(x)$$

and

$$\overline{f}(x) = \overline{f(x)}$$

where $x = (x_1, x_2, \ldots, x_n) \in B^n$. Here $f(0, 0, \ldots, 0) = 0$ and $f(1, 1, \ldots, 1) = 1$.

The rules for a Boolean algebra extend to the set of switching functions.

As an example for $n = 1$ we have the 4 switching functions, see Table 11.2.

Table 11.2: 4 switching functions for $n = 1$.

For $n = 1$	function	values		
variable value 0	0	0	1	1
variable value 1	0	1	0	1
function	0	x	$\overline{x}$	1
	zero function	identity	complement	unit function

If $n \geq 2$ then we have the special switching functions $(x_1, x_2, \ldots, x_n) \mapsto x_v$ the projection onto the vth coordinate. We denote these simply by x_v. If $z = (z_1, z_2, \ldots, z_n) \in B^n$ arbitrary but fixed we define

$$f_z(x) = \prod_{v=1}^{n} (x_v z_v + \overline{x_v}\, \overline{z_v}).$$

We have

$$f_z(x) = 1 \quad \text{if and only if} \quad x = z$$

and

$$f_z(x) = 0 \quad \text{if and only if} \quad x \neq z$$

because

$$x_v z_v + \overline{x_v}\, \overline{z_v} = 0 \quad \text{if and only if} \quad x_v \neq z_v.$$

This gives the following theorem.

Theorem 11.50. *Each switching function $f : B^n \to B$ has a unique representation of the form*

$$f(x) = \sum_{\substack{S \\ z \in B^n}} \lambda(z) f_z(x)$$

with $\lambda(z) = 0$ or 1. This is the disjunctive normal form.

The uniqueness of the representation follows directly from $\lambda(z) = f(z)$ for each $z \in B^n$.

11.7 Boolean Rings

Let M be a set and $\mathcal{P}(M)$ its power set. We construct a ring structure on $\mathcal{P}(M)$. For a discussion of rings see, for example, [13].

First we define multiplication by intersection. That is, if $A, B \in \mathcal{P}(M)$ then $AB = A \cap B$. For addition we use the *symmetric difference of two sets*. By this we mean

$$A \triangle B = (A \cup B) \setminus (A \cap B) = (A \cap B^c) \cup (A^c \cap B).$$

On $\mathcal{P}(M)$ we define for any two sets $A, B \in \mathcal{P}(M)$ their sum as

$$A + B = A \triangle B.$$

With these operations we get the following result.

Theorem 11.51. *Let M be a set and R = $\mathcal{P}(M)$. Then R with addition and multiplication as defined above forms a commutative ring with unity. The zero element is $\emptyset$ and the unity is M. The additive inverse of any subset A is A itself. The ring R is finite if and only if M is finite. If M = $\emptyset$ then R is the null ring {0}.*

Proof. The proof that R forms a ring under the given operations consists of verifying the ring axioms. This is very straightforward and we omit here the easy calculations. The other assertions follow directly from set properties.

   If M is finite then $\mathcal{P}(M)$ is also finite while if M is infinite its power set is also infinite.   □

   Any subring of $\mathcal{P}(M)$ is called a *ring of sets*. Recall that a nonempty subset of a ring forms a subring if it is closed under addition, multiplication and additive inverses. For subsets of $\mathcal{P}(M)$ to be a ring it is sufficient that this subset is closed under union, intersection and complement since symmetric difference is defined in terms of union, intersection and complement.

Lemma 11.52. *Let $R_1 \subset \mathcal{P}(M)$ be a nonempty collection of subsets of the set M. Then R_1 forms a subring of $\mathcal{P}(M)$, and hence a ring of sets, if it is closed under union, intersection and complement.*

   From Lemma 11.52 we get the following:

Lemma 11.53. *Let $R_1 \subset \mathcal{P}(M)$ be a ring of sets. Then R_1 forms a Boolean lattice under union, intersection and complement.*

   Let $A \in R = \mathcal{P}(M)$. Then $AA = A^2 = A \cap A = A$. Hence the square of any element in a ring of sets is itself. Further A is its own additive inverse since $A + A = 2A = \emptyset$. Hence for any $A \in R$ we have $2A = 0$ where R is any ring of sets.

   The property $A^2 = A$ for any A in a ring of sets we abstract to define a special type of ring.

Definition 11.54. A *Boolean ring* R is a ring where $x^2 = x$ for all $x \in R$.

   We now show that any Boolean ring is commutative and $2x = 0$ for any $x \in R$.

Theorem 11.55. *Let R be a Boolean ring. Then R is commutative and $2x = 0$ for any $x \in R$.*

Proof. If R consists of just the zero element 0 then the assertions are clear. Assume then that R has at least two elements. Let $x, y \in R$. Then $x^2 = x$ and $y^2 = y$. Consider $(x + x)^2 = x + x$. On the other hand,

$$(x + x)^2 = x^2 + x^2 + x^2 + x^2 = x + x + x + x.$$

This implies that $x + x = 0$ or $2x = 0$. This implies that $-x = x$ for any $x \in R$. Now consider $(x + y)^2 = x + y$ and get

$$(x + y)^2 = x^2 + xy + yx + y^2 = x + y \quad \Longrightarrow \quad xy + yx = 0 \text{ or } xy = -yx.$$

However, we have shown that $-yx = yx$, therefore $xy = yx$, and hence R is commutative. □

If R is a Boolean ring, then we define $x \leq y$ if $x = xy$. This relation is a partial order. It is reflexive because $x^2 = x$, it is antisymmetric since R is commutative, and finally it is transitive because

$$xz = xyz = xy = x \quad \text{for } x = xy \text{ and } y = yz.$$

Theorem 11.56. *Let $R = (R, +, \cdot, 0, 1)$ be a Boolean ring, and let $x \leq y$ if and only if $x = xy$. Then $(R, \leq)$ is a Boolean lattice. In this lattice we have:*
(a) *0 is the smallest element and 1 is the greatest element.*
(b) *$x \wedge y = xy$.*
(c) *$x \vee y = x + y + xy$.*
(d) *$\bar{x} = 1 + x$.*

Proof. We have already seen that $(R, \leq)$ is a partial order. The statement (a) is obvious.
(b) We have $xy \leq x$ and $xy \leq y$ since $xyx = xy = xyy$.
Now let $z \leq x, z \leq y$, that is, $z = zx$ and $z = zy$. It follows that $z = zy = zxy$, and therefore $z \leq xy$. Hence $x \wedge y = xy$, the greatest common lower element.
(c) We have

$$x(x + y + xy) = x^2 + xy + x^2y = x + 2xy = x$$

and analogously

$$y(x + y + xy) = y.$$

Therefore $x \leq x + y + xy$ and $y \leq x + y + xy$.
Now let $x \leq z$ and $y \leq z$. Then $x = xz$ and $y = yz$. Then

$$(x + y + xy)z = xz + yz + xyz = x + y + xy.$$

Hence $x+y+xy \leq z$ and therefore $x \vee y = x+y+xy$ the smallest common upper element.
(d) Now $x + (1 + x) = 1$ and $x(1 + x) = 0$. Therefore $\bar{x} = 1 + x$.
This shows that $(R, \leq)$ is a complementary lattice. The distributive law is satisfied because

$$(x \vee y) \wedge z = (x + y + xy)z = xz + yz + xyz = (x \wedge y) \vee (y \wedge z).$$

Hence $(R, \leq)$ is a Boolean lattice. □

Theorem 11.57. *Let* $(V, \le)$ *be a Boolean lattice with minimal element* 0 *and maximal element* 1. *We define addition and multiplication by*

(a) $x + y = (x \wedge \overline{y}) \vee (\overline{x} \wedge y)$,

(b) $xy = x \wedge y$.

Then $(V, +, \cdot, 0, 1)$ *is a Boolean ring and* $x \le y$ *if and only if* $x = xy$.

Proof. Immediately we have that $x + x = 0$, $x + 0 = x$, $xx = x$ and $x \cdot 1 = x$. Further $x + y = y + x$ and $xy = yx$, as well as $x(yz) = (xy)z$.

We show that addition is associative:

$$(x + y) + z = (x \wedge \overline{y} \wedge \overline{z}) \vee (\overline{x} \wedge y \wedge \overline{z}) \vee (\overline{x} \wedge \overline{y} \wedge z) \vee (x \wedge y \wedge z)$$
$$= x + (y + z).$$

Finally, we must show that

$$(x + y)z = xz + yz.$$

We have

$$(x + y)z = \big((x \wedge \overline{y}) \vee (\overline{x} \wedge y)\big) \wedge z = (x \wedge \overline{y} \wedge z) \vee (\overline{x} \wedge y \wedge z).$$

Further

$$xz + yz = (x \wedge z \wedge \overline{y \wedge z}) \vee (y \wedge z \wedge \overline{x \wedge z})$$
$$= \big(x \wedge z \wedge (\overline{y} \vee \overline{z})\big) \vee \big(y \wedge z \wedge (\overline{x} \wedge \overline{z})\big)$$
$$= (x \wedge \overline{y} \wedge z) \vee (\overline{x} \wedge y \wedge z).$$

Both calculations together prove that $(x + y) \cdot z = (xz + yz)$.

This proves the statement. □

We note that the concepts *Boolean lattice*, *Boolean algebra* and *Boolean ring* are entirely equivalent. This leads to the second version of Theorem 11.44.

Theorem 11.58 (Stone's theorem for Boolean rings). *Each finite Boolean ring R is ring-isomorphic to a power set ring* $\mathcal{P}(M)$ *for a set M.*

11.8 The General Theorem of Stone

We now use the theory of Boolean rings, which is equivalent to the theory of Boolean lattices, to prove the general theorem of Stone. We recall that $x \le y$ if and only if $x = xy$ via the mentioned equivalence. Let $R = (R, +, \cdot, 0, 1)$ be a Boolean ring. A subset $F \subset R$ is called a *filter* if the following four conditions are satisfied:

(a) $F \neq \emptyset$.

(b) $0 \notin F$.

(c) $x = xy$ implies that $y \in F$ for all $x \in F$ and $y \in R$.

(d) $xy \in F$ for all $x, y \in F$.

Remark 11.59.

(1) From (c) we have that $1 \in F$.

(2) If $0 \neq x \in R$ then $F_x = \{y \in R \mid x \leq y\}$ is a filter called a *principal filter generated by x*. If $\{F_i \mid I \in I\}$ is a family of filters with I a linearly ordered index set and if $F_i \subset F_j$ for all $i \leq j$ then $\bigcup_{i \in I} F_i$ is a filter. This means that in the set of all filters in R each chain $\{F_i \mid i \in I\}$ has a smallest upper element $\bigcup_{i \in I} F_i$. By Zorn's lemma, named after M. Zorn (1906–1993), each filter is contained in a maximal filter, that is, in a filter $U \subset R$ with the property that each filter F with $U \subset F \subset R$ is already U.

Zorn's lemma is equivalent to the axiom of choice (see [14]). Here we may take Zorn's lemma as an axiom. We denote the maximal filter as an *ultrafilter*.

Now filter $F \subset R$ cannot contain both x and $\bar{x} = 1 + x$ for otherwise $0 = x(1 + x)$ is in F, which is impossible.

This observation provides a nice characterization of ultrafilters.

Lemma 11.60. *A filter $F \subset R$ is an ultrafilter if and only if either $x \in F$ or $\bar{x} \in F$, $\bar{x} = 1 + x$, for each $x \in R$.*

Proof. Let $x \in R$ and let $F \subset R$ be a filter. As mentioned F cannot contain both x and $1 + x$.

Assume that F contains neither x nor $1 + x$. We show that F is not maximal. We define

$$F' = \{z \in R \mid xy = xyz \text{ for some } y \in F\}.$$

We have $F \subset F'$ and $x \in F'$ since F is not empty. Therefore $F \neq F'$.

We must show that F' is a filter. Assume that $0 \in F'$. Then $xy = xy \cdot 0 = 0$ for some $y \in F$ and therefore $y(1 + x) = y$ which gives $y \leq 1 + x = \bar{x} \in F$, which is impossible. Hence $0 \notin F'$.

Let $z \in F'$ and $z = zz'$. For some $y \in F$ we have the equation

$$xy = xyz = xy(zz') = (xyz)z' = xyz'$$

and then $z' \in F'$. Hence F' is a filter. $\square$

The set $\mathcal{U}$ of ultrafilters is therefore exactly the set of filters which contain either x or $1 + x$ for each $x \in R$.

We now allocate to each $a \in R$ a set $\rho(a) \subset \mathcal{U}$ of ultrafilters and we define

$$\rho(a) = \{U \in \mathcal{U} \mid a \in U\}.$$

This gives us the general Stone's theorem.

Theorem 11.61. *Let R be a Boolean ring and let $\mathcal{U}$ be the set of ultrafilters of R. The allocation $a \mapsto \rho(a)$ defines an injective map $R \to \mathcal{P}(\mathcal{U})$. In particular, the Boolean ring R is ring-isomorphic to a subring of $\mathcal{P}(\mathcal{U})$, and therefore each Boolean lattice is lattice-isomorphic to a Boolean set lattice.*

Proof. Directly we have $\rho(0) = \emptyset$ and $\rho(1) = \mathcal{U}$. We show that ρ is injective.

Let $a, b \in R$ with $a \neq b$. By symmetry we may assume that $a \neq ab$ because either $a \neq ab$ or $b \neq ba = ab$. Then $a(1 + ab) = a + ab \neq 0$.

Therefore there exists an ultrafilter U_c which contains the principal filter $\{y \in R \mid c \leq y\}$ where $c = a(1 + ab)$. The ultrafilter U_c contains a and $1 + ab$ but it cannot contain b because otherwise $ab \in U$ contradicting $1 + ab \in U_c$.

We finally have to show that

$$\rho(ab) = \rho(a) \cap \rho(b)$$

and

$$\rho(a + b) = \rho(a) \triangle \rho(b).$$

The ultrafilters which contain ab are exactly those which contain a and b. Hence we have $\rho(ab) = \rho(a) \cap \rho(b)$. We now consider an ultrafilter U which contains $a + b$.

Assume that U contains neither a nor b. Since U is an ultrafilter then $1 + a$ and $1 + b$ are contained in U by Lemma 11.60. Hence

$$(a + b)(1 + a)(1 + b) \in U.$$

However, this is not possible since

$$(a + b)(1 + a)(1 + b) = (a + b)(1 + a + b + ab) = a + a + ab + ab + b + ab + b + ab = 0.$$

This shows that $\rho(a + b) = \rho(a) \cup \rho(b)$.

Now assume that U contains both a and b. Then $ab \in U$. This also is impossible because

$$(a + b)ab = ab + ab = 0.$$

Hence $\rho(a + b) \subset \rho(a) \triangle \rho(b)$.

The last thing we must show is that

$$\rho(a) \triangle \rho(b) \subset \rho(a + b).$$

For this we start with an ultrafilter U which contains a and $1 + b$. We have to show that $a + b$ is in U. If not, then a, $1 + b$ and $1 + a + b$ are all in U. Again this is impossible since

$$a(1 + b)(1 + a + b) = (a + ab)(1 + a + b) = a + a + ab + ab + ab + ab = 0.$$

Altogether we have $\rho(a + b) = \rho(a) \triangle \rho(b)$, proving the theorem. □

We note that for finite Boolean rings the ultrafilters are exactly the filters generated by the atoms.

Hence Theorem 11.44 and Theorem 11.58 are indeed special cases of Theorem 11.61.

Exercises

1. Let B^n, $n \geq 1$, be the n-fold Cartesian product of $B = \{0, 1\}$, that is, the set of the n-digit 0–1-sequences $(a_1, a_2, \ldots, a_n)$ with $a_i = 0$ or 1 for $i = 1, 2, \ldots, n$.
   Show that B^n is partially ordered by

   $$(a_1, a_2, \ldots, a_n) \leq (b_1, b_2, \ldots, b_n) \quad \Leftrightarrow \quad a_i \leq b_i \text{ for } i = 1, 2, \ldots, n \quad \text{and}$$
   $$(a_1, a_2, \ldots, a_n) < (b_1, b_2, \ldots, b_n) \quad \Leftrightarrow \quad a_i \leq b_i \text{ for } i = 1, 2, \ldots, n \quad \text{and}$$
   $$a_i < b_i \text{ for at least one } i.$$

   Give the Hasse diagrams for B^1, B^2 and B^3.
2. Let $(M, \leq)$ be a finite partially ordered set.
   Show that M contains minimal elements and, if M contains exactly one minimal element, then this is the smallest element.
3. Let $(M, \leq)$ be a partial order
   (a) Show that $\sup(\emptyset)$ exists if and only if M has a smallest element, and then $\sup(\emptyset) =: 0$ is the smallest element.
   (b) Show that $\inf(\emptyset)$ exists if and only if M has a greatest element, and then $\inf(\emptyset) =: 1$ is the greatest element.
4. Let $(M, \leq)$ be a complete partial order. Show that $(M, \leq)$ has a uniquely determined smallest element, and this is $\sup(\emptyset) =: 0$.
5. Let $(M, \leq)$ be a complete partial order and $f : M \to M$ be continuous. Show that

   $$x_f = \sup\{f^i(\sup(\emptyset)) \mid i \geq 0\}$$

   is the uniquely determined smallest fixed point of f.
   (*Hint*: An element $x \in M$ is a fixed point of f if $f(x) = x$.
   Show first that

   $$\sup(\emptyset) \leq f(\sup(\emptyset)) \leq f^2(\sup(\emptyset)) \leq \cdots.$$

   This is the fixed point theorem of S. C. Kleen (1909–1994).)

6. Show the existence of an infinite set lattice which does not contain irreducible elements.

7. Let $(V, \leq)$ be a Boolean lattice and $a \in V$.

   Show the general statement that a is an atom if and only if a is irreducible.

8. (a) Let $(V, \leq)$ be a complete lattice, that is, a lattice and a complete partial order. Show that in V each subset D has a greatest lower bound

$$\inf(D) = \sup\{x \in V \mid x \leq y \text{ for all } y \in D\}.$$

   In particular, each complete lattice has a smallest and a greatest element.

   (b) Show that each finite lattice is complete.

9. Let $(V, \leq)$ be a complete lattice (see Exercise 8) and $f : V \rightarrow V$ be a lattice homomorphism.

   Then the set

$$P(f) = \{y \in V \mid f(y) = y\}$$

   of the fixed points of f is a complete sublattice. In particular, there exist uniquely determined smallest and greatest fixed points.

   This is the fixed point theorem of B. Knaster (1893–1980) and A. Tarski (1901–1983).

10. Show in detail that each switching function $f : B^n \rightarrow B$ has a unique representation of the form

$$f(x) = \sum_{\substack{S \\ z \in B^n}} \lambda(z) f_z(x), \quad \lambda(z) = 0 \text{ or } 1.$$

11. Let M be any nonempty set. In the power set $\mathcal{P}(M)$ we define addition and multiplication by

$$A + B = (A \cup B) \setminus (A \cap B) \quad \text{and}$$
$$A \cdot B = A \cap B.$$

   (a) Show that $R = \mathcal{P}(M)$ forms a commutative ring with unity.

   (b) Explain why we will not get a ring if we define operations in $\mathcal{P}(M)$ by

$$A + B = A \cup B \quad \text{and}$$
$$A \cdot B = A \cap B.$$

12. Let R be a ring such that

$$x^3 = x \quad \text{for all } x \in R.$$

   Show that R is commutative.

Bibliography

[1]   L. V. Ahlfors. *Complex Analysis*. McGraw-Hill, 1966.
[2]   K. I. Appel and W. Haken. *Every Planar Map is Four Colorable*. Contemporary Math., 98, 1989.
[3]   H. Bauer. *Wahrscheinlichkeitstheorie*. De Gruyter, 2002.
[4]   A. Beardon. *The Geometry of Discrete Groups*. Springer-Verlag, 1983.
[5]   S. Bosch. *Linear Algebra*. Springer-Verlag, 2008.
[6]   T. C. Brown. *A proof of Sperner's lemma via Hall's theorem*. Proc. Camb. Philos. Soc. 78, 387, 1975.
[7]   T. Camps, V. gr Rebel, and G. Rosenberger. *Einführung in die kombinatorische und die geometrische Gruppentheorie*. Heldermann, 2008.
[8]   C. Carstensen-Opitz, B. Fine, A. Moldenhauer, and G. Rosenberger. *Abstract Algebra*, 2nd edition. De Gruyter, 2019.
[9]   W. Cock. *In the Pursuit of the Traveling Salesman*. Princeton Univ. Press, 2014.
[10] V. Diekert, M. Kufleitner, and G. Rosenberger. *Elemente der Diskreten Mathematik*. De Gruyter Verlag, 2008.
[11] J. Elstrodt. *Maß- und Integrationstheorie*. Springer-Verlag, 2011.
[12] B. Fine, A. Gaglione, A. Moldenhauer, G. Rosenberger, and D. Spellman. *Algebra and Number Theory: A Selection of Highlights*. De Gruyter, 2017.
[13] B. Fine, A. Gaglione, and G. Rosenberger. *Introduction to Abstract Algebra*. John Hopkins University Press, 2013.
[14] P. R. Halmos. *Naïve Set Theory*. Springer, 1968.
[15] P. Haxell. *On forming committees*. The American Mathematical Monthly 118(9), 777–788, November 2011. Taylor & Francis, Ltd.
[16] G. A. Johnes and D. Singerman. *Complex Functions*. Cambridge University Press, 1987.
[17] W. Kaballo. *Einführung in die Analysis I*. Spektrum-Verlag, 1996.
[18] T. Kosky. *Catalan numbers with Application*. Oxford University Press, 2008.
[19] D. N. Kozlov. *Chromatic subdivision of a simplicial complex*. Homology, Homotopy and Applications 14(2), 197–209, 2021. Oxford University Press.
[20] D. N. Kozlov. *Organized Collapse: An Introduction to Discrete Morse Theory*. Graduate Studies in Mathematics, 207. AMS, 2020.
[21] R. C. Lyndon. *Groups and Geometry*. LMS Lecture Note Series, 101. Cambridge Univ. Press, 1986.
[22] R. C. Lyndon and P. Schupp. *Combinatorial Group Theory*. Springer, 1977.
[23] J. R. Munkres. *Elements of Algebraic Topology*. Addison-Wesley, 1984.
[24] E. M. Palmer. *The hidden algorithm of Ore's theorem on Hamiltonian cycles*. Computers & Mathematics with Applications 34, 113–119, 1997.
[25] R. L. E. Schwartzenberger. *N-dimensional Crystallography*. Pitman, 1980.
[26] M. Tenenbaum and H. Polland. *Ordinary Differential Equations*. Dover Books on Mathematics, 1986.

https://doi.org/10.1515/9783110740783-012

Index

https://doi.org/10.1515/9783110740783-013